# THE STUDY OF VEGETATION

# THE STUDY OF VEGETATION

a review of the developments in the various branches of
vegetation science with special attention to the Dutch
contributions compiled to celebrate the one-hundredth
meeting of the Commission for the Study of Vegetation
of the Royal Botanical Society of the Netherlands

edited for the Commission by M. J. A. Werger

23 March 1979

Dr. W. Junk bv Publishers, The Hague-Boston-London

ISBN 90 6193 594 6

# CONTENTS

# PREFACE

With a subtitle like the one attached to this volume one would suppose that a preface is hardly needed. However, the kind of jubilee celebrated with the publication of this book justifies a few words about its how and why. With forty-six years of age the Commission for the Study of Vegetation is one of the oldest of the Royal Botanical Society of the Netherlands and, measured in activities, it has been one of its most succesful ones, as it is the first to have organized one hundred symposia meetings. These meetings dealt with topics from the entire wide spectrum of branches of vegetation science as they come under this Commission. The list of lectures presented at these meetings as compiled by Vroman in the final chapter of this book, amounts to a total of 505 titles and adequately illustrates the variety of aspects of vegetation science represented.

· Initially the meetings of the Commission were held on Sundays, and from 1952 onwards when the Saturday was included in the week-end, that day became the day of the meetings, until a few years later, it changed to a week-day. First of all this clearly reflects the recognition the meetings gradually got from many of the authorities by which most of the participants to the meetings were employed. The authorities realized the value of these meetings, with their possibilities for acquintance and discussion with professional colleagues, for their staff for a proper fulfilment of the official duties, and they did no longer regard them as leisure activities. Thus, they allowed their staff to attend in office hours. Partly this change from the week-end to a week-day was also promoted by a change in proportion of the various groups of attendants: Until the early fifties a relatively large percentage of the participants were teachers who, obviously, could only attend at week-ends. Gradually their number became proportionally smaller, mainly because more and more people became professional biologists as more jobs became available for them, and as the level of education in the Netherlands rose. The proportional increase of professional biologists amongst the attendants together with the recognition of the value of the meetings by the authorities led to this change from spare-time meetings at week-ends to office-hours meetings during the week.

Apart from these changes, the number of meetings increased from once a year in the period from 1933 till 1940, to twice yearly until 1961, and since then nearly always three times a year. At present the meetings are generally recognized amongst vegetation scientists and their employers as offering a good forum to present one's points of view, in a lecture or in a discussion following on a lecture, and as a useful means to meet and discuss with one's direct colleagues or with those specializing in branches of vegetation science that differ from one's own field. And it looks as though things are not going to be worse in the near future, as it is still fairly easy to attract enough speakers to fill the programmes of the three yearly meetings, and as these meetings virtually consistenly are attended by about a hundred or more interested persons.

The (ex-)members of the Commission for the Study of Vegetation thought it attractive to invite for the hundredth meeting a number of recognized Dutch authorities from the various branches in vegetation science falling within the scope of the Commission to review the state of the art in their fields of specialization. They were asked to present brief summaries of the main historical trends of progress in their fields of science, to outline the present state of knowledge, and to sketch the probable or desirable future developments in these sciences. They were also advised to pay special attention throughout their reviews to the Dutch contributions in these respects. In this way it was hoped to commemorate the one-hundredth symposia meeting with an up-to-date review of vegetation science in most of its aspects and as seen from a Dutch perspective. As such a review was considered to be of interest to a wider public than the local community, the Commission decided to publish the lectures in an extended version in English. Special efforts by the authors and Dr. W. Junk Publishers made it possible to have this publication available at the day of the meeting.

Thus, the present volume was obtained, in which W.H. van Dobben sketches the significance of autecological research in the whole of vegetation science, as it allows the testing of hypotheses formulated in other fields of specialization, and thus can lead to a better understanding and a more effective management of field situations. In this way, van Dobben provides the base for the following two chapters in which ter Borg highlights the importance of a detailed study of various aspects of population biology, illustrated by several well-chosen examples, for an understanding of the dynamical aspects of a vegetation cover; and van den Bergh emphasizes upon competition as a fundamental factor in vegetation composition. In discussing some case studies, the latter author especially draws attention to the rather large discrepancies existing between the explanation of competition experiments under laboratory conditions and fluctuations in vegetation in actual field situations. Since the concepts, techniques and general theories of the floristic approach to vegetation science have been comprehensively reviewed by Westhoff & Van der Maarel in a recent volume of the Handbook of Vegetation Science, Westhoff, in his chapter, chose to emphasize particularly the Dutch contribution to the developments in phytosociology, and the advances in vegetation science made by Dutch scholars. The early developments of phytosociology in the Netherlands certainly deserve such special attention since the influences of both the 'Nordic approach' and the 'Braun-Blanquet approach' have been strong here in those early years. Some of the points made by Westhoff are well reflected in the list of lectures compiled by Vroman. The possibilities for and the relevance of the investigation of structural characters in vegetation science are lucidly explained by Barkman, who also points out the interrelations of this approach with the floristic one, and the significance of vegetation structure in several dynamic processes in the vegetation. In a useful outline, van der Maarel provides a fairly comprehensive, though brief, review of the progress made in multivariate analysis in various fields of vegetation science. Against this general background he clearly shows the state of this art in the Netherlands, thus also indicating the promising possibilities for developments in this field still lying ahead. The developments in palynology and the fruitful cooperation between palynologists and vegetation scientists sensu stricto for their mutual benefit as regards the development of theories are considered by Janssen. In that chapter it is also pointed out that the enormous flux in cultural measures has severely reduced the possibilities for a palynological reconstruction of the vegetation's history in the Netherlands. Thus, in indicating the importance of nature reserves, the link to Bakker's chapter is laid. Bakker shows very clearly that nature conservation has two main aspects, a legal one and a scientific one, and the latter aspect is virtually indentical to applied vegetation science. It is in his chapter that the significance of the practicallities so far obtained from theoretical vegetation science become very obvious, but it is also here where it becomes clear how much still remains to be done.

I believe that with this book the Commission for the Study of Vegetation properly fulfils its purpose: the book presents a modern survey of the advances in the various branches of vegetation science that come under the Commission, while it stimulates the discussion between close colleagues as well as between specialists (and persons interested) in fields that have diverged somewhat further. The realization of this book is due to the great readiness of its authors to prepare their manuscripts on relatively short notice, sticking to a very strict time-table, in the middle of their very busy summer and autumn seasons. I want to thank them most sincerely for their efforts. A word of thanks to Dr. W. Junk Publishers is also in place, as they have been extremely cooperative in producing this book neatly and within a very short period of time. I further like to thank all my (ex-)Commission colleagues for their spirit and support during the preparations of this book; three of them, Jo Willems, Jaap Wiegers and Ties Oomes, deserve special mentioning for their help in the planning stage of this project.

for the Commission for the Study of Vegetation

Marinus J.A. Werger

Division of Geobotany,
University of Nijmegen,
Toernooiveld, Nijmegen.

# AUTHORS

P.A. Bakker,
> Society for the Preservation of Nature Reserves in the Netherlands,
> Schaep en Burgh, Noordereinde 60,
> 's Graveland.

J.J. Barkman,
> Biological Station, Agricultural University, Kampsweg 27,
> Wijster (Dr.).

J.P. van den Bergh,
> Centre for Agrobiological Research, Bornsesteeg 65,
> Wageningen.

S.J. ter Borg,
> Biological Centre, Laboratory for Plant Ecology, University of Groningen,
> P.O. Box 14,
> Haren (Gr.).

W.H. van Dobben,
> Section Vegetation Science and Plant Ecology, Agricultural University,
> P.O. Box 8128,
> Wageningen.

C.R. Janssen,
> Laboratory of Palaeobotany and Palynology, Institute of Systematic
> Botany, University of Utrecht, Heidelberglaan 2,
> Utrecht (De Uithof).

E. van der Maarel,
> Division of Geobotany, University of Nijmegen, Toernooiveld,
> Nijmegen.

J. Vroman,
> Van Kempensingel 21,
> Woerden.

V. Westhoff,
> Division of Geobotany, University of Nijmegen, Toernooiveld,
> Nijmegen.

# 1. AUTECOLOGY AND VEGETATION SCIENCE

## W.H. VAN DOBBEN

# 1. AUTECOLOGY AND VEGETATION SCIENCE

## W.H. VAN DOBBEN

### 1. Introduction

In an ecological context the properties of organisms can be understood as a complex of adaptations to a specific way of living in a specific habitat. The proposition formulated in this way is in accord with the Darwinian approach to evolution which supposes a constant selective pressure in the direction of adaptation. It allows for the possibility that properties which are non-adaptive but do not reduce survival may be preserved. This forbids a dogmatic standpoint.

Such a concept of adaptation appears very enlightening and it can be considered a primary task of autecological research to develop it further. Adaptation may refer to form and function and to the physiology of organisms. It can be made plausible that life form and growth form of a plant species are useful adaptations. This approach leads to a functional morphology which could be called ecomorphology but this term is not in use. For investigations on properties such as water- and nutrient relations or reactions to light the term ecophysiology is applied. It is better to restrict these terms to investigations on the significance of form or vital functions for a special way of living. It is of course possible to end any report on physiological work with some reference to its significance within the context of survival in nature, but the term ecophysiology should only be used when the investigations are focussed upon special niche relationships. For example, investigations on the mechanism of photoperiodicity by way of phytochrome studies belong to pure plant physiology, whereas a study of the adaptive significance of photoperiodism for some species in a specific habitat should be called ecophysiology.

In the same way it is possible to distinguish pure morphology, such as investigations on tissue structures, and ecomorphology, such as studies of the significance of life and growth forms for survival. Comparison between species is very enlightening in this respect. Concepts such as xeromorphy and hygromorphy depend on it. In zoology this has led to the use of the terms comparative morphology and comparative physiology.

The use of such terms meets with the objection, that both form and function generally contribute in close cooperation to an adaptation and can hardly be studied separately. For example, drought resistance is the result of integrated morphological and physiological features. In other cases an adaptation is primarily physiological, such as the production of specific scents for the attraction of pollinating insects, or primarily morphological, as clinging roots for climbing plants. Besides, it should be acknowledged that no clinging root would ever emerge or function efficiently without suitable physiological processes. It follows that most classifications of adaptations in the sphere of ecophysiology or comparative morphology have a relative value only.

Analysis of growth presents a good example: All growth studies refer to a complex of physiological and morphological reactions. For example, growth rates of plants depend on the share of assimilating tissue in the total plant weight. In its turn this morphological relation depends on physiological reactions to water and nutrient supply and also temperature. When analysis of growth is presented as 'ecophysiology' one must admit that this does no justice to the morphological aspects of the growth process.

Many investigations in specific adaptations such as encountered during growth studies could quite satisfactorily be indicated as autecological research work sensu lato.

## 2. The evaluation of adaptations

A major difficulty in autecology is the evaluation of a particular property. Is it really an adaptation and if so, how important is it for survival? The value of frost resistance in a severe climate is clear enough and it is plausible that the possession of tendrils and climbing roots is a useful adaptation for climbing plants. But can the same be supposed for pinnate leaves or for the holes in the leaves of a *Philodendron*? Holes in higher leaves enable lower leaves (which indeed have less or no holes) to intercept some light so that they can stay alive as a stand-by when eventually the top of the plant is damaged. It is however not easy to prove such a supposition experimentally. Thorns and spines seem a useful defence against browsing. It is not by accident that woody undergrowth often bears thorns on branches within reach of herbivores. Holly (*Ilex*) forms spineless leaves on high branches. This supports the supposition about selective pressure because of browsing, for when there is no danger it is more profitable to form light intercepting tissue rather than spines. It is clear that we should be very cautious with this type of indirect argument and that it is difficult to measure the adaptive value of properties.

It has been supposed that differences in congeneric species as well as resemblances in species with remote descent point in the direction of adaptation. In these cases the terms divergence and convergence are used respectively.

A related point of view is the distinction between adaptive and descendant qualities. For example, a woody trunk is a useful adaptation for a macrophanerophyte. This holds for both dicotyledonous and for monocotyledonous tree species. A trunk can be built in several ways. In this case wood anatomy confirms the differences in descent. The fact that unrelated species form a functionally comparable organ as a trunk confirms the supposition that this is an important adaptation. In this case circumstantial evidence of this type is hardly required but it is given here to explain the argument.

In dicotyledonous tree species wood anatomy also shows systematic relationships but this is most clearly done by the flowers. It cannot be by chance that flowers are important in taxonomy. Flowers are conservative and reveal kinship, whereas vegetative parts are progressive and show adaptive properties eventually diverging from relatives and changing in the direction of unrelated groups.

This does not mean that flower forms are non-adaptive. It is the basic flower design that shows descendant qualities. Moreover, this design may have taken form as an adaptive feature of the ancestor of a plant family. However, such general considerations do not alter the fact that it remains difficult to recognize a quality with certainty as an adaptation, and still more to exactly evaluate its significance for survival.

## 3. The significance of vegetation science for autecological work

The object studied by vegetation science is the plant cover as a distinct multispecific entity and its relations to the abiotic environment, especially soil and climate. Description of field conditions dominates this type of work.

It is often claimed that descriptive vegetation studies only raise questions about plant ecology and never answer them. Only experimental work, by preference of a physiological design — so goes the argument —, can provide firm knowledge.

In the first place we must admit that our present knowledge about the ecology of wild species is mainly acquired by simple field observations, for example about life forms and growth forms, growth rhythm, pollination, dissemination and several other relationships with properties of the environment. When the argument is restricted to the classification and ordination of plant communities it must be considered that this type of work investigates relations between species combinations and habitat, for instance edaphic conditions, and in this way also contributes to the knowledge of the single species.

Detailed field observations especially can give important indications about autecology, for example the sequence of related species along a gradient such as the series *Ranunculus bulbosus* — *acer* — *repens* from dry to wet, *Plantago major* — *coronopus* — *media* — *lanceolata* from compacted to loose soil, or *Carex acutiformis* — *diandra* — *lasiocarpa* from eutrophic to oligotrophic conditions. This type of vegetation research provides the basis for comparative experimental work as indicated above. The same applies to observations on establishment of seedlings under field conditions, longevity of individual plants, succession studies and cyclic processes.

It is often necessary to test suppositions resulting from field observations by experiments. This is especially important where plants are concerned. The specific requirements of animals (birds, for example) can be studied more easily. The difference is that in animals with some luck we can assess by direct observation what they take from the environment, whereas the reactions to soil conditions of plants can only be studied by analysis of soil (or water) followed by experiments. A further difficulty is the importance of mutual interference in plants. This is often of a very complex nature, such as competition or the preconditioning of the soil profile by a vegetation cover for species of later succession stages.

Also in these cases vegetation science provides experiences which can stimulate research work of an autecological nature. The indicator value of individual species

5

for specific environmental factors as summarized by Ellenberg (1974) has been composed from such a combination of field observations completed by autecological investigations.

## 4. Autecology and the explanation of field situations

Autecological knowledge should in principle enable us to explain the occurrence or establishment of a species in a given habitat. In many cases this is possible to a certain extent. It is easiest when abiotic factors are decisive, such as pioneer situations. We can, for example, indicate the properties of *Senecio congestus* and *Ranunculus sceleratus* which enable them to establish and thrive as first pioneer species on bare and poorly aerated mud flats that are rich in nutrients. This biotope is predestined to be overgrown in a very short time. Plant species adapted to the bare situation must grow very quickly and produce large quantities of seed as early as possible. Annuals (or winter annuals) meet this condition. In *Senecio congestus* the exceptional growth rate is attained by a very high shoot-root ratio (possible because of a rich nourishment) and a low dry matter content (van Dobben 1967). One vigorous plant may produce 200.000 germinative seeds disseminated by wind (over large distances) and water. *Ranunculus sceleratus* is exclusively hydatochorous, and seed numbers of a well-developed plant may approximate one million.

Bakker (1960) has described how in the course of a rapid succession in a reclaimed Lake Yssel polder a *Senecio congestus* aspect can change within some months into an aspect of *Phragmitis australis*, *Scirpus maritimus*, or *Typha latifolia*. The decline of *Senecio congestus* could be explained by ecological characteristics of the species and the decrease of the available mineral nitrogen of the soil under influence of the first generation. *Senecio congestus* is dependend on a very high level of mineral nitrogen, in this case in the form of ammonium. The perennial species may emerge together with the annuals *Senecio congestus* and *Ranunculus sceleratus* but they stay back in the first year. This could be explained by a slower vegetative growth of these perennials caused by an early initiation of rhizomes. After the decline of the first generation annuals the perennials, having a more extensive root system, apparently succeed in taking up enough nitrogen produced by gradual mineralization of organic matter in the soil to enable good growth.

Generally the establishment of plant species in a habitat is more difficult to explain when later succession stages are at stake. In *Pinus* plantations in the northern dune district of the Netherlands which are about 100 years old, boreal species such as *Goodyera repens*, *Linnea borealis* and *Moneses uniflora* begin to appear. Apparantly soil conditions have changed under the influence of tree growth in such a way that essential demands of such species are met. We are, however, not able to specify these demands and to do so may require a very difficult type of research work. Just as we ask for such exact explanations of the sequence of establishment of species or explanations of densities and patterns we become aware of the limitations of our knowledge.

As concluded above it is very difficult to evaluate exactly the significance of an adaptation for survival of the single species and this holds even for qualities of which the function is quite clear. The preference of autecology for a comparative approach, however enlightening, reveals an inherent weakness, namely the difficulties met in quantifying the efficiency of vital functions in organisms. For every adaptation it is easy to draft a list of complications hampering its evaluation in hard figures. For instance, to quantify drought resistance in terms of water requirements it would at least be necessary to distinguish between growth stages in the plant. Moreover, drought resistance is not merely the ability to overcome water stress. Many species have organs for water storage and this gives them opportunities to survive dry periods. Here again the phase of development in which water is available is of importance.

Our considerations thus far mainly refer to the relations of species with the biotope. In closed vegetations, however, specific interrelations are as important and generally situations are more difficult to explain as the number of species involved increases.

It is generally supposed that by competition species can be prevented from establishing in a habitat where conditions otherwise would permit their growth and propagation. When a species is crowded out by another species with the same niche relationships but which performs more efficiently in a given habitat the term competetive exclusion is used. De Wit (1960) has shown that competition phenomena can be quantified when the ability of the species to claim space is decisive. Then it is even possible to predict the result of competition from spacing experiments with the same species.

The ability to claim space (depending on conditions of soil and climate) can be quantified as a 'crowding coefficient'. However, the calculation of a 'relative crowding coëfficient' predicting the result of competition between two species appears only possible when growth form and growth rhythm agree.

Still, competition for space belongs to those interspecific relations which are relatively easy to study with experimental methods, just as many adaptations which are induced by conditions of climate and soil. Generally the influence of other organisms in the same habitat, such as through pathogens, herbivory and animal activity in pollination and seed dispersal, is very difficult to evaluate. Many of these relations can only be studied in the field where standard conditions are unattainable. Such studies may indicate the importance of some specific relations but an interpretation in quantitative terms is only rarely possible. This situation explains why population studies must start from their own empirism; however knowledge of the autecology of the individual species contributes to a better understanding of events.

Our knowledge about establishment and ultimate status of a species in a given habitat is essentially empirical. Autecological data generally lead to explanations afterwards. Therefore, predictions are based more on experience than on insight.

The history of introductions of exotics is illustrative. It is safe to predict that a tropical species has no chance in an arctic tundra, but we cannot explain why exactly *Prunus serotina, Elodea* and *Azolla* species could establish themselves in temperate Europe. We can eventually show – on the basis of autecological data – that climate and soil (or water) conditions were favourable but the same may hold for many unsuccessful introductions. We may further consider that the habitats occupied in these cases (reafforested heathlands and ditches respectively) are artificial but such statements give part-explanations only.

Besides, it is impossible to consider the species as a homogeneous entity with the possibility to assess and quantify adaptations once and for all by investigations on a restricted number of specimens.

Provenance research shows substantial genetic differences between regional populations of every species (for some recent investigations see ter Borg (1972), Pegtel (1976) and van der Toorn (1972). This agrees perfectly with the evolution concept which supposes a selective pressure in the direction of adaptation. This pressure certainly will differ according to regional climate, soil and management conditions. The point of impact of this selection is the genetic diversity within the population which in itself offers an obstacle for the assessment of the properties of a species. Differences between regional populations or ecotypes may take the form of an ecocline (gradual transition) which again adds to the difficulties.

## 5. The significance of autecology for vegetation science

The first question we may ask is about the contribution autecology can make to the classification of vegetation.

Vegetation types can be characterized in several ways. For a rapid and global orientation in some remote part of the earth the distinction of 'formations' such as grassland, dwarf shrub heaths, or tropical rain forest gives a satisfactory first approach. Formations are determined by structural features such as life and growth forms. In detailed classifications of formations, such as designed by Schimper & Von Faber (Whittaker 1973) or Müller-Dombois & Ellenberg (1974) the environment to which the physiognomy of the vegetation is a response is taken into account, adding to the involvement of ecological points of view. For this type of classification, however, no knowledge of the autecology of the participating species is required other than available by direct observation. Even the identification of species is not necessary. A more detailed description of a formation must lead to the distinction of dominating species for different habitats within the formation's area, or the drafting of characteristic species combinations, for example according to the procedures of the Braun-Blanquet school.

Vegetation scientists working along these lines make some use of structural features of vegetations, especially in dubious cases. Braun-Blanquet 'classes' can be grouped into formations which confirms the notion that the vegetation structure of an area corresponds with its flora. However, floristic criteria very clearly dominate ecological ones in this classification. For example, the northern Atlantic salt marsh communities characterized by grasses, forbs and dwarf shrubs were united by Braun-Blanquet and Tüxen in a class *Salicornietea*. Now most authors distinguish a class *Asteretea tripolii* for the higher parts of the saltings and two classes, *Thero-Sa-*

*licornietea* and *Spartinetea*, for the floristically poor and open vegetation in lower parts of the same marshes. Both these classes are so poor in species that it is hardly possible to link them to other coena on floristic grounds, whereas their ecological affinities are obvious.

In this context it is interesting to compare the phytosociological classification of the vegetation of the Netherlands according to the Braun-Blanquet system, as summarized by Westhoff & den Held (1969), with a division in ecological groups designed by Arnolds & van der Meijden (1976). The latter authors place all species of medium and low parts of salt marshes in one group, whereas these species are divided by Westhoff & den Held over 5 classes (N.B. the highest rank in the phytosociological hierarchy!) This is no exception. Of the 37 ecological groups distinguished in the Netherlands 16 are divided on class level in the Braun-Blanquet system as applied by Westhoff & den Held. This illustrates the priority given to floristic criteria. Of course, the procedure can be defended by the observation that our floristic knowledge is far better developed than our ecological insight.

In cases where the occurrence of specific adaptations in a community is used for its classification it always concerns features which can be easily observed in the field.

Vegetation science, however, is more than the classification and ordination of communities. The relation of communities with the abiotic environment (or biotope) can be described but also submitted to investigations of diverse character. Characteristics like productivity or evaporation can be studied in a community as a whole and related to edaphic conditions. Further analysis in such directions, however, must lead sooner or later to autecological research work since the reactions of the community are largely determined by the properties of the participating species (see also Werger & van der Maarel 1978).

Now we can proceed further and demand explanations of how floristic composition and vegetation patterns are determined. Then we meet major obstacles.

The study of vegetation as a distinct entity is hampered by the fact that it is a component of the whole ecosystem, though, indeed, the productive part, the part that shapes the structure of the community and therefore dominates the physiognomy of the system. It is understandable, therefore, that description, productivity studies and investigations on the indicator value of species combinations for the properties of the system, are prominent in vegetation science. Community dynamics, however, are largely determined by interrelations with the soil and with other organisms and are submitted to the complexity of the 'web of life'. The possibility of explaining several phenomena via the autecology of separate plant species thus remains limited. There are clear cases of dominating species or species groups with comparable properties which have a decisive influence on soil conditions and, therefore, on the system as a whole. The acidifying effects of the litter of *Calluna vulgaris* are well established (see Gimingham 1972) and its promotion of podsolization is very likely. *Sphagnum* bogs are another example where the influence of distinct species is clear. The same holds for many features of forest ecosystems.

Vegetation science is very prominent in the applied sphere. Pure science demands explanations, applied science workable concepts based on insight where possible and on empirism when necessary. The management of vegetation types for several purposes is factually management of ecosystems and mainly a matter of experience. There is no doubt that the autecology of plant species can broaden the basis of real understanding. When management is focussed on one or a restricted number of species for their use or conservation the autecology of these species is of course of primary importance.

## References

Arnolds, E.J.M. & R. van der Meijden. 1976. Standaardlijst van de Nederlandse Flora 1975. Rijksherbarium Leiden.

Bakker, D. 1960. Senecio congestus (R.Br.) DC. in the Lake Yssel polders. Acta Bot. Neerl. 9: 235-259.

Borg, S.J. ter. 1972. Variability of Rhinanthus serotinus (Schonh.) Oberny in relation to the environment. Thesis Groningen.

Dobben, W.H. van. 1967. Physiology of growth in two Senecio species in relation to their ecological position. Jaarboek IBS 1967: 75-83.

Ellenberg, H. 1974. Zeigerwerten der Gefäszpflanzen Mitteleuropas. Scripta Geobotanica 9.

Gimingham, C.H. 1972. Ecology of Heathlands. London, Chapman & Hall.

Müller-Dombois, P. & H. Ellenberg. 1974. Aims and Methods of Vegetation Ecology. Wiley & Sons, New York-London.

Pegtel, D.M. 1976. On the ecology of two varieties of Sonchus arvensis L. Thesis, Groningen.

Toorn, J. van der. 1972. Variability of Phragmitis australis (Cav.) Trin. ex Steudel in relation to the environment. Thesis, Groningen.

Werger, M.J.A. & E. van der Maarel. 1978. Plant species and plant communities: some conclusions. In: E. van der Maarel & M.J.A. Werger (eds.), Plant species and plant communities. pp. 169-175. Junk, Den Haag.

Westhoff, V. & A.J. den Held. 1969. Plantengemeenschappen in Nederland. Thieme & Co, Zutphen.

Whittaker, R.H. (ed.) 1973. Handbook of vegetation science. Vol. 5: Ordination and Classification of Communities. Junk, Den Haag.

Wit, C.T. de. 1960. On competition. Pudoc, Wageningen.

# 2. SOME TOPICS IN PLANT POPULATION BIOLOGY

S.J. TER BORG

# 2. SOME TOPICS IN PLANT POPULATION BIOLOGY

## S.J. TER BORG

### 1. Introduction

Population biology is concerned with the study of populations. The most simple definition of a population says that it consists of all individuals of a species present in a certain locality at a certain time. This definition is quite different from the one that describes a population as a group of individuals with a common gene pool and barriers to external exchange (a 'Mendelian' population). At the start of a study usually little is known about the genetic relationships of the individuals. This is particularly so in plants, because of their variable reproductive systems. For instance, it has long been known that the species of *Taraxacum* are agamospermic, i.e. that seeds are formed without fusion of gametes; as a result lines of genetically identical individuals develop. Thus, according to the definition of a Mendelian population the number of *Taraxacum* populations in an area equals the number of lines. Since these cannot be easily distinguished a study necessarily includes several populations. And in fact matters are fairly simple in this genus since it is almost an all-or-none case. Problems increase when isolation is only partial, or when external isolating mechanisms are important, e.g. some types of pollinators (cf. p. 19). Therefore, the only possible approach at the start of a study is to use the simple, practical definition given first. It depends on the problems to be solved whether more has to be known about the genetic interactions within a population.

Population biology is concerned with the study of the processes which determine the size and composition of populations, and cause their establishment, growth and decline. It comprises population dynamics which is primarily concerned with numbers and their variation, and population genetics which studies the genetic processes. These fields have a separate tradition but the necessary integration of the two is increasing (e.g. Jain & Solbrig 1976, Freysen & Woldendorp 1978). Demography is mainly concerned with descriptive studies; it forms a part of population dynamics. The latter has a wider scope since it includes also analysis of causes.

Animal population biology began to build a tradition in the first decennia of this century, whereas plant population biology in its modern meaning developed later, from the forties onwards. In fact the basis for both had already been laid by Darwin. In a paper called 'A Darwinian approach to plant ecology', Harper (1967) illustrated this point with various quotations.

Botanists usually excuse the late development in this field by saying that plants are so difficult to count when compared with animals. However, difficulties with respect to counting are being solved (cf. p. 21). And, as stated by Antonovics (1976), plants have properties which allow easier research, as they are sessile unlike animals. Hence, individuals can be followed easily and mark-recapture techniques are superfluous except for seeds. Besides, the plant's environment can be fairly well defined. Moreover, propagation is relatively easy, certainly in species that can be cloned; and finally transplants allow comparisons under field conditions. There are serious limitations also since much of the essential processes in plant life occur underground. Part of the plant, the root system, is in the soil, but also complete individuals in the form of seeds, bulbs etc., can be underground, either in a dormant or an active state. Another complicating factor is the variability of the reproductive system.

Plant population biology developed later than vegetation science and later than autecology and ecophysiology but progress in these fields has evoked its growth. On the one hand it became clear that processes in a vegetation can be better understood if more is known about its components. On the other hand, the availability of detailed autecological and ecophysiological knowledge, and the increased development of genetics allowed and stimulated an increased attention to the role of variation within populations. Whereas autecology and ecophysiology usually treat variation in populations as a nuisance that has to be overcome by statistics, population biology considers variation as an important characteristic with its own adaptive significance (e.g. Allard et al. 1978).

Population biology requires a good knowledge of the autecology of the species studied, i.e. of its life cycle and the factors affecting it. Therefore, both fields often overlap and the border between them is difficult to draw. In order to define the scope of this paper a delineation had to be made, however, and it was decided to exclude studies concerned with ecotypic differentiation within species. This subject is treated by van Dobben (this volume).

Population biology can be an aim in itself. Populations, however, form part of two types of entities of higher order: they are part of a species and they are part of a vegetation. Therefore, the study of populations may be a basic step to the understanding of speciation, selection, niche differentiation etc.; it may also give an insight into the interrelations between the components of the vegetation, the community and the ecosystem. Moreover, there are applications in management of natural populations of plants and animals, e.g. in pest control.

I have not tried to review the whole field of plant population biology in this paper. Recent books are available on this subject (Harper 1977, Solbrig 1978). Instead I preferred to discuss some topics in more detail, with an emphasis on those to which members of the Royal Botanical Society of the Netherlands have made contributions. The first topic to be treated is dispersal as the early contributions to the meetings concern this subject (cf. p. 291 ff.). Besides, some recent developments evoked renewed attention. In the chapter on population dynamics Dutch studies in this field will be mentioned and reviewed within a framework of the international literature. The biennials have been chosen to be treated more extensively because they are a limited group of plants which has been studied in relatively great detail. One set of papers, on *Dipsacus sylvestris*, illustrates the advantages of a combined experimental and theoretical approach. The sections dealing with the manipulation of populations and with the relationship between population biology and vegetation science have been included to indicate ways to integrate both these fields of study, an aim of Dutch plant ecologists. Finally, some possibilities will be given of applying the results of population biology in weed control and management of nature reserves. By making this choice I had to omit almost completely two important subjects: studies on seeds and germination, and work on genetic variation within populations. As a consequence of excluding the latter, selection and evolution will hardly be mentioned. These subjects easily might have filled another two papers.

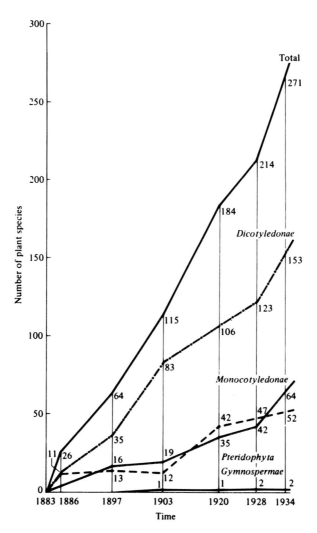

Fig. 1. Numbers of plant species recorded on the islands of the Krakatao group after a volcanic eruption in 1883 (After MacArthur and Wilson 1967, data from Docters van Leeuwen 1936).

15

## 2. Dispersal: time, chance and numbers

In order to establish a population somewhere propagules first must reach the area. Several Dutch botanists presented papers on transport and dispersal of seeds at various meetings organized by the Commission for the Study of Vegetation of the Royal Botanical Society (p. 291 ff.). Still earlier some studied the recolonization of the Krakatao islands in the former Dutch East Indies after a volcanic eruption that erased previous life (Backer 1929, Docters van Leeuwen 1936). Some of their data were compiled by MacArthur & Wilson (1967) in a graph (Fig. 1) which clearly

Fig. 2. Embankments in the Netherlands in the 20th century, their year of enclosure, and the main papers on their natural vegetation.

| | | | |
|---|---|---|---|
| 1. | Kroonpolders, Vlieland | 1910 – 1925 | de Vries (1961) |
| 2. | Proefpolder Andijk | 1927 | Kloos & de Leeuw (1928, 1930) |
| 3. | Wieringermeer | 1930 | Feekes (1936) |
| 4. | North-eastern Polder | 1942 | Feekes & Bakker (1954), |
| | | | Bakker & van der Zweep (1957) |
| 5. | Braakman | 1952 | Bakker & Boer (1952) |
| 6. | Eastern Flevoland | 1957 | Bakker (1957) |
| 7. | Veerse Meer | 1961 | Beeftink et al. (1971) |
| 8. | Southern Flevoland | 1968 | van der Toorn et al. (1969b) |
| 9. | Lauwerszeepolder | 1969 | Joenje (1978a, b) |

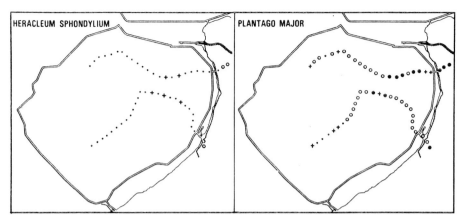

Fig. 3. The distribution of *Heracleum sphondylium* and *Plantago major* along road verges in Eastern Flevoland in 1967, i.e. 10 years after the enclosure. • absent, + rare, ○ occasional, ● frequent (From van der Toorn et al. 1969a).

showed that for recolonization by plants time is an important factor: after 50 years the number of species still increased continuously.

About 1930 when the Commission started its activities, engineers were enclosing the Zuiderzee and dyking and draining the first polders there. Huge stretches of land bare of any vegetation drained dry. The history of its early development was studied by Feekes (1936). As the engineers continued their work in later years botanists followed so that several accounts could be published on the vegetation in the Zuiderzee polders as well as in other embankments (Fig. 2). All of them have their own peculiarities: sandy or silty soils, saline or fresh water, enclosures in spring or autumn, rapid or gradual drying, etc. The records include wide ranges of data, and only few, particularly some on transport, will be mentioned here.

Feekes (1936) and Feekes & Bakker (1954) found that the dispersal of seeds into the polders takes place in three phases. The first set is carried with water and is present before the area is dry; the second set arrives by wind from the nearby 'old land'; and the third one is transported in various ways by man, e.g. when the new area is brought into cultivation. The latter group includes the species of road verges, whose migration was studied by van der Toorn et al. (1969a, Nip-van der Voort 1977; Fig. 3). Comparion of species distribution along a main road and one of its cross-roads revealed the effect of traffic (Table 1).

*Table 1.* Number of immigrant species in stretches of 100 m road verge along a main road and a cross-road, from the point where these part onwards (From v.d. Toorn et al. 1969b).

| | Distance to old land (km) | | | | | | | |
|---|---|---|---|---|---|---|---|---|
| | 5.5 | 5.6 | 5.7 | 5.8 | 5.9 | 6.0 | 6.1 | 6.2 |
| Main road | 13 | 19 | 17 | 20 | 20 | 14 | 16 | – |
| Cross-road | 12 | 11 | 4 | 10 | 5 | 7 | 8 | 6 |

17

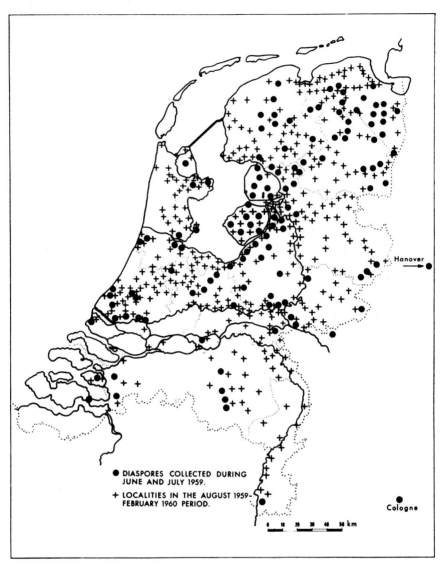

Fig. 4. The dissemination and establishment of *Senecio congestus* in the Netherlands after a dry summer, when the species had covered 10,000 ha in Eastern Flevoland (From Bakker 1960b).

Detailed data on transport were also collected by Bakker (1960a, b). Two sets will be discussed here. The first one concerns that on *Senecio congestus** and indicates the importance of chance. The species covered a 10,000 ha area in the Eastern Flevoland polder during the dry summer of 1959. During the period of seed

* Nomenclature follows that of the papers cited.

*Table 2.* Dissemination of diaspores (after Bakker 1960a).

| | Distance (km) between the place of collecting and the parent population | Number of diaspores collected | % plumes with achenes |
|---|---|---|---|
| *Cirsium arvense* | 0.01 | 2841 | 9.9 |
| | 1.00 | 3017 | 0.2 |
| | 2.00 | 2211 | 0.0 |
| *Tussilago farfara* | 0.01 | 462 | 100 |
| | 0.10 | 240 | 100 |
| | 2.00 | 30 | 100 |
| | 4.00 | 14 | 100 |

ripening convection currents and fairly strong winds carried the diaspores over long distances. It was even possible to follow the seeds by car for over 90 km as the pappus is large and groups of seeds are easily visible. A search of the country in the following autumn and winter showed that the species, which was rather rare previously, could be found in many places where growth conditions were satisfactory (Fig. 4; Bakker 1960b). In this species the dispersal agent was effective. This does not hold in all cases as was found when migration of *Tussilago farfara* and *Cirsium arvense* were studied closely. Though the pappus of both species was found at several kilometers' distance from the parent population, in *Cirsium arvense* the achenes had been lost by then. Further study learnt that this was probably due to the fact that in this species the achenes stick rather strongly to the receptacle while the plumes break off easily. In *Tussilago*, on the other hand, the achenes easily loosen from the receptacle while the plumes are strongly attached. So, while both species have a pappus on their seeds, its effectiveness differs widely (Table 2; Bakker 1960a).

When dispersal is discussed emphasis is usually on the capacities of species to cover distances. However, as also indicated by the *Tussilago* data, the large majority of seeds land within the neighbourhood of the parent plants. This phenomenon was discussed at length by Levin & Kerster (1974) in the context of gene flow by means of seeds or pollen. If dispersal distances of seeds or transport distances of pollen (estimated from flight distances of pollinating insects) are measured, strongly skewed curves are usually found. We observed this phenomenon in the course of a study on hybridization and isolation in two *Rhinanthus* species that are pollinated by bumblebees (Kwak 1977, 1978). In an area with an average distance of 8 cm between the inflorescences 90% of the pollinators' flights covered less than 30 cm, 9.3% between 30 and 100 cm, and only 0.4% flights of over 100 cm were observed (Fig. 5). With increasing plant distances (5 plants/m$^2$) only 5% of the flights covered more than 100 cm. The majority of the pollinators' movements apparently are strongly localized. The data on seed dispersal concern only a limited number of observations but yet it can be concluded that seed transport over more than 50 cm

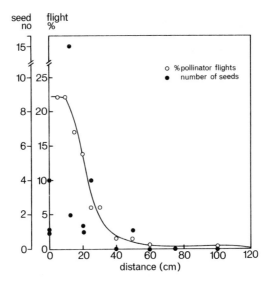

Fig. 5. Pollinator flight distances (o) and seed dispersal distance (•) in *Rhinanthus*. The % flight distance is based on a total of 760 observed flights. The data on seed dispersal concern the number of seeds recovered from 9 cm petri dishes after a period of 3-4 days (unpublished data from Floris Bennema & Harry Hummel).

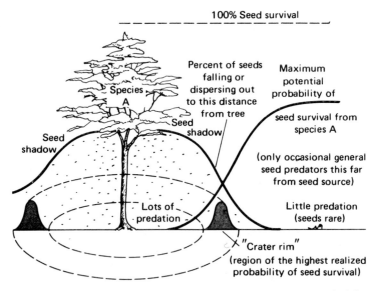

Fig. 6. Seed dispersal and post-dispersal predation on these seeds in a tropical forest. Tree species A has maximum realized survival where the decreasing 'seed shadow' curve (indicating seed quantity dispersed per unit distance) intersects the increasing survivorship curve well away from the concentrated predation around the parental source of seeds (From Emmel 1976, data from Janzen 1970).

20

is negligible. These data concern dispersal in the vegetation before mowing. Mowing and haymaking result in dispersal distances of several meters but their effectiveness is highly dependent on the number of seeds still present in the capsules by that time.

Around their parents seedlings find a habitat to which they are adapted, but their high numbers can give rise to high seedling densities so that competition is strong and many succumb. Another consequence may be the attraction of predators and parasites that kill many of them. According to Janzen (1970) this mechanism gives rise to the scattered distribution of some species, e.g. tropical forest trees (Fig. 6).

So we see in many cases high numbers of seeds being produced, the majority of which is not transported and doomed to death, or if transported, lands in unsuitable habitats. For the few that find a safe site this is a matter of chance to a high degree. This minority, however, is of vital importance for the species and leads to its persistence. The excess, seemingly an overproduction, appears to be necessary to guarantee the minimum number of survivors (see also p. 28).

Baas Becking (1936) formulated a statement, based on the work of the Dutch microbiologist Beyerinck, that is often cited in connection with matters of dispersal: 'Alles is overal, maar het milieu selecteert' (Everything is everywhere, but environment selects). This may hold true for organisms which are dispersed by airborne spores, like bacteria for which one world handbook for identification suffices (Buchanan & Gibbons 1974). However, the Krakatao data, data on the history of the flora and vegetation (e.g. van Zeist 1970), and the observations discussed above, make clear that arrival of many plant species is a chance process that takes much time. Therefore, one should be careful to apply this dictum to all groups of organisms too easily.

## 3. Fluctuations in existing populations

### 3.1 *Introduction*

A problem at the basis of plant population studies is the plasticity of plants and the delimitation of individuals. Counting of numbers is relatively easy in groups with discrete individuals like most therophytes and trees, but apart from such data some quantitative measure is required to estimate size, be it biomass, seed production, or another appropriate measure. In clone forming plants with tillers, suckers or other means of vegetative growth it is often difficult to establish the numbers of individuals so that measuring their size and other properties is even more difficult.

Harper (1977, 1978) argues that for plant population studies a basic morphological unit should be — and always can be! — chosen, that is no more variable than animals or their limbs. The size of a plant is determined by the number of these 'modular units'; plant size is so variable because this number can vary widely. Examples of modular units are a leaf with its axillary bud, a tiller (in grasses), a ramet (in plants like *Carex arenaria*), or a short branch developed in one season (in

*Table 3.* 'Age groups' in coenopopulations of perennial polycarpic herbaceous plants in natural coenoses (after Rabotnov 1969, 1978).

1. *The period of dormancy.*
   a. viable seeds in the soil or on its surface
   b. plants in the dormant state
2. *The virginal period*, from germination to flowering and fructification.
   a. seedling
   b. juvenile
   c. immature (intermediate between 2b and 2d)
   d. mature virginal (individuals that have not yet flowered, but have the appearance of mature plants)
3. *The generative period*, covering reproduction by seeds.
   a. initial phase of increasing vegetative and generative vigour
      – with generative shoots
      – without generative shoots
   b. maximal vigour representing the period of life culmination
      – with generative shoots
      – without generative shoots
   c. the onset of senescene with decreasing vegetative and generative vigour
      – with generative shoots
      – without generative shoots
4. *The senile period*, when due to senescence plants lose the ability to reproduce by seeds.

trees). It may be argued that population studies using single leaves as basic units go into too much detail. However, they allow one to establish a link with growth analysis practices as well as with plant morphology (cf. Hallé & Oldeman 1970, Oldeman 1977, Harper 1979).

Another unit which can be distinguished in plant populations is the 'genet'. It indicates the sum of modular units which are derived from one zygote (Kays & Harper 1974). Sometimes it concerns an easily recognizable individual, like in many monocarpic plants and trees, in other species it may include a number of ramets distributed over a large area.

A second point which complicates the study of plant populations is the habit of plants to grow in patterns. A regular or random distribution can rarely be assumed, hence mathematical models can become extremely complicated. Simulation studies, even if based on simplified systems, are complex and often result in models that are either a caricature or far from reality (Schaffer & Leigh 1976).

For the construction of life tables and their analysis with the methods developed in animal and human demography (e.g. Varley et al. 1973) the age of individuals has to be known. Again this usually presents no problems in therophytes and trees but in longer-lived perennial herbaceous plants direct age determinations are only possible in a limited number of cases. In some species it was found feasible to distinguish growth rings in subterranean parts, e.g. in *Spergularia media* (Sterk 1968, 1969) and *Chamaenerion angustifolium* (van Andel 1974, 1975); a number of other examples, mainly from Russian authors, is mentioned by Harper (1977). Harberd (1961) and Rozema (1978a, b) succeeded in establishing plant ages by measuring the yearly growth of clones. It was found that genets can become very old (ages of the order of several centuries were measured), while individual ramets have shorter lives (of the order of years).

While in the western world a tradition developed of measuring and counting plants in relation to size and age, in Russia a separate school developed where populations are characterized by the numbers of plants belonging to certain 'life states' or 'age groups', i.e. sets of individuals with the same morphological or developmental status (Rabotnov 1969, 1978). Table 3 presents the various age groups distinguished in herbaceous perennials. Use of both, age or status, was compared by Werner; she found the latter to be more satisfactory (cf. p. 35).

## 3.2 Descriptive studies

For a study of a population a basic knowledge of the autecology of the species involved is required. Apart from that questions should be asked about:
— population size (number of individuals, genets, ramets, or other modular units),
— the composition of the population, i.e. the distribution of individuals over size and age classes,
— the proportion of individuals belonging to different phases of the life cycle (c.q. age groups),
— the number of individuals that reproduce, and the number of progeny per parent (be it seeds, seedlings or tillers),
— intra-population variation with respect to morphological and physiological characteristics of the type of reproductive system,
— the variation of all these characteristics in space and time,
— and their significance for the phenomena observed in the population.

Some population properties can be determined by direct measurements, e.g. the total number of plants or modular units and their distribution over size classes. Basic studies in this field have been done by Japanese workers. Koyama & Kira (1956) studied the intra-population variation of various characters (weight, height, etc.) relative to plant density; Yoda et al. (1963) developed the 3/2 power law, relating log mean density with log mean plant weight. Joenje's data on almost pure populations of *Salicornia* spp. and *Spergularia marina* seem to be in conflict with this law (Joenje 1978a, Fig. 7). However, these species, particularly *Salicornia*, have a growth form such that even at the highest densities and at maximal production soil cover never exceeds 40%. For that reason the data might support a hypothesis saying that competition for light is the major factor at the basis of this law, rather than root competition.

These and other studies in this field (cf. Harper 1977) usually consider monocultures, either natural or experimental ones. Russian authors studied the structure of natural populations in mixtures. They determined the numbers of individuals in the age groups mentioned in Table 3, and constructed spectra. Based on these data they distinguished three main types of 'coenopopulations' (i.e. populations including all phases of the life cycle, plants as well as seeds) in the course of a succession process (Rabotnov 1969):

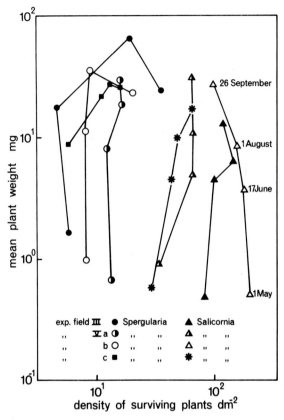

Fig. 7. Mean plant weight against mean density of surviving plants. Data on *Salicornia* spp. and *Spergularia marina* in the Lauwerszeepolder (From Joenje 1978a).

1. Populations of the invasive type. Species which are in the process of ecesis and do not yet complete the life cycle;

2. Populations of the normal type. Species which complete the life cycle from seedlings up to the formation of viable seeds;

3. Populations of the regressive type. Species which in the present coenose have lost their ability to reproduce themselves by seeds.

The properties described thus far can be determined by direct measurements but information about several other characteristics, e.g. turn-over processes, requires repeated mapping of individual plants. The classic studies in this direction were performed by Tamm (1956, 1972a, b), who charted perennial herbs in open woodlands and meadows in southern Sweden. One of his graphs is given as an example (Fig. 8). It demonstrates survival, flowering, mortality, vegetative propagation and seedling establishment of *Anemone hepatica* in a sq.m woodland plot over a period of 14 years. From Tamm's data conclusions can be drawn which have been confirmed in several studies since:

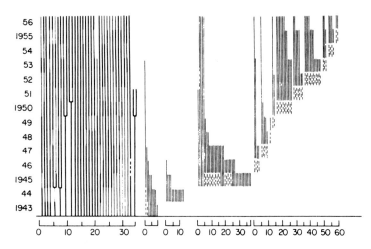

Fig. 8. The behaviour of *Anemone hepatica* in a 1 sq. m woodland plot. Each vertical line represents one individual, ramified if the plant ramified itself. Flowering is indicated with a heavy line, broken lines indicate that the plant was not observed that year (From Tamm 1956).

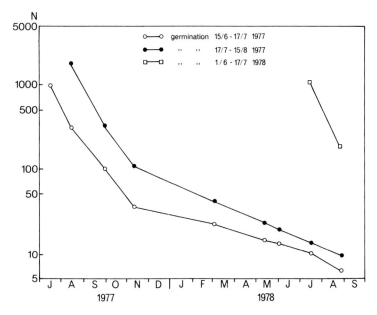

Fig. 9. Survivorship curves of three cohorts of seedlings of *Taraxacum* in a grassland in the Krimpenerwaard (Data from P.J. van Loenhoud. They form part of an extensive study on the systematics and ecology of *Taraxacum* spp.; cf. Sterk 1978a, b).

25

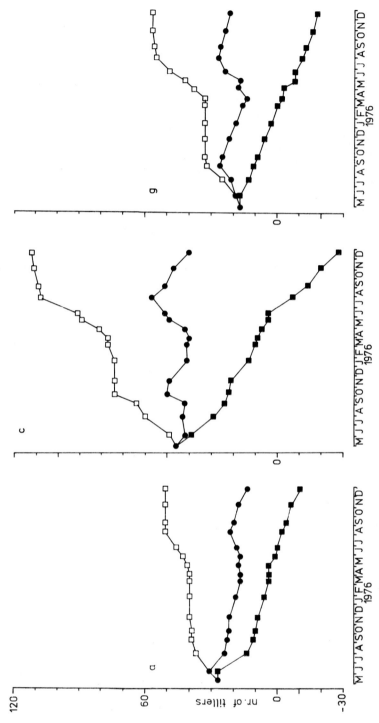

Fig. 10. The turn-over of tillers of *Ammophila arenaria* in three plots along a transect in a dune system on Anglesey (Wales). □ births and ■ deaths (both cumulative) and ● total tiller number; (a) first dune ridge; (c) second dune ridge, (g) old grey dune (From Huiskes 1977).

1. Plant species that flower within one or two years after germination when they grow at wide spacing, may require several years to reach this stage under competitive conditions.

2. Individual plants do not flower every year.

3. As long as environmental conditions do not change the decline of established plants is surprisingly linear when numbers are plotted on a semi-log basis (which means that during periods of equal length a fixed percentage of the individuals present disappears). This allows calculating the 'half-life' and hence predictions on longevity. While Tamm had mapped his populations only once a year, later workers did so at more frequent intervals. The following points emerged from these studies and seem to hold nearly always for herbaceous perennials:

1. Linear decline occurs in established plants but mortality is often higher in the early phase of the life cycle (Fig. 9). In other words, survivorship curves of plants usually conform to the Deevey type II or III curve, i.e. they are concave or linear (Deevey 1947). This appears to hold usually for annuals too, but an extreme case of a Deevey type I curve (convex, i.e. mortality concentrated at the end of the life cycle) was observed in *Vulpia*, a dune annual (Watkinson & Harper 1978).

2. In the course of the year there may be a period of faster decline of numbers. This is sometimes the season with adverse conditions (cold winter or dry summer) but often, when competitive environments are concerned, it is the time when growth is fastest, i.e. usually spring and early summer.

3. Even when population size is relatively constant, high numbers of births and deaths may occur. This population turn-over varies between species (*Ranunculus* spp., Sarukhán 1974) as well as within species when different sites are compared (Fig. 10).

## 3.3 Analysis of causes

When descriptive studies have provided information on a population, concerning its size and composition and the variation of these with time, the question arises as to which factors determine these properties. In the following attention will be focussed mainly on questions with respect to population size.

In principle a population is able to grow exponentially, in practice its numbers are kept in check. What factors govern its size? Are there internal factors inherent to the species or external ones? If the latter applies, are they biotic or abiotic? In plants, are these factors primarily other plants, or on the contrary, animals and microorganisms? It is known that all of these factors may affect plant distribution and that they can influence all phases of the life cycle but it is often difficult to prove that they are acting. An example forms the discussion on the effectiveness of allelopathy (Whittaker 1970). Another illustration is found in a paper by Ernst (1978), who disputes the view that differences between physiological and ecological optimum curves are mainly due to interspecific competition. He mentions several causes which may be involved, e.g. fluctuation of factors, interactions, and intraspecific variation. The biotic factors should be added to this list, as is illustrated by

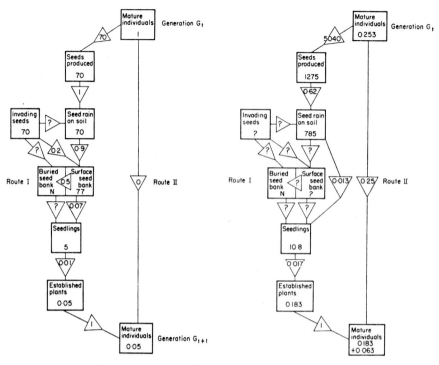

Fig. 11. Life tables for *Poa annua* in an open habitat (left, data from Mortimer 1974) and *Senecio jacobaea* (right, data from van der Meijden 1971). In squares: number of individuals, in triangles: seed production and survival probabilities. Route I: generative reproduction; route II vegetative survival (lacking in ephemerals and annuals) (From Sagar & Mortimer 1976).

Huffaker's work on *Hypericum perforatum* (Huffaker 1964), and by a study of the effects of root- and stemboring larvae on *Phragmites australis* in either wet or dry habitats (van der Toorn & Mook 1977).

However, identification of the environmental factors is one thing, quantifying and determining their relative importance under field conditions is another. A first step may be made by indicating in which phase of the life cycle major reductions occur. Sagar & Mortimer (1976) did so by constructing life table diagrams for a number of species from which they inferred where population control occurs (Fig. 11). They used data from van der Meijden (1971), who had presented them in tabular form (Table 4, included to give an example of this way of presenting data on life history.)

It is important to note that a factor causing a high loss is not necessarily the key factor regulating population size. The latter may be controlled by a certain factor, and it may be irrelevant to which factor the remaining losses are due (e.g. Williamson 1972). This point can be illustrated with van der Meijden's data. It can be gathered from Fig. 11 that from a seed rain of 785 seeds 1.3% produced a seedling. One might ask why the other 775 seeds failed to do so. In a later paper van der

*Table 4.* A life table for the seeds produced by 112 parent plants of *Senecio jacobaea* in a dune area near Leyden (From van der Meijden 1971).

|  |  | Loss | Loss per stage (%) |
| --- | --- | --- | --- |
| Number of capitula 1969 | 23,437 |  |  |
| Incompletely developed |  | 4,858 | 20.7 |
| Damaged |  | 14 | 0.1 |
| Predated by *Pegohylemyia* |  | 492 | 2.1 |
| Total loss |  | 5,364 | 22.9 |
| Number of capitula left | 18,073 |  |  |
| (Mean number of seeds per capitulum: 69.7) |  |  |  |
| Total seed production | 1,260,000 |  |  |
| Dispersed by wind out of area |  | 475,000 | 37.7 |
| Number of seeds fallen in area | 785,000 |  |  |
| Sterile |  | 93,000 | 11.8 |
| Fertile | 692,000 |  |  |
| Fertile seeds not germinated |  | 681,200 | 98.4 |
| Fertile seeds germinated |  |  |  |
| in 1969 | 2,890 |  |  |
| in 1970 (up till August): | 7,940 |  |  |
| Total germinated | 10,800 = 0.9% of total seed production |  |  |

Meijden & van der Waals-Kooi (1979) showed that germination and establishment is limited to those places where some disturbance occurs so that seeds become covered with a thin layer of sand; a high degree of disturbance, causing a sand cover of over 4 mm, reduces germination. The strict requirements are only met in a few 'safe sites' and, therefore, the number of these determines the number of seedlings. As long as the parent plants produce enough seeds to fill all safe sites it is irrelevant which factors cause the loss of the other seeds, be it wind, insects, fungi, etc. The only group that still has an important function are those seeds that become dormant, e.g. by sand cover, and can give rise to later recruitments.

Another point to be noted is that the interpretation of fluctuations may require information over a long period. This is illustrated by the discussions of Sterk (1976a, b) and de Wilde (1976) on the interpretation of a decline in the number of flowering of *Anacamptis pyramidalis* in a population in a dune area near Wijk aan Zee (Fig. 12). De Wilde attributed this to recent extensive withdrawal of water, but Sterk, basing his views on available data covering a period of c. 35 years, pointed out that in this population the decline falls within the normal pattern of fluctuations (Fig. 13). He found similar fluctuations in the *Anthyllis vulneraria* population nearby, most of which could be attributed to climatic factors.

Whatever the actual causes of the fluctuations in the *Anacamptis* population may have been, one thing became clear: long ranges of data can be indispensable. In some cases information apparently exists, as is illustrated by Willems' data (1978) on numbers and plant size in populations of *Orchis mascula* over a 10 year period. A search for similar data might be useful. Yet, additional studies are required including observations on non-flowering individuals. However, the present-day poli-

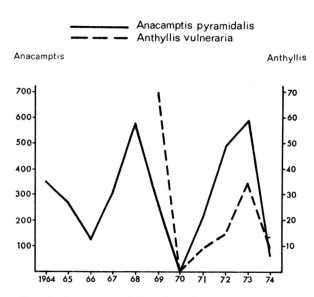

Fig. 12. Fluctuations in the number of flowering plants in North Holland dune areas. For *Anacamptis* the total number of flowering plants in the population, for *Anthyllis* the average number of flowering plants per sq. m is given (From Sterk 1976a).

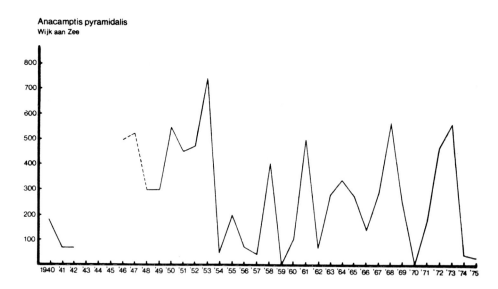

Fig. 13. Fluctuations in the number of flowering plants in a population of *Anacamptis pyramidalis*, 1940-1975 (From Sterk 1976b).

30

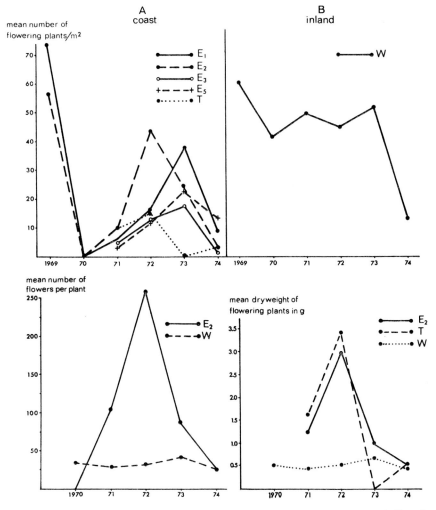

Fig. 14. Fluctuations in some parameters of coastal and inland populations of *Anthyllis vulneraria*. E and T: coastal populations; W: inland population (From Sterk 1975).

cy with short term grants does not give much hope that the gap will be filled soon, if ever.

In his study of *Anthyllis*, Sterk compared a number of dune populations with one in a limestone grassland in South Limburg which was found to be more stable (Fig. 14). According to Sterk this might be due to the higher habitat stability in Limburg compared to that in the dune area (Sterk 1975). A rather abstract environmental characteristic, habitat instability, is indicated here as a cause of population fluctuations. Instability in itself is of course not the real factor, but humidity, temperature, parasitic attack, rabbit grazing, and others, as mentioned by Sterk.

However, the degree of variation may affect plant characters, e.g. reproductive effort, as becomes clear from the discussion about r- and K-species (Solbrig & Simpson 1974, Moore 1976).

Although by these and similar case studies much becomes known about environmental factors and their effects (e.g. *Salicornia*, Beeftink 1978) we are far from being able to describe, quantify and predict all processes in a natural population, be it for only one species. Perhaps this is too high a goal. Meanwhile two quite different approaches are possible: studies of simplified systems, or the design of theoretical models. Both can point to the consequences of certain assumptions or, to quote Lewontin (1967), they can indicate 'the limits of the possibilities'. Examples of the first approach can be found in the work on agricultural systems and that on competition (cf. de Wit 1960, van den Bergh, this volume). As for the second: theoretical studies on various subjects have been published in recent years. Apart from those mentioned elsewhere in this paper there are, e.g., Cohen (1966, 1968) about the relation of the optimal dormancy ratio to habitat instability, Bullock (1976) on population growth and seed dispersal, and Fresco (1979) on the analysis of species amplitudes.

Some possibilities of combining field work and theoretical studies will be illustrated by a discussion on biennial plant species in the next section.

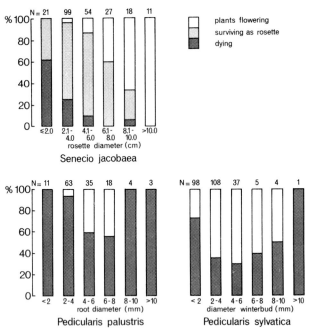

Fig. 15. Relation of plant size to survival and flowering in the following year.
*Senecio jacobaea*: rosette diameter (Data from van der Meijden & van der Waals-Kooi 1979).
*Pedicularis palustris*: diameter of thickened root just below winterbud (Unpublished data Johan Brouwer).
*Pedicularis sylvatica*: diameter of winterbud (Unpublished data Henk Everts).

## 3.4 Biennials

Biennials are supposed to complete their life cycle in two years but in field conditions this seems to be more the exception than the rule. This may be connected to the consideration that a species behaving like a biennial must meet rather strict conditions, as will be shown below, and hence built-in securities increasing the chance of seed production are required. In some species variation between an annual and a biennial life cycle was observed (e.g. *Cirsium vulgare*, Bakker, pers. comm.), in others biennial and monocarpic perennial individuals were found to occur (see below). Sometimes genotypic variation forms the basis of the variation (e.g. *Linum catharticum*, Zijlstra 1974), more often environmental conditions seem to induce the variation. Many biennial species which behave as monocarpic perennials do so because flowering is delayed until the rosettes have reached a certain minimum size (e.g. *Cirsium palustre*, Linkola 1935; *Dipsacus sylvestris*, Werner 1975a; *Senecio jacobaea*, van der Meijden & van der Waals-Kooi 1979; Fig. 15). Delayed flowering of the latter species may also be a response to defoliation during the period of flowering in the previous year by caterpillars of the Cinnabar moth (*Tyria jacobaeae*) (van der Meijden et al. l.c.).

*Table 5.* Colonization success of *Dipsacus sylvestris* in relation to the vegetational composition of 8 experimental fields (From Werner 1977).

| Grasses | Herbaceous dicots | Shading by shrubs | |
|---|---|---|---|
| | | Nil | Heavy |
| Very dense | Nil | Field K: Germination very low (20% ± 6 SD per quadrat); 96% mortality of seedlings in 1st year. Population died out without reproducing. | |
| Moderate levels | Moderate levels | Fields A, B, L, M: Mean germination levels from 24% to 57%; lower values, in fields with more grass. Survival of seedlings and growth rates of rosettes high (mean diameter in August of year 1: A, 12.4 cm; B, 9.2 cm; L, 15.4 cm; M, 13.1 cm). Flowering plants produced in the 2nd (Fields A, L, M) or 3rd (Field B) year. | Field C: Germination low (24% ± 7 SD per quadrat). Growth rate of rosettes moderately high (mean diameter in August of year 1:7.1 cm) Flowering plants produced by the 3rd year. |
| Low levels | High levels | Field D: Germination moderate level (43% ± 5 SD per quadrat). High survival of 1st year seedlings (15%). Growth of rosettes very slow (mean diameter in August of year 1:2.7 cm). Flowering plants produced by the 4th year. | Field J: Germination high (58% ± 4 SD) but 97% of the seedlings died within a year. Population died out without reproducing. |

*Table 6.* Matrix of transition coefficients; classification according to stage. The numbers are the means of the values of all but one field calculated from the data presented by Werner and Caswell 1977.

| stage: | 1 | 2 | 3 | 4 | 5 | 6 | 7 |
|---|---|---|---|---|---|---|---|
| 1. seeds | – | – | – | – | – | – | 678 |
| 2. dead or dormant seeds, year 1 | .617 | – | – | – | – | – | – |
| 3. dead or dormant seeds, year 2 | – | .954 | – | – | – | – | – |
| 4. small rosettes (< 2.5 cm) | .015 | .027 | .010 | .023 | – | – | – |
| 5. medium rosettes (2.5 – 18.9 cm) | .053 | .007 | .000 | .169 | .256 | – | – |
| 6. large rosettes (≥ 19.0 cm) | .002 | .001 | .000 | .036 | .284 | .121 | -- |
| 7. flowering plants | – | – | – | .000 | .013 | .658 | – |

In *Pedicularis palustris* and *P. sylvatica*, two hemiparasitic species that behave as strict biennials in the populations we studied, we expected to find a similar relation: increased percentage of surviving and flowering plants with increased size of the overwintering organs (winterbud and thickened roots). Instead of this an optimum curve was observed (Fig. 15). This conclusion is based on a limited number of observations, but it seems to be reliable because the phenomenon was observed in two species. Repeated observations and a study of causal factors are nevertheless required.

*Dipsacus sylvestris* was studied extensively by Werner c.s. Seeds of the species were introduced into plots in eight types of abandoned fields and their fates were followed for 5 years (Werner 1977). Table 5 summarizes the results. Grass litter was found to have a profound effect on seedling establishment (see also Werner 1975b); competition was mainly with other herbaceous dicots, less with grasses; shading by low shrubs (*Rhus typhina*) affected seedling growth. A combination of the two latter effects was fatal: plants did not reach the flowering stage in plot J. A further analysis of the *Dipsacus* data was carried out by applying the Leslie matrix method (Williamson 1972, Harper 1977). It was possible to derive from the field data the average chance a seed has either to stay dormant or to develop into a small, a medium sized or a large rosette. Similarly the rate of transition between rosettes of

*Table 7.* The 'state vectors' entered into the matrix, one set according to age, another according to stage (size) (From Werner & Caswell 1977).

| Age classification | Stage classification |
|---|---|
| (1) seeds | (1) seeds |
| (2) dead or dormant seeds, year 1 | (2) dead or dormant seeds, year 1 |
| (3) dead or dormant seeds, year 2 | (3) dead or dormant seeds, year 2 |
| (4) rosettes, year 1 | (4) small rosettes (< 2.5 cm) |
| (5) rosettes, year 2 | (5) medium rosettes (2.5 – 18.9 cm) |
| (6) rosettes, year 3 | (6) large rosettes (≥ 19.0 cm) |
| (7) rosettes, year 4 | (7) flowering plants |
| (8) flowering plants | |

various sizes and flowering plants could be calculated, etc. (Werner & Caswell 1977; Table 6). From a comparison of matrices based on age and on stage (Table 7) it was learnt that the latter give more satisfactory predictions. This supports the point that age determination is often not essential in plant population studies (cf. p. 23). The data in the matrix, together with information about the composition of the population at the start of the study (in this case 8 x 100 seeds) allow calculation of the population development. For simplicity the averaged transition coefficients were given in Table 6, but Caswell & Werner (1978) used the separate data and hence were able to calculate population growth in the eight separate plots (Fig. 16, left).

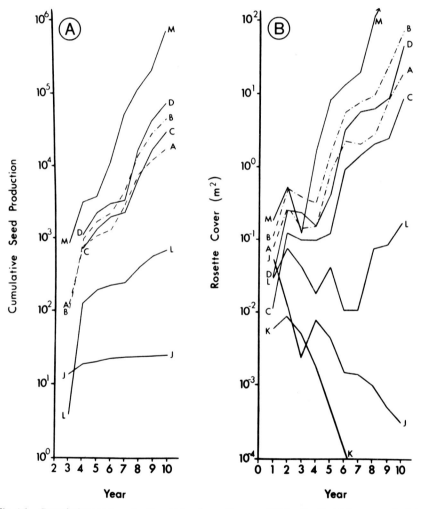

Fig. 16. Cumulative seed production (A) and rosette cover (B) for the first 10 years, calculated from population data from each of eight fields, starting from an initial number of 100 seeds of *Dipsacus sylvestris* (From Caswell & Werner 1978).

*Table 8.* Matrix of the sensitivity of the population growth rate to changes of the transition coefficients. The numbers are the means of the values of all but one of the experimental fields (From Caswell and Werner 1978).

| stage: | 1 | 2 | 3 | 4 | 5 | 6 | 7 |
|---|---|---|---|---|---|---|---|
| 1. Seeds | – | – | – | – | – | – | .00067 |
| 2. Dead or dormant, year 1 | .067 | – | – | – | – | – | – |
| 3. Dead or dormant, year 2 | – | .003 | – | – | – | – | – |
| 4. Small rosette | 1.896 | 0.702 | 0.446 | 0.32 | – | – | – |
| 5. Medium rosette | 7.260 | 2.677 | 1.537 | 0.113 | 0.219 | – | – |
| 6. Large rosette | 43.170 | 19.010 | 18.541 | 1.137 | 2.185 | 0.242 | – |
| 7. Flowering plant | (103.930)[a] | (45.547)[a] | (45.011)[a] | 2.826 | 5.380 | 0.595 | (0.241)[b] |

[a] Hypothetical; changes *Dipsacus sylvestris* to an annual life history
[b] Hypothetical; changes *Dipsacus sylvestris* to a perennial life history.

From other experiments predictions were derived about rosette growth and the moment at which intraspecific density effects were to be expected: this appeared to be within 5-20 years after seed introduction. This information was used to calculate rosette growth over the first 10 years (Fig. 16, right). Though a rigid test of comparison was not given the descriptions suggest a good fit of the calculated data with the field observations. The authors were also able to calculate the degree to which a population benefits from seed germination in the second year instead of the first. They found that this has a profound effect on rosette cover and its drop in the first few years (Fig. 16, right) Exclusion of delayed germination intensified this drop and therefore could be supposed to enlarge the risk of elimination in this period. The latter conclusion could be made since the authors succeeded in predicting the effects of changes of the transition coefficients in the matrix on population growth rate (Table 8). It may be supposed that similarly the effect on other population parameters can be calculated and thus an estimate of the effects of various factors which influence coefficients can be obtained.

Werner & Caswell's data added a quantitative example to the discussion on the relative advantages and disadvantages of certain life histories. There exist several theoretical studies on this subject which were reviewed by Stearns (1976); later additions, especially concerning biennials, came from Hart (1977) and van der Meijden & van der Waals-Kooi (1979). Amongst other things, it was predicted that the biennial strategy is superior to the annual under those conditions where germination and survival of seeds are relatively low but rosettes have a high growth potential. This was found to be the case in *Dipsacus*. On the other hand the high cost of maintaining the vegetative parts in relation to reproductive output was found to favour the biennial or monocarpic perennial life histories of *Dipsacus* over a perennial polycarpic one. Calculations predicted that in only one of the eight habitats studied, where a biennial *Dipsacus* did not persist, the relations between survival and growth were such that either an annual or a perennial population might have survived (population J in Table 5 and Fig. 16). In 7 of the 8 fields an annual or a perennial would fare worse than a biennial.

These studies on biennials, like many others, explain processes observed in the past. Some, like those of Werner c.s., give also quantitative models that allow predictions. However, the question to what degree predictions are fulfilled by comparing them with field observations, has not yet been answered.

## 3.5 *Manipulations with populations*

Another method to study natural populations is to change their size and composition. In studies of animal populations this proved to be a highly successful approach (e.g. Kluyver 1971). In plants plasticity is a complicating factor but nevertheless, the method deserves more attention than it has received thus far. It may offer information on questions with respect to the degree at which the population processes are governed by endogeneous factors, i.e. by interactions between the components of the population. For instance, net population size is usually relatively constant because the numbers of births and deaths equal each other; are births possible because space came available after the death of old individuals or is the death of the latter induced by the arrival of the newcomers?

Several types of manipulations with populations can be distinguished:
— introduction of seeds or transplants in an area where the species does not occur;
— changes in population size by introducing or removing seeds or seedlings in an existing population;
— changes in population structure by reducing or enlarging the proportion of certain population components.

Sowing seeds outside the area where the species occurs can have various effects, as is illustrated by the study of *Dipsacus* (Table 5). More examples are available in the literature but this approach will not be discussed in more detail since it does not usually concern population processes.

Addition of seeds to an existing population did not have much effect in several cases mentioned (Sagar 1960, Holt 1972), whereas in others it influenced population size (Hawthorn & Cavers 1976, Watkinson & Harper 1978). Watkinson & Harper's study concerned *Vulpia fasciculata*, a winter annual of sand dunes. The size of natural populations was changed either by seeding or by weeding. Both influenced the final population size and the total seed yield. The authors found density dependent birth rate and density independent death rate. Their remark that studies of more species will be required before generalizations will be possible is fully confirmed by our data on the summer annual *Rhinanthus serotinus*, a grassland hemiparasite (Leemburg-van der Graaf & ter Borg in prep.). A demographic study of eleven 0.25 m$^2$ meadowplots suggested density dependence of both birth and death rate but the variable carrying capacities of the plots may have interacted here. In the experimental garden a range of densities was established by weeding the seedlings from a turf of mixed grasses where they had established spontaneously in large numbers. It may be concluded from Fig. 17 that mortality was higher at higher densities, while the mean number of fruits per plant, i.e. birth rate, was lower; the total fruit production per unit area was almost independent of density. So, while in *Vulpia* only plasticity effects were observed, plasticity as well as mortality were found to act in *Rhinanthus*.

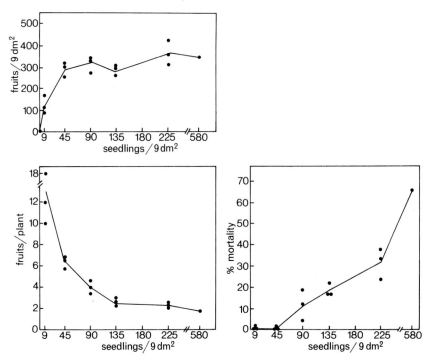

Fig. 17. The effect of density on mortality and fruit production per plant and per unit area in *Rhinanthus serotinus* in 30 x 30 sq. cm plots (unpublished data from Piet Drent).

Changing the structure of populations of annual species may be done by removing the individuals of a certain size class or, in case of a polymorphic species, removing a particular morph. Examples of this type of interferences seem to be lacking as yet. In perennial species more possibilities are available since seeds or seedlings and ramets can be varied separately or in combination. Examples of the latter were given by Putwain, Machin & Harper (1968) who studied *Rumex acetosella* and by Hawthorn & Cavers (1976) who worked with *Plantago rugelii* and *P. major*.

If the experiments are carried out in a successional environment knowledge of the position of the species in the successional series will enhance the interpretation of the data. Horn (1972) distinguished three main types of succession patterns: (1) all species in the area are able to fill openings after chronic or patchy disturbance, (2) obligatory succession, and (3) competitive hierarchy (Fig. 18). Particularly in the case of obligatory succession seed introduction is highly time-dependent, or better, development-dependent. This can be illustrated by data presented by Beeftink et al. (1971) who studied succession on the Middelplaten after closure of the Veerse Meer (Fig. 22). Many seedlings of *Phragmites australis* were present during the first years and to prevent the species from dominating the vegetation the area was weeded. This had to be done for four years only; afterwards no more seedlings were observed despite continued immigration. Establishment had become impossible by then, either because of the competition of the vegetation which had developed meanwhile or because of the changes in the habitat. During these years the soil had dried and germination conditions for reeds, – soil covered with a thin layer of water –, did not occur any longer.

Failures and successes of seed introduction can also be regarded against the background of the population structure which may change with the progress of succession (p. 24). The following results can be predicted for populations of herbaceous perennials of the three types distinguished: (1) The introduction of seeds will be most effective in populations of the invasive type, particularly in the species which require some time to build up a population. (2) Seed production is probably not a limiting factor in populations of the normal type so that introduction of surplus seed cannot be very effective. (3) In populations of the regressive type the effect is highly dependent on the causes of the regression and cannot be easily predicted.

Interpretation of the results of manipulations may be strongly facilitated and improved if the experiments can be combined with a demographic study. If a matrix like the one presented in Table 8 is available, the effects of various manipulations can be predicted and those treatments excluded which are not likely to have much effect. On the other hand, these manipulations are a way to test the predictions of the matrix.

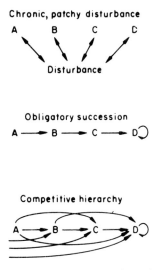

Fig. 18. Patterns of succession. A, B, C and D represent hypothetical species. Where disturbance is chronic and patchy, any species is likely to invade an opening that results from the death of any other species. An obligatory succession results if later species require preparation of their environment by earlier species. In an ideal competitive hierarchy each later species can outcompete earlier ones, but can also invade in their absence (From Horn 1972).

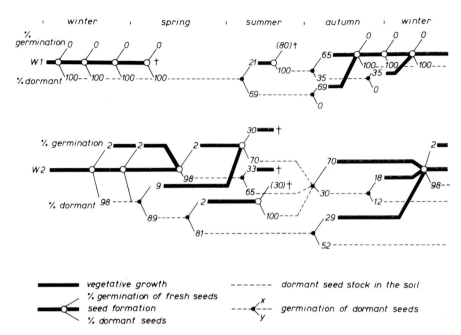

Fig. 19. Theoretical reconstruction of the course of development during one year of two strains of *Stellaria media*, based on field observations and experimental data (From van der Vegte 1978).

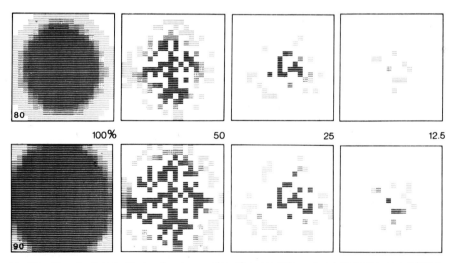

Fig. 20. The simulated effects of various ratios of susceptible and resistent lines in a mixture on spread of a disease; situation after 80 (upper row) and 90 days (lower row). The susceptible line accounts for (from left to right) 100, 50, 25 and 12,5% of the mixture respectively (From Kampmeijer & Zadoks 1978).

40

Thus far no attention has been given to the possibility of a genetic basis for the variations observed in populations. Since Turesson's publications on ecotypic variation in species much has become known about genetically determined differences *between* populations (Heslop-Harrison 1964; for some Dutch contributions see also van Dobben, this volume), but far less is known about genetic variation *within* plant populations and its adaptive significance. The information in this field is increasing however (Antonovics 1976, Allard 1978, Harper 1978).

Plants have properties that make them unsuitable for an automatic application of the rules developed in population genetics, as they are sessile, plastic, and have variable reproductive systems. The papers by Bradshaw (1972) and Levin & Wilson (1978) indicate that this has various unexpected consequences which will not be further treated here. These authors also show that much will have to be done before the processes in plant populations can be really understood.

The Dutch contributions on variation within plant populations will be briefly summarized here. The first to be mentioned comes from Feekes (1936) who during his study of the pioneer vegetation in the Wieringermeerpolder, was surprised by the wide morphological variation he observed in several pioneer species, e.g. *Salicornia herbacea, Spergularia salina* and *Aster tripolium*. The phenomenon was observed again in all later polders. In such areas huge numbers of plants grow at fairly wide spacing, allowing rare genotypes to establish themselves. In fact, a large scale sowing experiment is carried out under natural conditions. The variation has now been observed repeatedly and various biosystematic accounts on these and other pioneer species have been published but as yet the knowledge about the ecological function of the various morphs is limited (for a short review see Joenje 1978a).

A study on a pioneer species of inland habitats, *Stellaria media*, was presented by van der Vegte (1978). This species is usually supposed to be an ephemeral but closer observations showed that in fact different sub-populations occur. Van der Vegte studied two of those which grow mixed and are morphologically similar but have strongly different life histories and germination characteristics. Field observations and germination experiments under controlled conditions enabled him to construct diagrams of both strategies (Fig. 19). The work nicely illustrates the care that is needed in population biological work, as one can easily presume to be studying one population while in fact sub-populations are present (cf. p. 13).

Genetic variation with respect to disease resistance may affect the spread of pathogens, as shown by the models presented by Zadoks & Kampmeyer (1977). They simulated the effect of varying ratios of susceptible and resistant lines in a mixture (Fig. 20) and found that the introduction of a low proportion of resistant plants can relatively strongly reduce the spread of an infestation. The work was started as part of a project on the yellow stripe rust in wheat and several other crop diseases, but it is extended now to include wild species and to test whether the principle works in natural populations. The perennial salt marsh species *Limonium vulgare* has been chosen as an example. It occurs in extensive populations which can be heavily infested with a rust, *Uromyces limonii*, and by mildew, *Erysiphe communis*. In spite of this its populations persist. The hypothesis that this is due to genetic variation with respect to disease resistence is being tested at the moment.

Studies of the isoenzyme variation of *Chamaenerion angustifolium* and *Plantago* spp. have been started recently; results are not yet available. Both studies fall within the framework of multidisciplinary projects with a wider scope. The first is concerned with species succession in woodland clearings (e.g. van Andel & Vera 1977), the other considers grassland perennials (Kuiper et al. 1977).

## 4. Relations of population biology to other fields of research

### 4.1 *Relations between vegetation science and plant population biology*

Vegetation consists of populations belonging to various — rarely one — plant species. Hence, a study of the component populations will broaden our understanding of the patterns and processes in the vegetation. This statement might be attacked by those who adhere to the holistic school in vegetation science but I shall not go into this discussion here.

Nobody will dispute that plant populations are influenced by the surrounding species, either by competitive relations, allelopathy, parasitism, etc. The effect of a single species on the surrounding vegetation and its role in the whole complex is less evident. It can be attempted to study this subject by direct methods as, e.g., used by Putwain & Harper (1970), who studied the position of *Rumex acetosa* and *Rumex acetosella* in a grassland vegetation. They tried to compare the niches of both species and their overlap with other plants and plant groups. However, the necessary removals were done with herbicides, leaving a mass of dead and dying material. Therefore, the results should be treated with some care.

Gančev (1963) followed the changes in the vegetation in grasslands that became infested with the hemiparasites *Rhinanthus rumelicus* and *R. wagneri*. He compared the composition of the vegetation inside and outside the 'weeded' areas and concluded that *Rhinanthus* affects mainly the boreal and tussocky species; the xerophytes of southern origin and species with rootstocks were less affected. The original grasslands changed into steppe-like meadows, the same type of vegetation that develops under grazing, but in the latter case the process takes a far longer period of time.

When using Gančev's method one can never be sure that the observed changes are brought about by the species studied. This uncertainty can be avoided when seeds are introduced in experimental plots, or seedlings removed, and the vegetation of these plots is compared with controls. Werner (1977) applied the first method when she measured dry weight production of the species in the plots where she had introduced *Dipsacus sylvestris*. She found that *Dipsacus* competed with the herbaceous dicots, not with the grasses.

Both methods, sowing and weeding, were applied in studies on the effect of the hemiparasite *Rhinanthus serotinus* on the surrounding vegetation (Rabotnov 1959,

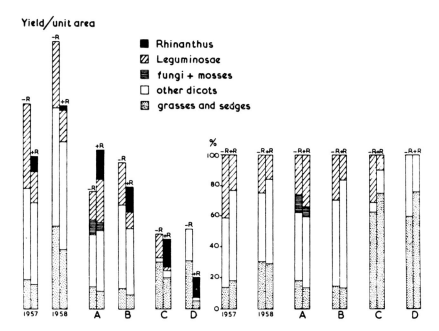

Fig. 21. The effect of *Rhinanthus serotinus* on yield and composition of a mixed vegetation. Left absolute weights, right relative weights of component groups. Sizes of the plots are in the order of 1 – 6 sq. m (From ter Borg & Bastiaans 1973).

Krylova 1963, ter Borg & Bastiaans 1973). Fig. 21 shows that the proportion of the *Leguminosae* generally decreased but the effects on other dicots and grasses varied. The variation may be due to the small scale of these experiments, but it may also have been a result of the different roles of the various plant groups in the vegetation types in which the experiments were carried out.

Allen & Forman (1976) studied the effects of removals of various species in 'old fields'. They found that species composition and community structure (vertical layering and horizontal pattern) influence the effects, whereas there is only partly a correlation with the degree of dominance of the removed species.

Dominance of a species is sometimes evident: the vegetation under an isolated tree or shrub differs from the vegetation in the rest of the area. Also in herbaceous vegetation some species may function as 'trendsetters', e.g. *Molinia caerulea* or *Holcus lanatus*, when growing in dense tussocks. They are the 'violents' of the Russian authors, who distinguish two more groups which may become dominant in an area, viz. the 'patients', which tolerate marginal and extreme conditions (cf. 'stress tolerants', Grime 1977, 1978), and the 'explerents', which respond to decreasing competition and fill openings (cf. r-species). Only a minor proportion of all species belong to these groups, all others are in the fourth group, the 'additors' (Rabotnov 1975). Our understanding of the processes in the vegetation will be broadened primarily by a study of the life histories of the 'violents', as is demon-

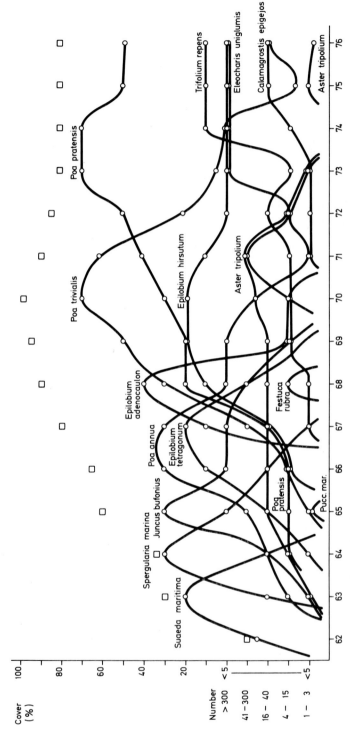

Fig. 22. Succession of species on the Middelplaten, which came dry after closure of the Veerse Meer (From Beeftink et al. 1971 and additional data; provided by the authors).

44

strated clearly by papers on cyclic succession. Watt (1947) found that processes in the low and open turf in the Brecklands are determined by development, growth and decline of the tussocks of *Festuca ovina*, while he (1955) and Stoutjesdijk (1959) showed that the life history of *Calluna vulgaris* is decisive for the vegetation of a dry heath. A study of the other species, the additors, may indicate in which way their life histories are adapted to the current conditions.

The foregoing concerned contributions of population studies to vegetation science, the reverse will be discussed next.

As is illustrated in Section 3.3 (*Anacamptis*) the interpretation of population fluctuations sometimes requires the availability of data over long periods of time, longer than are usually provided by demographic studies. Therefore, repeated observations of the vegetation in permanent quadrats can be very valuable, as for instance those presented by Beeftink et al. (1972, 1978, Fig. 22) and Londo (1971, 1978), though they give only net results and do not offer any information about, e.g., turn-over processes in the vegetation. However, they provide additional information when demographic data, which offer more details but cover short periods, have to be interpretated. Ranges of data over longer periods can indicate the normal rate of fluctuations and may sometimes answer the question whether a population is in a phase of growth or decline (Leemburg & ter Borg in prep.). Long ranges of data appear to be rare, however, certainly from habitats where the conditions have been constant. Some chance occurrence or some change of environmental conditions seems to be the usual motive to start repeated observations. The papers mentioned above do not deviate in this respect. Ranges of data from areas where changes were slight and due to natural succession processes are therefore invaluable. Vegetation science may be of great help to population biology when long ranges can be supplied, either from changed or from unchanged habitats.

## 4.2 *Biological weed control*

Like population biology and particularly population dynamics, weed control is generally concerned with one or a few taxa and the regulation of their numbers. Therefore, connections are clear and cross-references in the literature are many. This is not the place to give a review of the biological control of weeds. Most of us are acquainted with the famous textbook examples: control by insects of *Opuntia* spp. in Australia and of *Hypericum perforatum* in the U.S. (Huffaker 1964); and control of *Chondrilla juncea* by the rust *Puccinia chondrillina* (Wapshere, cited in Harper 1977). Much more work is going on in this field, however. A recent development is to stimulate population growth of pests and pathogens which are already present in the region of a weed. This prevents the use of introduced pests which may have dangerous side effects and, therefore, need time consuming, scrupulous screening (van Zon 1977). Two other methods of weed control, using herbivores and other plants, were applied in our country. Herbivorous fish, – grass carp –, is used to control waterweeds (van Zon & Zonderwijk 1976), and in fact all those cases where grazers are used in managing nature reserves may be said to belong to this category.

The principle of weed control by competition with other plant species was applied by Bakker, who advised sowing reeds (*Phragmites australis*) in polder areas. It prevented infestation by more troublesome species, e.g. *Tussilago farfara, Polygonum* spp., etc. (Bakker 1960c, Bakker et al. 1960). The method was successfully applied in Eastern and Southern Flevoland. A similar principle is applied when floating waterplants are used to control submerged species (Zonderwijk & van Zon 1976).

An exchange of knowledge between specialists on biological control and those studying population dynamics of plants might be useful. Biological control as the applied branch of population biology has a longer tradition and may have quantitative data available as well as data on longer time series. The study of population dynamics of plants as such, is still in its infancy and tries to reveal the mechanisms and principles in the regulation of plant populations. Integration may become urgent in the near future, since chemical control of weeds has already resulted in the first resistant weed strains (Zonderwijk, pers. comm.).

### 4.3 *Management of nature reserves*

The management of nature reserves is aimed primarily at the conservation of complete ecosystems. Its questions generally lie at another level of organisation and not at the population level; they should be answered by those working in that field (cf. P.A. Bakker, this volume). However, the study of populations may add insight into matters of species diversity and thus be useful. Moreover, there are sometimes problems at the species c.q. population level that are of direct importance, e.g. control of weeds, or population regulation of desired species.

Little attention is given to the biological control of weeds in nature reserves in our country. However, if this technique is applied anywhere, it should be in nature reserves. Control measures based on various management techniques (weeding, mowing, etc.) are used with different degrees of success, but a study of biological control by insects and pathogens might give better long-term results. Development of this branch of applied science could possibly provide answers to questions as to why, for instance, the 'forest-pest', *Prunus serotina*, is a serious pest here, while in Canada the species is planted as the favourite windshed and does not spread (van Zon 1977).

It will be clear from the preceding paragraphs that a detailed study will be necessary if there are problems to be solved concerning the regulation of the numbers of desired species. The discussion will not be repeated here. With respect to the presence and absence of species in a particular area a few additional remarks will be made. They are based partly on the gains of population dynamics, partly on those of biogeography.

In island biogeography a close relation was found to exist between the numbers of species present on an island and its size, its shape, its age, its distance to a source of immigrants and the size and composition of this source (MacArthur & Wilson 1967). Some of these relations can be explained as resulting from the chances of extinction of populations or the chances of immigration of propagules. From these studies it follows that extinctions of local populations are regularly occurring natural phenomena. In some species they are frequent, even in seemingly stable habitats (e.g. van der Meijden 1978), in others the chances of extinction are low. For all it holds that as soon as the zero line has been passed re-establishment is impossible, except by immigration. This is no problem as long as a source of seeds is nearby, but as shown on page 20 seed transport occurs mainly over short distances. Long distance dispersal often is subject to chance processes and, therefore, takes time. If the populations are widely distributed it will require a long time.

Nature reserves can be considered as islands in an ocean of other habitats (Wilson & Willis 1975). Within a few decennia man reduced the extent of various habitats and increased the distance between them (Fig. 23), thus bringing about a higher rate of population extinction and a lower rate of immigration. This resulted in a lower species diversity. It is often rather difficult to prevent extinctions, since such action requires a good knowledge of the factors governing the distribution and size of species populations. And even if such knowledge is available, it may yet be impossible to fulfill the species' requirements, since its decrease may be due to conditions which cannot be changed. On the other hand, it certainly is relatively easy to increase the rate of immigration simply by introducing seeds. An orthodox and rigorous refusal of introductions therefore seems to be wrong (see also Bradshaw 1977). However, because of intra- and interspecific variation (cf. p. 41), great care should be taken with respect to the propagules used and the places of introduction. Careful booking of the provenances of seeds and transplants, and of the results of introductions is a minimum requirement when these techniques are applied in nature reserves.

## 5. Concluding remarks

Some subjects chosen from the past and the present of plant population biology, particularly as it is carried out in the Netherlands, have been discussed in the preceding paragraphs. What about the future in this field of research?

Currently there is an increased interest in this branch of plant ecology, certainly in our country. Many studies are under way. Most of these, though not all, have been mentioned in the foregoing. Their results, which may be expected within the next few years, will provide a good amount of demographic data. These will allow the testing of models and theories (e.g. those on r- and K-species, on the role of various life histories, dispersal mechanisms, reproductive systems, dormancy, etc.). Since most of these studies concern projects of a multidisciplinary nature, they will produce data on the effects of a broad spectrum of environmental factors, as well as on the variation within populations and its genetic basis. One point seems to receive too little attention thus far: experiments with populations in natural habitats (see pp. 37 and 42). Such data might offer more information on the populations themselves and can form a link with studies at the vegetation level.

More basic information will probably reveal more mechanisms affecting size and

48

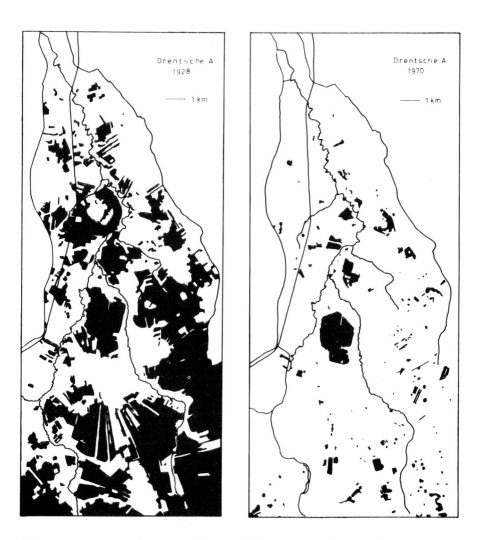

Fig. 23. The decrease of the area of heath land, moors, sanddunes, etc. in North-eastern Drenthe, 1853 – 1900 – 1928 – 1970. (Data J.P. Bakker; they form part of a study on the effect of man on flora, vegetation and landscape in Drenthe, cf. Bakker 1976).

composition of plant populations and may stimulate the generation of theories, hypotheses and generalizations. The latter will have to be watched with care since the past years have shown that every species studied so far is adapted to the environment in its own way, and that the real understanding of its presence, abundance and absence requires a special study in every single case.

## Acknowledgements

My sincere thanks are due to all those colleagues who – on the first request – provided me with all I asked for, not only reprints, but also manuscripts, figures, and unpublished data. Their cooperation was very stimulating.

The comments of prof. dr. D. Bakker, dr. W. Joenje and drs. Manja Kwak (Groningen), dr. A.A. Sterk (Amsterdam) and drs. E. van der Meijden (Leyden) led to many improvements of the manuscript. The data on *Rhinanthus* and *Pedicularis* were taken from reports on projects carried out by Floris Bennema, Johan Brouwer, Piet Drent, Henk Everts, Harry Hummel and Jelle Norder. Mr. E. Leeuwinga prepared the figures, Mrs. Anita Severijnse and Mrs. L. Goss corrected parts of the English text and Aida Sumual typed various versions of the manuscript. Their help is gratefully acknowledged.

## References

Allard, R.W., R.D. Miller & A.L. Kahler. 1978. The relations between degree of environmental heterogeneity and genetic polymorphism. In: A.H.J. Freysen & J.W. Woldendorp (eds.), Structure and functioning of plant populations. pp. 49-73. North-Holland Publ. Co., Amsterdam.

Allen, E.B. & R.T.T. Forman. 1976. Plant species removals and old-field community structure and stability. Ecology 57: 1233-1243.

Andel, J. van. 1974. An ecological study on Chamaenerion angustifolium (L.) Scop. Ph. D. thesis Free Univ. Amsterdam.

Andel, J. van. 1975. A study on the population dynamics of the perennial plant species Chamaenerion angustifolium (L.) Scop. Oecologia 19: 329-337.

Andel, J. van. & F. Vera. 1977. Reproductive allocation in Senecio sylvaticus and Chamaenerion augustifolium in relation to mineral nutrition. J. Ecol. 65: 747-758.

Antonovics, J. 1976. The input from population genetics: 'The new ecological genetics'. Syst. Bot. 1: 233-245.

Baas Becking, L.G.M. 1934. Geobiologie, of inleiding tot de milieukunde. Van Stockum en Zn., Den Haag.

Backer, C.A. 1929. The problem of Krakatao as seen by a botanist. Martinus Nijhoff, The Hague.

Bakker, D. 1957. Oostelijk Flevoland raakt begroeid. Levende Natuur 60: 305-310.

Bakker, D. 1960a. A comparative life-history study of Cirsium arvense (L.) Scop. and Tussilago farfara L., the most troublesome weeds in the newly reclaimed polders of the former Zuiderzee. In: J.L. Harper (ed.), The biology of weeds. Symp. Brit. Ecol. Soc. 1: 205-222. Blackwell, Oxford.

Bakker, D. 1960b. Senecio congestus (R.Br.) DC. in the Lake IJsselpolders. Acta Bot. Neerl. 9: 235-259.

Bakker, D. 1960c. Het botanisch onderzoek in de IJsselmeerpolders. Vakbl. v. Biol. 40: 63-79.

Bakker, D. & A.C. Boer. 1952. De vegetatie van de Braakman in 1952. Van Zee tot Land 19: 1-15.

Bakker, D., J.J. Jonker & H. Smits. 1960. Land reclamation in Holland; bringing the polders into production. Span 3: 143-151.

Bakker, D. & W. van der Zweep. 1957. Plant migration studies near the former island of Urk in the Netherlands. Acta Bot. Neerl. 6: 60-73.

Bakker, J.P. 1976. Botanisch onderzoek ten behoeve van natuurtechnisch beheer in het Stroom-dallandschap Drentsche A. Natuur en Landschap 1: 1-12.

Beeftink, W.G. 1978. Structure and dynamics of the salt marsh ecosystem; progress report. Verh. Kon. Ned. Akad. Wetensch. afd. Natuurkunde, 2e reeks, 71: 113-133.

Beeftink, W.G., M.C. Daane & W. de Munck. 1971. Tien jaar botanisch oecologische verkenningen langs het Veersche Meer. Natuur en Landschap 25: 50-63.

Beeftink, W.G., M.C. Daane, W. de Munck & J. Nieuwenhuize. 1978. Aspects of population dynamics in Halimione portulacoides communities. Vegetatio 36: 31-43.

Borg, S.J. ter & J.C. Bastiaans. 1973. Host-parasite relations in Rhinanthus serotinus. 1. The effect of growth conditions and host; a preliminary report. Symp. Parasitic Weeds Eur. Weed Res. Counc. pp. 236-246. Malta Univ. Press, Valeta.

Bradshaw, A.D. 1972. Some of the evolutionary consequences of being a plant. Evol. Biol. 5: 25-48.

Bradshaw, A.D. 1977. Conservation problems in the future. Proc. Roy. Soc. London B, 197: 77-96.

Buchanan, R.E. & N.E. Gibbons. 1974. Bergey's manual of determinative bacteriology (8th ed.) Williams & Wilkins Co., Baltimore.

Bullock, S. 1976. Consequences of limited seed dispersal within simulated annual populations. Oecologia 24: 247-256.

Caswell, H. & P.A. Werner. 1978. Transient behaviour and life history analysis of teasel (Dipsacus sylvestris Huds.). Ecology 59: 53-66.

Cohen, D. 1966. Optimising reproduction in a randomly varying environment. J. Theor. Biol. 12: 119-129.

Cohen, D. 1968. A general model for optimal reproduction in a randomly varying environment. J. Ecol. 56: 219-228.

Deevey, E.S. 1947. Life tables for natural populations of animals. Quart. Rev. Biol. 22: 283-314.

Docters van Leeuwen, W.M. 1936. Krakatau, 1883 to 1933. Ann. Jard. Bot. Buitenzorg. 56-57: 1-506.

Emmel, T.C. 1976. Population Biology. Harper & Row, New York.

Ernst, W. 1978. Discrepancy between ecological and physiological optima of plant species. A re-interpretation. Oecol. Plant. 13: 175-189.

Feekes, W. 1936. De ontwikkeling van de natuurlijke vegetatie in de Wieringermeerpolder, de eerste groote droogmakerij van de Zuiderzee. Ned. Kruidk. Arch. 46: 1-294.

Feekes, W. & D. Bakker. 1954. De ontwikkeling van de natuurlijke vegetatie in de Noordoost-polder. Van Zee tot Land 6: 1-92.

Fresco, L.F.M. 1979. A non-experimental analysis of species amplitudes and of competition. (in prep.)

Freysen, A.H.J. & J.W. Woldendorp. 1978. Structure and functioning of plant populations. Verh. Kon. Ned. Akad. Wetensch. afd. Natuurk., 2e reeks, dl. 70. 323 pp. North-Holland Publ. Co., Amsterdam.

Gančev, I. 1973. Influence of Rhinanthus on the composition and dynamics of meadow communities. Bull. de L'inst. Bot. Sofia 23: 88-118 (in Bulgarian, with English summary).

Grime, J.P. 1977. Evidence for the existence of three primary strategies in plants and its relevance to ecological and evolutionary theory. Amer. Nat. 111: 1169-1194.

Grime, J.P. 1978. Interpretation of small-scale patterns in the distribution of plant species in space and time. In: A.H.J. Freysen & J.W. Woldendorp (eds.), Structure and functioning of plant populations. pp. 101-124. North-Holland Publ. Co., Amsterdam.

Hallé, F. & R.A.A. Oldemann. 1970. Essai sur l'architecture et la dynamique de croissance des arbres tropicaux. Masson, Paris.

Harberd, D.J. 1961. Observations on population structure and longevity of Festuca rubra L. New Phyt. 60: 184-206.

Harper, J.L. 1967. A Darwinian approach to plant ecology. J. Ecol. 55: 247-270.

Harper, J.L. 1977. Population biology of plants. Academic Press, Oxford.

51

Harper, J.L. 1978. The demography of plants with clonal growth. In: A.H.J. Freysen & J.W. Woldendorp (eds.), Structure and functioning of plant populations. pp. 27-48. North-Holland Publ. Co., Amsterdam.

Harper, J.L. 1979. In: Population dynamics. Symp. Brit. Ecol. Soc. Blackwell, Oxford (in press).

Harper, J.L. & J. White. 1974. The demography of plants. Ann. Rev. Ecol. Syst. 5: 419-463.

Hart, R. 1977. Why are biennials so few? Amer. Nat. 111: 792-799.

Hawthorn, W.R. & P.B. Cavers. 1976. Population dynamics of the perennial herbs Plantago major L. and P. rugelii Decne. J. Ecol. 64: 511-529.

Heslop-Harrison, J. 1964. Forty years of genecology. Adv. Ecol. Res. 2: 159-247.

Holt, B.R. 1972. Effect of arrival time on recruitment, mortality, and reproduction in successional plant populations. Ecology 53: 668-673.

Horn, H.S. 1976. Succession. In: R.M. May (ed.) Theoretical ecology. pp. 187-204. Blackwell, Oxford.

Huffaker, C.B. 1964. Fundamentals of Biological Weed Control. In: P. De Bach & E.L. Schlesinger (eds.), Biological control of insect pests and weeds. pp. 74-117. Chapman & Hall, London.

Huiskes, A.H.L. 1977. The population dynamics of Ammophila arenaria (L.) Link. Ph.D. thesis Univ. Wales.

Jain, S. & O.T. Solbrig. 1976. Plant population biology at the cross-roads (a symposium sponsored by the Amer. Soc. for Plant taxonomists, and the Systematic section of the Bot. Soc. of America) Syst. Bot. 1: 202-323.

Janzen, D.H. 1970. Herbivores and the number of tree species in tropical forests. Amer. Nat. 104: 501-528.

Joenje, W. 1978a. Plant colonization and succession on embanked sand flats; a case study in the Lauwerszeepolder. Ph.D. thesis Univ. Groningen.

Joenje, W. 1978b. Migration and colonization by vascular plants in a new polder. Vegetatio 38: 95-102.

Joenje, W. 1978c. Plant succession and nature conservation of newly embanked tidal flats in the Lauwerszeepolder. In: R.L. Jefferies & A.J. Davy (eds.), Ecological processes in coastal environments. pp. 617-634. Blackwell, Oxford.

Kampmeijer, P. & J.C. Zadoks. 1978. EPIMUL, a similator of foci and epidemics in mixtures of resistent and susceptible plants, mosaics and multilines. PUDOC, Wageningen.

Kays, S. & J.L. Harper. 1974. The regulation of plant and tiller density in a grass sward. J. Ecol. 62: 97-105.

Kloos, A.W. & W.C. de Leeuw. 1928. De spontane vegetatie van den proefpolder te Andijk in 1928. Ned. Kruidk. Arch. 1928: 149-161.

Kloos, A.W. & W.C. de Leeuw. 1930. De vegetatie van den proefpolder te Andijk in 1929. Ned. Kruidk. Arch. 1930: 113-118.

Kluyver, H.N. 1971. Regulation of numbers in populations of Great Tits (Parus m. major). In: P.J. den Boer & G.R. Gradwell (eds.), Dynamics of Populations. pp. 507-523. Pudoc, Wageningen.

Koyama, H. & T. Kira. 1956. Intraspecific competition among higher plants. VIII. Frequency distribution of individual plant weight as affected by the interaction between plants. J. Inst. Polytech. Osaka City Univ. 7: 73-94.

Krylova, N.P. 1963. Increasing stability of legumes in swards of flood meadows. Bull. Soc. Imperial Naturalist Moscow 68: 72-83.

Kuiper, P.J.C., J.W. van Delden & J.W. Woldendorp. 1977. Vergelijkend onderzoek naar demografische, fysiologische en genetische eigenschappen van plantensoorten in relatie tot hun standplaats in graslanden. BION, aanvraag zwaartepuntsubsidie (Unpubl.).

Kwak, M.M. 1977. Pollination ecology of five hemiparasitic, large flowered Rhinanthoideae with special reference to the pollination behaviour of nectar thieving, short-tongued bumblebees. Acta Bot. Neerl. 26: 97-107.

Kwak, M.M. 1978. Pollination, hybridization and ethological isolation of Rhinanthus minor and Rhinanthus serotinus (Rhinanthoideae) by bumblebees (Bombus Latr.) Taxon 27: 145-158.

Leemburg-van der Graaf, C.A. & S.J. ter Borg. 1979. Population dynamics of Rhinanthus. (In prep.)

Levin, D.A. & H.W. Kerster. 1974. Gene flow in seed plants. Evol. Biol. 7: 139-220.

Levin, D.A. & J.B. Wilson. 1978. The genetic implications of ecological adaptations in plants. In: A.H.J. Freysen & J.W. Woldendorp (eds.), Structure and functioning of plant populations. pp. 75-100. North-Holland Publ. Co., Amsterdam.

Lewontin, R.C. 1967. Introduction. In: R.C. Lewontin (ed.), Population biology and evolution. pp. 1-4. Syracuse Univ. Press, Syracuse.

Linkola, K. 1935. Ueber die Dauer und Jahresklassenverhältnisse des Jugendstadiums bei einigen Wiesenstauden. Acta Forest. Fenn. 42: 5-53.

Londo, G. 1971. Patroon en proces in duinvallei vegetaties langs een gegraven meer in de Kennemerduinen. Ph.D. thesis Univ. Nijmegen.

Londo, G. 1978. Over het gedrag in ruimte en tijd van Taraxacum en Plantago. Gorteria 9: 174-178.

MacArthur, R.H. & E.O. Wilson. 1967. The theory of island biogeography. Princeton Univ. Press, Princeton, N.J.

Meijden, E. van der. 1971. Senecio and Tyria (Callimorpha) in a Dutch dune area: a study on an interaction between a monophagous consumer and its host plant. In: P.J. den Boer & G.R. Gradwell (eds.), Dynamics of Populations. pp. 390-404. Pudoc, Wageningen.

Meijden, E. van der. 1978. Interactions between the Cinnabar Moth and Tansy Ragwort. Proc. 4th. Int. Symp. Biol. Contr. Weeds (1976) pp. 159-162. Gainesville, Florida.

Meijden, E. van der & R.E. van der Waals-Kooi. 1979. Population ecology of Senecio jacobaea. I. Reproductive strategy and the biennial habit. J. Ecol. 67 (1) (in press).

Moore, P.D. 1976. r, K and evolution. Nature 262: 351-352.

Mortimer, A.M. 1974. Studies of germination and establishment of selected species with special reference to the fates of the seeds. Ph.D. thesis Univ. Wales.

Nip-van der Voort, H. 1977. Ontwikkeling van de bermflora in de IJsselmeerpolders. Contactbl. v. Oecol. 13: 28-30.

Oldeman, R.A.A. 1977. Bouwplan en groeikringlopen van het regenwoud. Contactbl. v. Oecol. 13: 65-70.

Putwain, P.D. & J.L. Harper. 1970. Studies in the dynamics of plant populations III. The influence of associated species on populations of Rumex acetosa L. and R. acetosella L. in grassland. J. Ecol. 58: 251-264.

Putwain, P.D., D. Machin & J.L. Harper. 1968. Studies in the dynamics of plant populations II. Components and regulation of a natural population of Rumex acetosella L. J. Ecol. 56: 421-431.

Rabotnov, T.A. 1959. The effect of Rhinanthus major Ehrh. upon the crops and the composition of the floodland herbage. Bull. M.O.-va Isp. Prirody, Otd. Biol. 64: 105-107 (Russian).

Rabotnov, T.A. 1969. On coenopopulations of perennial herbaceous plants in natural coenoses. Vegetatio 19: 87-95.

Rabotnov, T.A. 1975. On phytocoenotypes. Phytocoenologia 2: 66-72.

Rabotnov, T.A. 1978. On coenopopulations of plants reproducing by seeds. In: A.H.J. Freysen and J.W. Woldendorp (eds.), Structure and functioning of plant populations. pp. 1-26. North-Holland Publ. Co., Amsterdam.

Rozema, J. 1978a. On the ecology of some halophytes from a beach plain in the Netherlands. Ph.D. thesis Free Univ. Amsterdam.

Rozema, J. 1978b. Population dynamics and ecophysiological adaptations of some coastal Juncaceae and Gramineae. In: R.L. Jefferies & A.J. Davy (eds.), Ecological processes in coastal environments. pp. 229-241. Blackwell, Oxford.

Sagar, G.R. 1960. Factors affecting the germination and early establishment of Plantains. (Plantago lanceolata, P. media and P. major.) In: J.L. Harper (ed.), The biology of weeds. Symp. Brit. Ecol. Soc. 1: 236-245. Blackwell, Oxford.

Sagar, G.R. & A.M. Mortimer. 1976. An approach to the study of the population dynamics of plants with special reference to weeds. Appl. Biol. 1: 1-45.

Sarukhán, J. 1974. Studies on plant demography: Ranunculus repens L., R. bulbosus L. and R. acris L. II Reproductive strategies and seed population dynamics. J. Ecol. 62: 151-177.

Schaffer, W.M. & E.G. Leigh. 1976. The perspective role of mathematical theory in plant ecology. Syst. Bot. 1: 209-232.

Solbrig, O. (ed.). 1978. Demography and dynamics of plant populations. Botanical Monographs 15. Blackwell, Oxford.

Solbrig, O. & B.B. Simpson. 1974. Components of regulation of a population of dandelions in Michigan. J. Ecol. 62: 473-486.

Stearns, S.C. 1976. Life history tactics: A review. Quart. Rev. Biol. 51: 3-47.
Sterk, A.A. 1968. Een studie van de variabiliteit van Spergularia media en Spergularia marina van Nederland. Ph.D. thesis Univ. Utrecht.
Sterk, A.A. 1969. Biosystematic studies on Spergularia media and S. marina in the Netherlands. III. The variability of S. media and S. marina in relation to the environment. Acta Bot. Neerl. 18: 561-577.
Sterk, A.A. 1975. Demographic studies of Anthyllis vulneraria L. in the Netherlands. Acta Bot. Neerl. 24: 315-337.
Sterk, A.A. 1976a. Jaarlijkse registratie van aantallen individuen van zeer zeldzame plantensoorten en de populatiedynamica. Gorteria 8: 1-11.
Sterk, A.A. 1976b. Anacamptis pyramidalis bij Wijk aan Zee. Gorteria 8: 81-85.
Sterk, A.A. 1978a. Inleiding tot de biosystematiek van het geslacht Taraxacum. Contactbl. v. Oecol. 14: 43-47.
Sterk, A.A. 1978b. Taraxacum onderzoek van de vakgroep Bijzondere Plantkunde, Universiteit van Amsterdam. Danseria 13: 15-20.
Stoutjesdijk, Ph. 1959. Heaths and inland dunes of the Veluwe. Wentia 2: 1-96.
Tamm, C.O. 1956. Further observations on the survival and flowering of some perennial herbs. I. Oikos 7: 273-292.
Tamm, C.O. 1972. Survival and flowering of some perennial herbs. II. The behaviour of some orchids on permanent plots. Oikos 23: 23-28.
Tamm, C.O. 1972. Survival and flowering of some perennial herbs. III. The behaviour of Primula veris on permanent plots. Oikos 23: 159-166.
Toorn, J. van der, M. Brandsma, W.B. Bates & M.G. Penny. 1969a. De vegetatie van Zuidelijk Flevoland in 1968. Levende Natuur 72: 56-62.
Toorn, J. van der, B. Donougho & M. Brandsma. 1969b. Verspreiding van wegbermplanten in Oostelijk Flevoland. Gorteria 4: 151-160.
Toorn, J. van der & J.H. Mook. 1977. Influence of environmental factors on the performance of reed vegetations. Verh. Kon. Ned. Akad. Wetensch., afd. Natuurkunde, 2e reeks, 69: 15-20.
Varley, G.C., G.R. Gradwell & M.P. Hassell. 1973. Insect population ecology. Blackwell, Oxford.
Vegte, F.W. van der. 1978. Population differentiation and germination ecology in Stellaria media (L.) Vill. Oecologia 32: 232-246.
Vries, V. de. 1961. Vegetatiestudie op de westpunt van Vlieland. Ph.D. thesis Univ. Amsterdam.
Watkinson, A.R. & J.L. Harper. 1978. The demography of a sand dune annual: Vulpia fasciculata. I. The natural regulation of populations. J. Ecol. 66: 15-34.
Watt, A.S. 1947. Pattern and process in the plant community. J. Ecol. 35: 1-22.
Watt, A.S. 1955. Bracken versus heather, a study in plant sociology. J. Ecol. 43: 490-506.
Werner, P.A. 1975a. Predictions of fate from rosette size in teasel. Oecologia 20: 197-201.
Werner, P.A. 1975b. The effects of plant litter on germination in teasel, Dipsacus sylvestris Huds. Amer. Midl. Nat. 94: 470-476.
Werner, P.A. 1977. Colonization success of a 'biennial' plant species: experimental field studies of species cohabitation and replacement. Ecology 58: 840-849.
Werner, P.A. & H. Caswell. 1977. Population growth rates and age vs. stage distribution models for teasel (Dipsacus sylvestris Huds.). Ecology 58: 1103-1111.
Whittaker, R.H. 1970. The biochemical ecology of higher plants. In: E. Sondheimer & J.B. Simeone (eds.), Chemical ecology. pp. 43-70. Academic Press, New York.
Wilde, W.J.J.O. de. 1976. Enkele opmerkingen over de verarming van de duinflora bij Wijk aan Zee. Gorteria 8: 49-51.
Willems, J.H. 1978. Populatiebiologisch onderzoek aan Orchis mascula (L.) L. op enkele groeiplaatsen in Zuid-Limburg. Gorteria 9: 71-80.
Williamson, M. 1972. The analysis of biological populations. Edward Arnold, London.
Wilson, E.O. & E.O. Willis. 1975. Applied biogeography. In: M.L. Cody & J.M. Diamond (eds.), Ecology and evolution of communities, pp. 522-536. Belknap Press, Harvard Univ., Cambridge Mass.
Wit, C.T. de. 1960. On competition. Versl. Landbouwk. Onderz. 66: 1-82.
Yoda, K., T. Kira, H. Ogawa & K. Hozumi. 1963. Self-thinning in overcrowded pure stands under cultivated and natural conditions. J. Biol. Osaka City Univ. 14: 107-129.

Zadoks, J.C. & P. Kampmeyer. 1977. The role of crop populations and their deployment, illustrated by means of a simulator, Epimul 76. Ann. New York Acad. Sc. 287: 164-190.

Zeist, W. van. 1970. Betrekkingen tussen palynologie en vegetatiekunde. In: H.J. Venema, H. Doing & I.S. Zonneveld (eds.), Vegetatiekunde als synthetische wetenschap. Misc. papers Landbouwhogeschool, Wageningen 5: 127-140.

Zon, J.C.J. van. 1977. Verslag van het 'IV International Symposium on Biological Control of Weeds', in Gainesville (Florida) en van een aansluitende studiereis in de V.S. CABO verslag nr. 14 (Unpubl.).

Zon, J.C.J. van & P. Zonderwijk. 1976. Profiel van de graskarper in Nederland. Vakbl. v; Biol. 56: 282-285.

Zonderwijk, P. & J.C.J. van Zon. 1976. Waterplanten als bondgenoten bij het onderhoud. Waterschapsbelangen 2: 21-23.

Zijlstra, G. 1974. Variabiliteit bij Linum catharticum L. in Nederland. Contactbl. v. Oecol. 10: 10-14.

# 3. CHANGES IN THE COMPOSITION OF MIXED POPULATIONS OF GRASSLAND SPECIES

## J.P. VAN DEN BERGH

# 3. CHANGES IN THE COMPOSITION OF MIXED POPULATIONS OF GRASSLAND SPECIES

## J.P. VAN DEN BERGH

### 1. Introduction

Knowledge about the dynamics of plant populations in the vegetation is important with respect to:
(1)  characterization of the environment via the vegetation;
(2)  determining the effect of changing environmental conditions on the botanical composition of the vegetation; and
(3)  management of the vegetation.
Mass proportions and the presence or absence of species in the vegetation give distinct indications on the properties of the environment. Seasonal aspects may have considerable effects not only on these mass proportions but also with respect to the occurrence of certain plant species, at least with respect to the above-ground observations of plants (thus not taking into account the underground presence and the presence of seeds or the like). Without any knowledge about the dynamics of the populations composing the vegetation during the year and during a sequence of years, characterization of the environment by means of the vegetation is hardly possible, if not entirely impossible.

In environmental controle there is a need for standard plots which can be used in monitoring vegetation considered to be influenced by human activities. The above-mentioned clearly indicates that also on (standard) plots where human activities are supposed to be minimal, the botanical composition may still fluctuate. Until sufficient information is obtained on the extent and the nature of these fluctuations, the use of standard plots to determine changes of the environment will be problematic.

The introduction of special measures to neutralize a disturbance in the equilibrium between the populations of the vegetation requires much knowledge about the factors regulating the populations and affecting their interactions. An accurate description of the dynamics of plant populations is essential for obtaining this knowledge.

The major part of long term observations on the dynamics of plant populations has been carried out on permanent quadrats, which have been laid out in communities in succession (pioneer vegetation), or in communities consisting predominantly of annual species (extreme conditions). The environments of such communities, because of their nature, are subject to many changes or fluctuations. Therefore, these communities are not suited to study the dynamics of plant populations as the dynamics of the environment itself may prevail in such situations, and one is not able to obtain any reasonable estimate of the extent of the impact of these environmental dynamics. Moreover, in almost all studies the dynamics of one population only has been described, so that the interference of the surrounding populations can not be taken into consideration.

Fortunately there are a few studies for which the objections mentioned above do not hold. On some old grassland communities the botanical composition has been analysed during a great number of years. It will be shown that even in these so-called stable communities the plant populations may fluctuate to a considerable extent. The pattern of fluctuations may differ for the various populations and is highly dependent on the fertilizer level. Before discussing this subject, the interactions between plant species will be dealt with in general. A theory has been developed which enables us to distinguish different ways of interference and to investigate the possibility of an equilibrium.

## 2. Interference between plant populations

In studying the complex process of interference between plant populations in a plant community, it seems reasonable to begin with simple experiments, in which two species in different mixtures grow in a homogeneous environment. As the extreme of one-sided mixtures, monocultures should also be included in these experiments. To avoid the effect of different plant densities replacement series are applied, i.e. a certain amount of one species (expressed in viable seeds, or tillers, or grammes of biomass, etc.) is replaced by a certain amount of the other species, so that the total plant density remains the same. When the relative plant frequency or relative yield ($r$) is defined by the quotient of the amount of a species in the mixture ($O$) and that in the monoculture ($M$), for a replacement series with the species a and b the following holds (de Wit 1960):

$$O_a/M_a + O_b/M_b = r_a + r_b = 1.$$

In the majority of the experimental mixtures discussed in the literature (excluding legume-nonlegume mixtures), this sum of the relative yields (Relative Yield Total = RYT) stays about equal to 1 for all the successive harvests (Trenbath 1974), which means that the species are growing in the same space (de Wit & van den Bergh 1965).

### 2.1 *Populations growing in the same space: RYT = 1*

An experiment under controlled conditions in a climatic chamber is discussed first. Tillers of *Anthoxanthum odoratum* (A) and *Lolium perenne* (L) were planted according to a replacement series in pots with well-mixed soil and harvested every 3 weeks. In the upper left replacement diagram of Fig. 1 the absolute yields of the separate species of each mixture at the second harvest are plotted against the relative yield of species A of the relevant mixture, also of the second harvest ($^2r_A$). Straight lines can be drawn through the points of each species, hence $^2r_A + {}^2r_L = 1$.

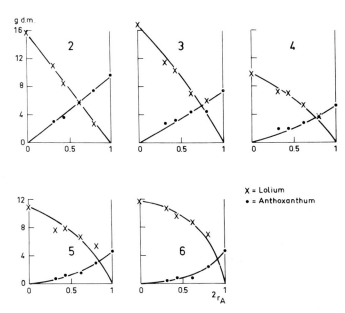

Fig. 1. Replacement diagrams of the dry-matter yields per pot of *Lolium perenne* and *Anthoxanthum odoratum* at five successive harvests plotted against the relative yields of *Anthoxanthum odoratum* at the second harvest ($^2 r_A$). After van den Bergh (1968).

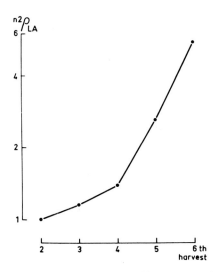

Fig. 2. Relative replacement rates (on a log scale) with respect to the second harvest presented as a course line for *Lolium perenne* (L) with respect to *Anthoxanthum odoratum* (A) for five successive harvests ($^{n2} \rho_{LA}$) of the same experiment.

61

This means that the yields of dry matter of the second harvest also behave according to the replacement principle.

The replacement diagrams of the following harvests show that the yields of species L in the mixtures continue to increase with respect to its yield in monoculture (convex curves) and those of species A continue to decrease to the same extent with respect to its monoculture (concave curves), resulting in RYT = 1. Under these conditions in the long run species A will disappear completely from the mixture and monocultures of species L will remain. The rate at which this process takes place is reflected in Fig. 2 in which the relative replacement rate of species L with respect to species A

$$^{n2}\rho_{LA} = \frac{^{n}r_L/^{2}r_L}{^{n}r_A/^{2}r_A} \tag{1}$$

for the successive harvests (n) with respect to the second harvest are plotted logarithmically. As the slope of this course line is a measure for the replacement rate, Fig. 2 shows that species A disappears from the mixture at an increasing rate.

A second example concerns a field experiment with *Lolium perenne* (L) and *Dactylis glomerata* (D). During 4 years 7 different mixtures and monocultures were harvested 4 times a year. On the vertical axis of the replacement diagrams in Fig. 3 the absolute yields of the third harvest of 4 successive years are plotted against the relative yields of species D of the third harvest in 1965 ($^{3}r_D$). Since RYT = 1 during the whole experimental period, the species are growing in the same space. In the second year species D shows already a distinct advantage over species L (convex curve for species D and concave curve for species L). In the course of the fourth year species L is suppressed by species D to such an extent that the mixtures have practically changed into monocultures of species D. The course line of species L with respect to species D (Fig. 4) clearly shows that the rate at which species D suppresses species L may vary considerably, dependent on the season and the year. In the months June, July, August and September species D is very aggressive (sharply declining course line), except in the very dry year 1967.

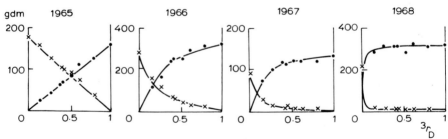

Fig. 3. Replacement diagrams with dry-matter yields per plot of *Lolium perenne* (X) and *Dactylis glomerata* (●) at the third harvest of four successive years plotted against the relative yields of *Dactylis glomerata* at the third harvest in 1965 ($^{3}r_D$). After van den Bergh & Elberse (1970).

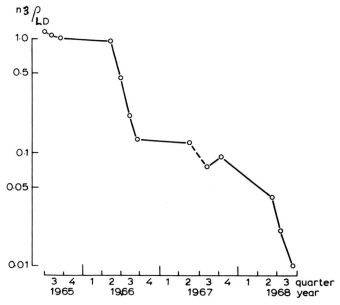

Fig. 4. Relative replacement rates (on a log scale) with respect to the third harvest in 1965 ($n^3 \rho_{LD}$) presented as a course line for *Lolium perenne* (L) with respect to *Dactylis glomerata* (D) for the various harvests in four successive years.

As a third example Fig. 5 reflects the course lines of a pot experiment with *Dactylis glomerata* (D) and *Agrostis tenuis* (A) at various pH values of the soil. This figure shows that the major effect of pH occurs before the first harvest, which is due to poor tillering of the seedlings of species D at the lowest pH. After this pH no

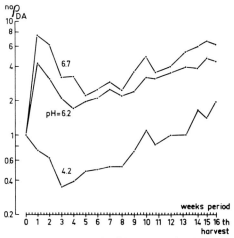

Fig. 5. Relative replacement rates of *Dactylis glomerata* (D) with respect to *Agrostis tenuis* (A) at three pH values of the soil for 16 successive harvests. After van den Bergh (1968).

63

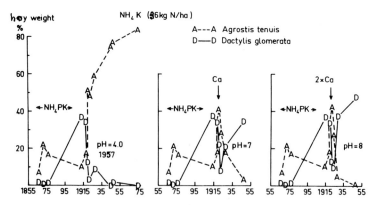

Fig. 6. Hay weight percentage of *Agrostis tenuis* (A) and *Dactylis glomerata* (D) during 120 years on plot 18 of the Park Glass Plots at Rothamsted. From left to right unlimed, moderately limed (4.4 tons CaO ha$^{-1}$ every 4 years) and heavily limed (7.6 tons CaO ha$^{-1}$ every 4 years) from 1920 onwards; pH values in 1957: 4.0, 7.0 and 8.0 resp. From 1865-1904 fertilized with a low dose of complete mixed mineral manure, thereafter 96 kg N ha$^{-1}$ as sulphate of ammonia and 225 kg K$_2$O, 15 kg Na and 11 kg Mg as sulphates. After Williams (1978).

longer affects competition, which is demonstrated by the parallelism of the course lines. In the long run species D suppresses species A at all pH values.

This contrasts with the results of plot 18 of the 120-year old experimental field at Rothamsted (Fig. 6, see also page 76). Fifty years after liming started in 1920, on the unlimed part of this plot (pH = 4) species D has practically disappeared and species A is present in the mixture with 83% hay weight, whereas on the moderately (pH = 7) and heavily limed parts (pH = 8) of this plot species A has practically disappeared and species D is the major species.

The explanation for the different results of the pot experiment and the field experiment probably is that in the pot experiment the plants could only reproduce vegetatively (4 weekly harvests), whereas in the field experiment the species can reproduce also via seed due to the hayfield treatment. The pot experiment clearly shows that species D is hampered mainly in the establishment phase by a low pH.

In the preceding examples the curves in the replacement diagrams can be described by the following hyperbolic functions

$$^n r_a = \frac{^{nm} \rho_{ab} \cdot {}^m r_a}{^{nm} \rho_{ab} \cdot {}^m r_a + {}^m r_b} \quad \text{and} \quad {}^n r_b = \frac{^m r_b}{^{nm} \rho_{ab} \cdot {}^m r_a + {}^m r_b} \tag{2}$$

which means that the species are growing in the same space and that $\rho$ is independent of the composition of the mixture (de Wit & van den Bergh 1965). This independence is clearly demonstrated in the ratio diagram in which the proportion of the relative yields of the species at the (n + 1) harvest are plotted against that at the nth harvest on a double log scale (Fig. 7). From equation (1) it follows that

64

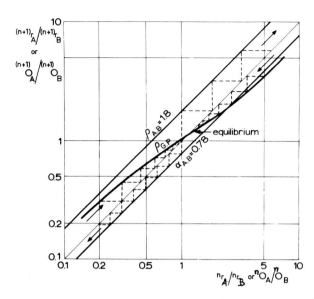

Fig. 7. Ratio diagram of the relative or absolute yield ratio of species A and B at harvest (n + 1) plotted against this ratio at harvest n. For explanation see text.

$\log \left( (n+1)_{r_a}/(n+1)_{r_b} \right) = \log {}^{(n+1)n}\rho_{ab} + \log \left( {}^n r_a/{}^n r_b \right)$ , so that the observations must lie around a straight line parallel to the diagonal when the relative replacement rate is constant. The distance of this ratio line to the diagonal determines the number of harvests or seasons (number of 'steps of the dotted staircase') necessary for a given shift in the composition of the mixture. This means that when the conditions do not change, in due course one of the species will predominate and so there will be no coexistence. The homogeneity of the environment, which diminishes the possibility of species maintaining themselves in their preferential habitat, or the extreme conditions (low pH) probably will have caused the predominance of only one species in the experiments discussed.

## 2.2 Annual species versus perennial species with mainly vegetative (re)production

The important difference between annual and perennial grassland plants with respect to population dynamics is that the plant parts with which the perennial continues growth in the next growing season, are the very parts lost by the annual (stubble, underground parts). This implies that with the perennial species the space already occupied is redistributed every year, whereas with annuals the empty space has to be occupied again by seedlings.

A difference should be made between competitive ability and reproduction via seed. In a situation as shown in Fig. 8, the species with the lowest production in monoculture (A) performs better (indicated by distance p) than when it is compe-

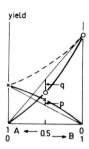

Fig. 8.  Replacement diagram of species A and B. For explanation see text.

ting with itself, and the highest yielding species in monoculture (B) performs worse (indicated by distance q) than when it is competing with itself. The relative replacement rate (a measure for competitive ability) $\rho_{AB} = 1.8 > 1$, which means that species A is stronger than species B. When conditions do not change ($\rho$ remains constant), ultimately species B will be replaced completely by species A (Fig. 7).

This reasoning only holds for species with mainly vegetative (re)production, since differences in reproduction via seed are not taken into account. With annuals this is taken into account when, instead of the double quotient of the relative yields of dry matter, the double quotient of the absolute number of seeds of the mixture is introduced: the relative reproductive rate $\alpha_{AB} = \dfrac{(n+1)O_A / {}^nO_A}{(n+1)O_B / {}^nO_B}$ (de Wit 1960).

In the example of Fig. 8 $\alpha_{AB} = 0.78 < 1$, which means that if the species are sown at the same ratio in the next year, as they were harvested in the previous year, in the long run with conditions unaltered, species A will be replaced by species B (Fig. 7 ). In spite of the greater competitive ability of the annual A (convex curve), it still loses because of its low generative reproduction ($M_A \ll M_B$). This phenomenom was first described by Montgomery (1912).

2.3 *Populations growing in partly different spaces (niche differentiation):* $1 < RYT < 2$

When in the replacement diagrams the one curve is not convex to the same extent as the other is concave, equation (2) does not apply and the sum of the relative yields (RYT) does not equal 1. In some experiments the following equation can be applied

$$ {}^nr_a = \frac{{}^{nm}k_{ab} \cdot {}^mr_a}{{}^{nm}k_{ab} \cdot {}^mr_a + {}^mr_b} \quad \text{and} \quad {}^nr_b = \frac{{}^{nm}k_{ba} \cdot {}^mr_b}{{}^{nm}k_{ba} \cdot {}^mr_b + {}^mr_a} \tag{3} $$

in which the relative crowding coefficients $k_{ab}$ and $k_{ba}$ (a measure for the curvature of the curves) are independent of the composition of the mixture (de Wit 1960).

66

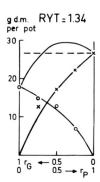

g d.m.  RYT = 1.34
per pot

Fig. 9.  Replacement diagram of *Glycine javanica* (G = o) and *Panicum maximum* (P = x). After de Wit, Tow & Ennik (1966).

A well-known example of this is a mixture of a legume with a grass. In Fig. 9 the curves of *Glycine javanica* (G) and *Panicum maximum* (P) can be described with equation (3) and the RYT of the 50/50 mixture is equal to 1.34. In this case the species clearly do not grow in the same space: *Glycine* has an additional 'nitrogen space' at its disposal because of its symbiosis with air nitrogen binding bacteria in its root nodules. The relative replacement rate $\rho$ is dependent on the composition of the mixture in such a way that the species perform relatively better as their proportion in the mixture decreases. In the ratio diagram (Fig. 7) this is shown by the curve intersecting the diagonal. Left from the intersection *Glycine* replaces *Panicum* (much grass in the mixture depletes the N supply in the soil → poor grass growth and decreased competition for light → increased proportion of the legume in the mixture) and right from the intersection *Panicum* replaces *Glycine* (much legume in the mixture fixes much air nitrogen → increased N supply of the soil → better grass growth and increased competition for light by the taller grass → increased proportion of grass in the mixture) and a stable equilibrium is attained at the ratio of intersection.

2.4  *Special cases*

In the (theoretical) case that the species do not interfere with each other at all (they grow in an entirely different space), both curves in the replacement diagram will be identical to the known asymptotic density curves and the maximum RYT value will be equal to 2 (Fig. 10a). Two species with a different growing period, as e.g. *Crocus* (spring) and *Dahlia* (late summer), probably will hardly interfere with each other in a mixture (different spaces in time) and may form a stable equilibrium.

Fig. 10 shows a sequence of possibilities in which the effect of interference varies from positive via indifferent to negative. When the relative yield > 1, there is a stimulating effect of species b on a (Fig. 10f), but this need not be reciprocal (supporting plant and climber, parasite and host plant, etc.). In these cases the

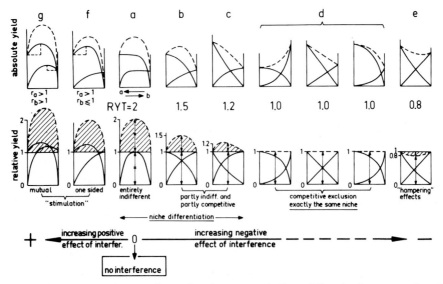

Fig. 10. Ways of interference (for explanation see text). Dotted lines in the upper series of diagrams represent the absolute total yields and in the lower series of diagrams the RYT values of the mixtures. After van den Bergh & Braakhekke (1978).

RYT may even be greater than 2 (Raininko 1968). RYT < 1 (Fig. 10e) may indicate hampering effects. In this case toxic substances may be excreted or a disease may occur by which the carrier species is not damaged but the neighbouring species is indeed infected (Sandfaer 1970).

Cases are known in which the relative crowding coefficients of equation (3) are dependent on the composition of the mixture. The curves in the replacement diagrams may be S shaped in that case. Ennik (1970) observed e.g. in the second year after sowing that the composition of the mixtures of *Lolium perenne* (L) and *Trifolium repens* (T) is independent of the proportion sown (Fig. 11). The curves are each other's reflection and this means that despite RYT = 1 an equilibrium may

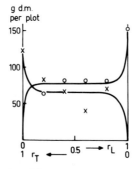

Fig. 11. Replacement diagram of *Trifolium repens* (T = x) and *Lolium perenne* (L = o). After Ennik (1970).

68

establish. This clearly shows that it is not sufficient to determine the RYT of the 50/50 mixture only to be able to discuss the nature of the mutual interference of species; it is necessary to analyse a complete scala of different mixtures in interference studies (see also Braakhekke 1979).

## 3. Species density and dry matter production

Al-Mufti et al. (1977) show a graph in which species density is plotted against maximum standing crop plus litter. Between 4 and 7 tons. $ha^{-1}$ species density is maximal, which means a high degree of niche differentiation, hence a low degree of competition. Above 7 tons. $ha^{-1}$ some tall growing species suppress most of the other species, hence a high degree of competition. Below 4 tons. $ha^{-1}$ conditions are probably so extreme that only specialized species can survive, hence the other species disappear not because of competition, but because of the unfavourableness of the environment.

In Fig. 12 the numbers of phanerogams per sample on 3 hayed plots of a 20-year old fertilizer experiment on permanent grassland (see also page 70) are plotted against time. These results agree well with the general picture presented above. The numbers of species on the unmanured plot and on the PK-plot stay high. The average dry matter production during the last 6 years is 4.7 and 6.0 tons $ha^{-1}$, respectively, which is according to Al-Mufti et al. (1977) within the range that allows a high degree of niche differentiation. The number of species on the NPK-plot with an average yield of 8.5 tons. $ha^{-1}$ steadily decreases but it took 20 years for the number of species to drop just below 10. In this case competition for light by the tall growing grasses is the dominating factor causing the much lower species density.

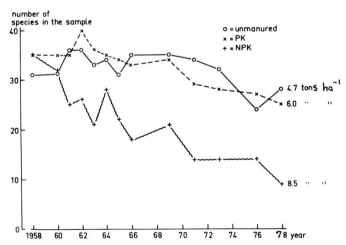

Fig. 12. Total number of phanerogam species in 50 samples of 25 sq. cm taken from plots of 16 × 2.5 sq. m (hayed plots at Wageningen, see page 70) plotted per year of observation. On the right hand side the average dry-matter yield in tons. $ha^{-1}$ during the last six years is given.

69

## 4. Dynamics of plant populations in old grassland communities

As already mentioned in the introduction there are very few examples of long duration observations on the botanical composition of old grassland.

### 4.1 *The 20-year old experiment at Wageningen (1958-1978)*

A long term experiment (Elberse 1966) has been set out 20 years ago on an old extensively grazed alternate pasture (one cut of hay every two years). It is located near Wageningen (the Netherlands) on a heavy river clay soil (clay fraction $< 20 \mu =$ 50%, organic matter 20%). Some other soil characteristics are given in Table 1. Three plots of 0.18 ha are grazed by young cattle and 12 plots of 16 x 2.5 sq. m are hayed at the end of June and a second cut is harvested in October. The fertilizer treatments of the grazed plots are O, PK, NPK (P = 40 kg $P_2O_5$ ha$^{-1}$, K = 60 kg $K_2O$ ha$^{-1}$, N = 60 kg N ha$^{-1}$). The treatments of the hayed plots (in duplo) are O, P, K, PK, NPK, Ca (P = 120 kg $P_2O_5$ ha$^{-1}$, K = 400 kg $K_2O$ ha$^{-1}$, N = 100 kg N ha$^{-1}$ in spring + 60 kg N ha$^{-1}$ after haying, Ca = 1000 kg CaO ha$^{-1}$). The fertilizers used are superphosphate, potassium sulphate, ammonium nitrate and limestone.

Besides determining the yield, the botanical composition is analysed by the 25 sq. cm-frequency method of de Vries (1937). On each plot 50 samples of 25 sq. cm are taken at random in May and the presence of each species in each sample is recorded. From this a frequency percentage (F%) is calculated. At the start of the experiment the species density was 58 in an area of 0.05 ha. The main species were (in order of decreasing F%): *Festuca rubra, Agrostis (tenuis + stolonifera), Anthoxanthum odoratum, Holcus lanatus, Rumex acetosa, Trisetum flavescens, Alopecurus pratensis* and *Lolium perenne.*

### 4.1.1 Unmanured plots

In fact, on none of the unmanured plots the original treatment (alternate use) is continued because the plots are either grazed or hayed. In 20 years the pH-KCl of the grazed plot decreased from 4.8 to 4.4 and that of the hayed plot from 4.3 to 3.8 (Table 1). The phosphorus content did not change, and the potassium content of the grazed plot remained the same, whereas that of the hayed plot decreased to 20%.

*Table 1.* Some soil characteristics of the experiment at Wageningen in different years.

| treatment year | pH-KCL grazed 0 | PK | NPK | pH-KCL hayed 0 | PK | NPK | P-AL grazed 0 | PK | NPK | P-AL hayed 0 | PK | NPK | K‰ grazed 0 | PK | NPK | K‰ hayed 0 | PK | NPK |
|---|---|---|---|---|---|---|---|---|---|---|---|---|---|---|---|---|---|---|
| 1958 | | — 4.8 — | | | — 4.3 — | | | — 6 — | | | — 6 — | | | — 25 — | | | — 30 — | |
| 1963 | 4.6 | 4.6 | 4.7 | 4.2 | 4.1 | 4.3 | 8 | 11 | 11 | 8 | 31 | 25 | 22 | 25 | 24 | 16 | 60 | 41 |
| 1969 | 4.4 | 4.4 | 4.6 | 4.0 | 4.0 | 4.0 | 9 | 18 | 21 | 9 | 60 | 45 | 24 | 31 | 26 | 18 | 86 | 62 |
| 1976 | 4.4 | 4.3 | 4.5 | 3.8 | 3.9 | 3.8 | 9 | 23 | 26 | 9 | 81 | 57 | 25 | 33 | 33 | 20 | 82 | 48 |

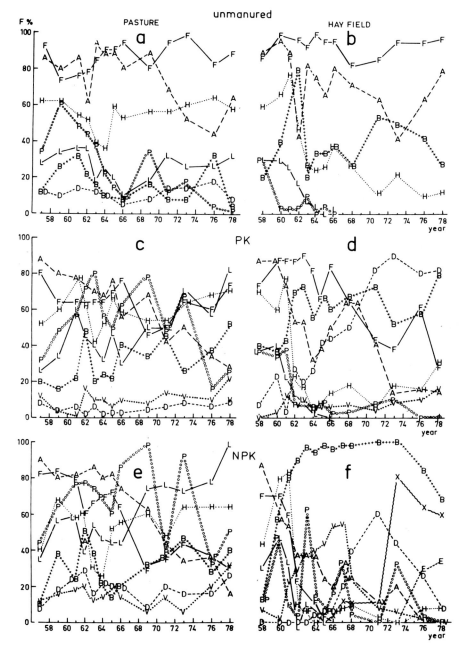

Fig. 13. Frequency percentages of some grass species on different plots of the experiment at Wageningen (see text) plotted against time.

| | | | | |
|---|---|---|---|---|
| A | = | *Agrostis (tenuis + stolonifera)* | H | = | *Holcus lanatus* |
| B | = | *Alopecurus pratensis* | L | = | *Lolium perenne* |
| D | = | *Dactylis glomerata* | P | = | *Poa trivialis* |
| E | = | *Elytrigia repens (= Agropyron repens)* | V | = | *Poa pratensis* |
| F | = | *Festuca rubra* | X | = | *Arrhenatherum* |

In Figs. 13a and b the frequency percentage of some grasses is plotted against time. The most striking feature is the dynamic behaviour of the populations, especially on the hayed plot. On this plot *Lolium perenne, Poa trivialis* and *Trifolium repens* (not presented) already have disappeared from the sample after a few years. Other species like *Alopecurus pratensis* and *Holcus lanatus* showed a sharply rising F% during the first years and a sudden decrease shortly afterwards. After some ten years of hay treatment *Holcus lanatus* and *Trisetum flavescens* (not presented) had changed to low frequency species (below F% = 10).

According to the flutuation pattern of the F% the species can be classified in 4 groups. The first group of species, like *Agrostis* spp., *Anthoxanthum odoratum* (not presented), *Alopecurus pratensis, Festuca rubra* and *Dactylis glomerata* did not show a special trend, though the F% is more or less fluctuating.

A second group of species showed a declining F%, such as *Lolium perenne, Poa trivialis* and *Holcus lanatus* on the hayed plot, and *Poa trivialis* on the grazed plot. The rapid disappearance of *Lolium perenne* on the hayed plot is in good agreement with the results of Kruijne et al. (1967): the relative average frequency of *Lolium perenne* decreased accordingly as grazing intensity decreases.

A third group consisted of species that have the tendency to increase. No species of this group are present on the unmanured plots.

To a fourth group belong some species that showed a remarkable rise-and-fall curve of long duration. On all the hayed plots (except for the NPK treatment) the F% of *Plantago lanceolata* (Fig. 14) increased after 1966 attaining a maximum within 5 to 7 years and falling back to its original level in 1978. On the grazed unmanured plot the F% of *Plantago lanceolata* fluctuated around 5, whereas on the grazed manured plots this species disappeared within 10 years.

## 4.1.2 PK plots

Comparing Figs. 13a-b and 13c-d it becomes obvious that species may change from one group to another, depending on the fertilizer treatment. Most marked on the grazed PK plot were the increasing F% of *Lolium perenne*, the decreasing F% of *Agrostis* and the widely fluctuating F% of *Poa trivialis*, whereas on the hayed plot *Dactylis glomerata* increased considerably. It is likely that the increase of the productive grass species *Lolium perenne* and *Dactylis glomerata* is caused by the improved fertility: in 20 years P-Al on the grazed and hayed PK plots increased from 6 to 23% and 81% respectively, whereas potassium increased on the grazed PK plot from 25 to 33°/oo and on the hayed PK plot from 30 to 82°/oo (Table 1).

## 4.1.3 NPK plot

The F% curves of the grazed NPK plot closely resemble those of the grazed PK plot (compare Figs. 13e and 13c): e.g., *Lolium perenne* showed an increase and *Agrostis* showed a decrease. The F% of *Poa trivialis* on the grazed NPK plot fluctuated even more than that on the grazed PK plot. Very striking are the synchronous ups and downs of this species on these two plots.

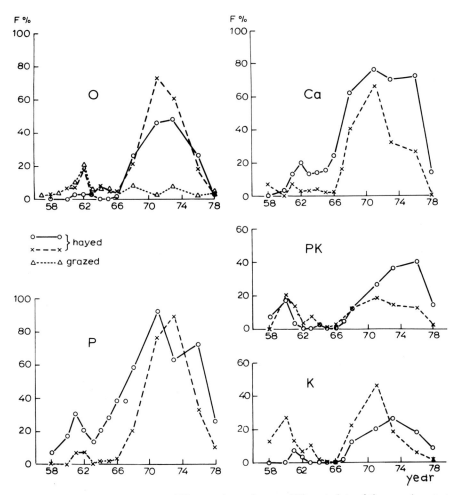

Fig. 14. Frequency percentages of *Plantago lanceolata* on different plots of the experiment at Wageningen plotted against time (the hayed plots are replicated).

On the hayed NPK plot *Alopecurus pratensis* attained a very high F% in 4 years time (Fig. 13f) and this is maintained for many years. Other species like *Poa pratensis*, *Dactylis glomerata* and *Arrhenatherum elatius* (and perhaps *Elytrigia repens* in the near future) showed one after another very distinct rise-and-fall curves. On the replicate (not presented) the same phenomenon occurred, but instead of *Arrhenatherum elatius*, *Anthriscus silvestris* rose very rapidly after 1971. This highly dynamic character of the vegetation with a succession of tall growing grasses is the result of the increasing fertility by the application of complete fertilizers (Table 1).

4.1.4  Discussion

It may be wondered why some populations show such wide annual fluctuations in F%. A few possible reasons are: (1) sampling errors; (2) external effects (changing weather conditions, diseases, pests, competition, etc.); or (3) internal effects (endogenous processes in the population, such as periodicity in the extent of flowering).

The first reason seems unlikely, because there are also species with quite different growth habits (tussock- and sod-forming) that did not show these wide annual fluctuations, like *Festuca rubra, Dactylis glomerata, Lolium perenne, Poa pratensis* and *Plantago lanceolata*.

It is generally accepted that weather conditions may have a considerable effect on the botanical composition. However, contrary to these expectations no correlation was found between changes in F% and the local monthly observations on precipitation, air temperature and evaporation surplus.

It seems unlikely that diseases and pests should have caused these fluctuations: firstly, because as such it was never observed in the field, and secondly, because it may be assumed that the species density in this field is a good buffer against explosive developments of diseases and pests (Harper 1977).

Competition may also change the botanical composition; however, this process in a perennial sward will doubtlessly take place at a much more gradual rate than indicated by the F%-curves.

It is generally accepted that in a perennial sward the species also reproduce via seed, but to what extent this takes place or how this type of reproduction is related quantitatively to the vegetative one, is as yet unknown. A striking phenomenon is that in some years certain species flower abundantly, while in other years they are not particularly noticeable. This periodicity in flowering could have serious consequences on the dynamics of a population (Sterk 1975). The time of putting the cows to pasture and the time of cutting with respect to the development stage of the individuals in the populations will vary from year to year and with it the amounts of viable seeds. However, as long as it is not known to what extent germination and establishment of these seeds markedly affect the F% the latter supposition remains speculative.

The striking behaviour of *Plantago lanceolata* on the hayed plot, where a complete rise-and-fall curve was developed within 12 years (Fig. 14), might be explained by assuming an abnormally high seed production having occurred in 1967. Delayed germination, caused by an extended dormancy of part of the annual seed crop (see also Sterk 1975), may be the reason why the increase of the F% is so gradual (5 to 7 years). Why this population returned to its original level in 1978 is obscure.

Of all the plots the grazed unmanured plot showed the least changes in F%. Of the species present with an average F% > 10 only *Poa trivialis, Trisetum flavescens* and *Cynosurus cristatus* (the latter two species are not presented) showed a distinct decrease of F%.

On the hayed NPK plot the dynamics of the productive species successively replacing each other was extremely great. Why *Poa pratensis* and *Dactylis glomerata* showed complete rise-and-fall curve, whereas *Alopecurus pratensis* was able to be present with a high F% for a very long time, is obscure. Neither can the question be answered, whether the decrease in *Poa pratensis* is due to the increasing competition of *Dactylis glomerata* or perhaps to endogenous factors in the population of *Poa pratensis*. The non-synchronous trend of these rise-and-fall curves on the replicate (not presented) render it unlikely that weather conditions have determined this process.

The unstability of these eutrophicating hayed plots is clearly demonstrated by the fact that in 1973 *Arrhenatherum elatius* on the one replicate, and *Anthriscus silvestris* on the other, suddenly formed a major contribution to the botanical composition.

## 4.2 *The 120-year old Park Grass Plots at Rothamsted (1856-1976)*

The famous Park Grass Plots give much information about the changes in the botanical composition in percentage contribution to hay weight (hayed at the end of June) since 1858 (recent report by Williams 1978). The years of observation, however, are very irregularly distributed throughout the period of 120 years with some intervals of more than 25 years (1877-1903 and 1948-1975). Therefore, these data only give an incomplete picture of the dynamics in a grassland community. Moreover, weight percentages are less stable than frequency percentages, because of their dependency on the season (de Vries 1940).

When the experiment started, the 'Park' had already been under grass for at least a century. Prior to the experiment the land was manured with farmyard manure, road scrapings and the like, and sometimes with guano or other purchased manures. Until 1872 the aftermath was always grazed by sheep, thereafter the grass crop was carried off the field.

### 4.2.1 Unmanured plot (plot 3)

In Fig. 15a the weight percentages of only 7 species are given. From 1921 to 1926 and from 1936 to 1940 yearly observations were made. The fluctuations from year to year were even more pronounced than those discussed on pages 70-76, making it very difficult to point out certain tendencies. Nevertheless, some populations showed a clear trend. *Lolium perenne*, for instance, in the long run almost disappeared (traces only), while *Holcus lanatus* was able to maintain itself with very low percentages. Other species, like *Briza media* and *Leontodon hispidus*, hardly present in the first 20 years, rose much later to some 20% hay weight. Also the small contributions of *Festuca rubra* (8%) and of *Agrostis tenuis* (7%) in 1858 compared to those in 1976 (32% and 23%, respectively) clearly demonstrated that, despite the constant management, in the long run considerable changes may occur.

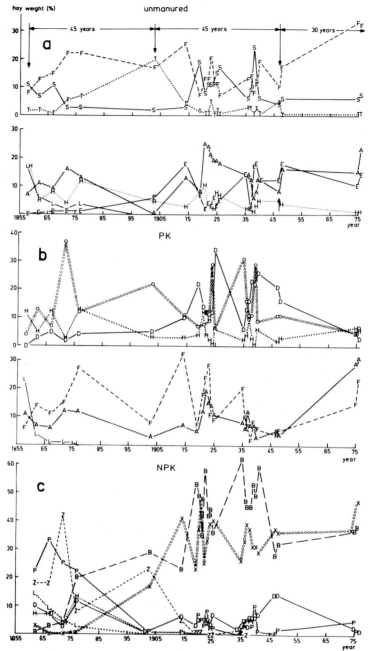

Fig. 15. Hay weight percentages of some species on three different plots of the Rothamsted Park Grass Plots plotted against time. After Williams (1978).

A = *Agrostis tenuis*
B = *Alopecurus pratensis*
D = *Dactylis glomerata*
E = *Leontodon hispidus*
F = *Festuca rubra*
H = *Holcus lanatus*
L = *Lolium perenne*

P = *Poa trivialis*
S = *Plantago lanceolata*
T = *Briza media*
Z = *Bromus mollis*
O = *Lathyrus pratensis*
X = *Arrhenatherum elatius*

## 4.2.2 PK plot (plot 7)

This plot received 35 kg $P_2O_5$ as superphosphate and 225 kg $K_2O$, 15 kg Na and 11 kg Mg as sulphates per ha per year. The four most important species on the PK plot were *Festuca rubra* and *Agrostis tenuis* (as on the unmanured plot), and *Dactylis glomerata* and *Lathyrus pratensis* (Fig. 15b). The yearly fluctuations have been tremendous. For example, in the years 1923-'24-'25 *Festuca rubra, Lathyrus pratensis* and *Dactylis glomerata* successively reached peaks of about 30% hay weight, whereas in the non-peak years of this period their weight percentages amounted to about 10. Another remarkable observation is the weight percentage of *Agrostis tenuis* and *Festuca rubra* in 1976 (together 54%), which is unique in the 120 years period.

## 4.2.3 NPK plot (plot 14)

The fertilizers applied were the same as in the PK plot plus 96 kg N ha$^{-1}$ as sodium nitrate.

It has taken *Alopecurus pratensis* and *Arrhenatherum elatius* 60 years before they definitely were the leading species, leaving the other species far behind (Fig. 15c). During the second period of 60 years these two dominating grasses have fluctuated in opposite directions, resulting in a very dynamic equilibrium.

## 4.2.4 Discussion

Here too, the tremendous fluctuations in the botanical composition attract attention. In a few extreme cases a possible relationship is pointed out between a drought period of long duration and the increase or decrease in certain species (Williams 1978), but in general the relationships between climate and changes in the botanical composition remain obscure. It is difficult to indicate particular trends because of the very great gaps in the years of observation. Nevertheless, even on the unmanured plot in the long run some populations decreased and others increased. The species that were of any importance 120 years ago are still there, though sometimes in traces only. It is demonstrated that on the NPK plot certain definite changes needed 60 years to take place. The rather low fertilizer level in comparison with that of the hayed NPK plot at Wageningen obviously caused merely very slow changes.

## 4.3 *The 34-year old micro plots near Cambridge (1936-1970)*

During 34 years Watt (1960, 1971) made yearly (the last 12 years monthly or even bimonthly) observations on the dynamics of a grass-heath community within an area of 10 x 160 sq. cm, subdivided into 1024 squares of $1\frac{1}{4}$ x $1\frac{1}{4}$ sq. cm. This rather open vegetation (on average half of the area consisted of bare soil) showed a very dynamic picture.

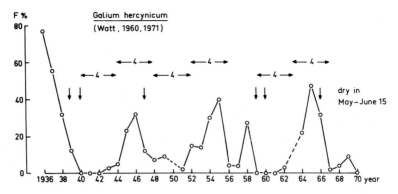

Fig. 16. Frequency percentages of *Galium hercynicum* in the first half of July during 34 successive years on a micro plot in a grass-heath community. After Watt (1960, 1971).

One of the most spectacular results of his study is given in Fig. 16. The F% of *Galium hercynicum* in the first half of July showed very distinct rise-and-fall curves within cycles of 2 x 4 years (except for the anomaly around 1958). Watt explained the periodic break-downs of this population by the occurrence of drought periods in May-June 15 (indicated in Fig. 16). However, in 1956 and 1967 there was no such dramatic drought period and the population still collapsed. Perhaps an endogenous rhythm of the population, like periodically increased and decreased seed production, may play a role but this has not been recorded systematically.

## 5. Conclusions and suggestions

From three long term experiments (20, 34 and 120 years) on old grassland communities the following conclusions may be drawn.
1. The fluctuations in plant populations from year to year may be considerable. This may be partly due to seasonal effects; however, frequency percentages are supposed to be rather constant during the season.
2. Except for some extreme drought periods, the relationship between these fluctuations and weather is obscure.
3. On the unmanured hayed plots at Wageningen (Fig. 13b) and at Rothamsted (Fig. 15a), as well as on the micro plot near Cambridge (Fig. 16) populations occur showing a definite trend (increasers, decreasers or rise-and-fall populations). This may partly be the result of a changed management at the beginning of the experiment (pasture → pure hayfield, lightly fertilized → unmanured), and partly of factors still obscure. Especially the pronounced rise-and-fall curves of *Plantago lanceolata* (Fig. 14) and *Galium hercynicum* (Fig. 16) suggest an endogenous mechanism. One of the possibilities is periodicity in the seed production, but unfortunately no data were collected on this aspect. By starting comparable experimental plots some years after another, populations of differing age can be compared. In

this way populations of one species showing a rise-and-fall curve can be compared in the rising and falling stage at the same time.

4. Of all the plots discussed, the grazed unmanured plot at Wageningen (Fig. 13a) shows the least fluctuating curves. This may point in the same direction: with grazing seed production is suppressed and therefore periodic changes in seed production and in number of established seedlings may not be manifested.

5. The successive rise-and-fall curves of the productive grasses *Poa pratensis, Dactylis glomerata, Arrhenatherum elatius* and *Elytrigia repens* (?) on the hayed NPK plot at Wageningen (Fig. 13f) probably are of a different nature than those of *Plantago lanceolata* (Fig. 14) and *Galium hercynicum* (Fig. 16) because on the former plot soil fertility is steadily increasing. Here it looks more like a succession of populations in which competition plays an important role: the first population is suppressed by a second one and the second population in its turn by the next one, etc. The vegetation is still on the move! On the other hand, on the NPK plot at Rothamsted after 60 years a very dynamic equilibrium was established between *Alopecurus pratensis* and *Arrhenatherum elatius* (Fig. 15c). Whether this equilibrium was caused by niche differentiation, or by host-parasite relationships, or by climatic fluctuations is unknown. It should be noted that on this plot probably the same kind of successive rise-and-fall curves occurred during the first 20 years, but here it concerned *Poa trivialis, Bromus mollis* and *Alopecurus pratensis*. The much smaller fertilizer dose is most likely responsible for the slow start of the exacting species *Alopecurus pratensis* (compare Fig. 15c with Fig. 13f).

6. The examples presented have shown that in addition to very rapid fluctuations (from year to year), long lasting fluctuations (about 8 years) and very slow trends (over 20 years) may occur in old grassland communities. For management of the vegetation it is important to know whether these fluctuations and trends are due to cyclic processes or to permanent changes. To gain insight on this aspect, it is necessary to make observations on, e.g., periodicity in the extent of flowering, seed production, germination and establishment, mortality, etc. Besides these endogenous cyclic processes, of course, external cyclic processes may affect the population, like diseases and pests (overground as well as underground), fluctuations in the amount of litter, etc. When these processes are better understood permanent changes can be distinguished and counter measures may be taken.

7. Competition experiments with sown or planted mixtures and monocultures are very usefull to get more insight in the way populations interfere, and in the mechanisms by which equilibria may become established.

In general these experiments will not be of predictive value with regard to the dynamics of populations in old perennial communities, because of the very complicated nature of the interactions between plant populations and environmental factors in these experiments. However, with regard to the dynamics of populations in annual communities and very simple mixtures of perennials models have been developed (Torssell & Nicholls 1978).

# References

Al-Mufti, M.M., C.L. Sydes, S.B. Furness, J.P. Grime & S.R. Band. 1977. A quantitative analyses of shoot phenology and dominance in herbaceous vegetation. J. Ecol. 65: 759-791.

Bergh, J.P. van den. 1968. An analysis of yields of grasses in mixed and pure stands. Agric. Res. Rep. 714: 1-71.

Bergh, J.P. van den & W.G. Braakhekke. 1978. Coexistence of plant species by niche differentiation. In: A.H.J. Freysen & J.W. Woldendorp (ed.), Structure and functioning of plant populations. Verh. Kon. Ned. Akad. Wet., Afd. Natuurk. 2e reeks, 70: 125-138. North-Holland Publ. Co., Amsterdam.

Bergh, J.P. van den & W.Th. Elberse. 1970. Yields of monocultures and mixtures of two grass species differing in growth habit. J. appl. Ecol. 7: 311-320.

Braakhekke, W.G. 1979. Coexistence of plant species. Thesis, Wageningen (in prep.).

Elberse, W.Th. 1966. Invloed van gebruik en bemesting op botanische samenstelling en produktie van verwaarloosd grasland. I.B.S. (Wageningen), Versl. No. 40: 1-12.

Ennik, G.C. 1970. White clover grass relationships: competition effects in laboratory and field. In: J. Lowe (ed.), White clover research. Brit. Grassl. Soc., Occ. Symp. No. 6: 165-174. Belfast.

Harper, J.L. 1977. Population biology of plants. Academic Press, Oxford.

Kruijne, A.A., D.M. de Vries & H. Mooi. 1967. Bijdrage tot de oecologie van de Nederlandse graslandplanten. Versl. landbouwk. Onderz. 696: 1-65.

Montgomery, E.G. 1912. Competition in cereals. Bull. Nebr. agr. exp. Sta. XXIV, art. V: 1-22.

Raininko, K. 1968. The effects of nitrogen fertilization, irrigation and number of harvestings upon leys established with various seed mixtures. Suom. maatal. Seur. Julk. 112: 1-137.

Sandfaer, J. 1970. An analysis of competition between some Barley varieties. Danish Atomic Energy Comm., Risö Rep. 230: 1-114.

Sterk, A.A. 1975. Demographic studies of Anthyllis vulneraria L. in the Netherlands. Acta Bot. Neerl. 24: 315-337.

Torssell, B.W.R. & A.O. Nicholls. 1978. Population dynamics in species mixtures. In: J.R. Wilson (ed.), Plant relations in pastures. CSIRO Symp. Brisbane. pp. 217-232.

Trenbath, B.R. 1974. Biomass productivity of mixtures. Adv. Agron. 26: 177-210.

Vries, D.M. de. 1937. Methods of determining the botanical composition of hayfields and pastures. Rep. 4th Intern. Grassl. Congr., Aberystwyth. pp. 474-480.

Vries, D.M. de. 1940. Verslag van een vergelijkend onderzoek van een drietal methoden van botanisch graslandonderzoek, in verband met de grootte der seizoensverschillen in samenstelling der graszode. Versl. landbouwk. Onderz. No. 46(6)A: 313-341.

Watt, A.S. 1960. Population changes in acidiphilous grass-heath in Breckland, 1936-57. J. Ecol. 48: 605-629.

Watt, A.S. 1971. Factors controlling the floristic composition of some plant communities in Breckland. In: E. Duffey & A.S. Watt (eds.), The scientific management of animal and plant communities for conservation. Brit. Ecol. Soc. Symp. No. 11: 137-152. Blackwell, Oxford.

Williams, E.D. 1978. Botanical composition of the Park Grass Plots at Rothamsted, 1856-1976. Rothamsted Exp. Sta., Harpenden. pp. 1-61.

Wit, C.T. de. 1960. On competition. Versl. landbouwk. Onderz. 66(8): 1-82.

Wit, C.T. de & J.P. van den Bergh. 1965. Competition between herbage plants. Neth. J. agric. Sci. 13: 212-221.

Wit, C.T. de, P.G. Tow & G.C. Ennik. 1966. Competition between legumes and grasses. Agric. Res. Rep. 687: 1-30.

# 4. PHYTOSOCIOLOGY IN THE NETHERLANDS: HISTORY, PRESENT STATE, FUTURE

## V. WESTHOFF

1. Introduction
2. Historical development from c. 1925-c. 1950
3. Description and classification of vegetation from c. 1950 onwards: symmorphology and syntaxonomy
   3.1. Survey based on some landscape units
       3.1.1. Coastal dunes, salt marshes, marine tidal belt
       3.1.2. Eutrophic wetlands and freshwater tidal delta
       3.1.3. Heaths, moorland, pools and bog
       3.1.4. Woodland and scrub
       3.1.5. Grassland
   3.2. Methodology; general concepts
4. Synecology and vegetation dynamics
5. Dutch studies abroad
   5.1. Europe
   5.2. Outside Europe
6. Evaluation; trends of future development
   References

# 4. PHYTOSOCIOLOGY IN THE NETHERLANDS: HISTORY, PRESENT STATE, FUTURE

## V. WESTHOFF

Dedicated to D.M. de Vries, the great pioneer of vegetation science in the Netherlands.

## 1. Introduction

To the development of phytosociological research in the Netherlands in the years 1925-1950 both the 'Nordic approach' and the 'Braun-Blanquet approach' contributed significantly. Proceeding from rivalry to cooperation and finally to integration, their co-existence had a major impact on the comprehensiveness and the special character of Dutch phytosociology. From 1950 onwards vegetation science showed a more uniform orientation and emphasis fell on the concepts of the 'Braun-Blanquet approach'. As this approach has been discussed at length in a recent review (Westhoff & van der Maarel 1973, 1978) it was decided to restrict this paper to a detailed review of the Dutch contributions to phytosociological theory, methods and achievements thus obtained. Additionally, an evaluation will be given and trends of future development of vegetation science will be discussed briefly.

The term 'phytosociology' is used here as a synonym for 'vegetation science' and 'phytosociology'. It can be defined according to Braun-Blanquet (1964, p. 2, transl. in Westhoff 1970, p. 22): 'Phytosociology is studying all phenomena and effects regarding the social life of plants'.

The common Anglo-American understanding, using the designation 'vegetation ecology' or even 'ecology' 'tout court', is rejected (see Westhoff 1970); this usage is neither practical nor logical (Egler 1942, 1951, Major 1958, 1961). 'Ecology' in the sense of 'study of vegetation' is a term both too wide and too narrow. Too wide, because 'ecology' is concerned with plant communities as well as with individual organisms or taxa. Hence, the definition of 'ecology' by Odum (1963) as 'the structure and function of ecosystems' has to be rejected. The term 'ecology' is too narrow, because 'ecology' in Europe as well as in America has been used generally in the sense of 'study of the relations between living beings and their environment', whereas phytosociology is dealing with much more aspects than the particular problem of the relation between the community and its environment.

According to the seven approaches enumerated in the empirical system of biological sciences by Tschulok (1910), phytosociology can be divided into seven approaches too (Du Rietz 1921, Braun-Blanquet 1928, 1951, 1964, Westhoff 1970). Without entering on some minor variations of opinion we arrive at the following subdivision:

(1) Symmorphology: sub-science studying the specific floristic composition and structure of phytocoenoses;

(2) Synecology: sub-science studying the interrelations of phytocoenoses and of coena as well as their interrelations with the environment;

(3) Syndynamics, vegetation dynamics or (narrower) succession study: subscience studying the development of phytocoenoses out of and into other phytocoenoses and exploring the laws which determine their arising and their decay;

(4) Synepiontology: sub-science dealing with the development of a given coenon as a type in the course of history;

(5) Synchronology: sub-science studying the evolution of vegetation, by preference with palynological methods;

(6) Synchorology: sub-science studying the arrangement of phytocoenoses and coena in space, as well as their occurrence and their distribution over the earth and their geographic differentiation;

(7) Syntaxonomy or synsystematics: sub-science studying the delimitation of vegetation units, and their hierarchical classification.

On the base of this survey it may be clear that it is incorrect to consider the terms 'synecology' and 'study of vegetation' as synonymous, as has been done by Schröter & Kirchner (1902), van der Klaauw (1936), Bremekamp (1962), Daubenmire (1968), and others.

The fifth aspect — that of synchronology — also may be considered as a subscience of palaeobotany (palynology), historical geobotany, or historical phytogeography. In the Netherlands the sub-sciences of phytosociology and palynology have been closely cooperating from the beginning through their organizational junction into one of the oldest sections, or Commissions, of the Royal Botanical Society of the Netherlands. The 'Commission for the Study of Vegetation' which was formed in 1933, be it under an other name (see Vroman, this volume), has organized symposia-meetings, increasing in number from one to three per year, in which both phytosociologists and palynologists participated. This integrated approach has proved to be a fortunate one. One of its results has been the interpretation of palynological data with syntaxonomic concepts, contributing to a more detailed and adequate understanding of human impact as a major factor in vegetation history, at least since Roman times. The subject of palynology is dealt with by C.R. Janssen in this volume.

Considering the six other sub-divisions of phytosociology, it may be stated that they all have been dealt with in the Netherlands, however not all of them with equal accent. Much work has been done on (1) symmorphology and syntaxonomy, (2) synecology, particularly in its aspect of competition research (see P. van den Bergh, this volume), (3) syndynamics, and (4) synchorology, mainly as vegetation mapping. It has to be stressed, however, that particularly the more elaborate vegetation monographs deal with several themes, aiming at an integrated study of symmorphology and syntaxonomy as well as that of synecology, synchorology and syndynamics; examples (in chronological order) are the studies by van Dieren (1934a), Feekes (1936), Meyer Drees (1936), Diemont (1938), Adriani (1945), Westhoff (1947), Sissingh (1950, 1952), Barkman (1958a), Maas (1959), Boerboom

84

(1960), Zonneveld (1960), Beeftink (1962, 1965, 1966, 1968, 1977), Segal (1969), Werger (1973), van Gils (1978).

Moreover, phytosociological research has evolved into some specialized, partly applied, directions, viz.: (1) application in grassland husbandry, mainly by D.M. de Vries and his school (see 3.1.5.); (2) study of autecology with vegetational methods, mostly aiming at a more adequate description of geographical variation in the ecological demand of species (e.g. Adriani & van der Maarel 1978, Blom 1974, Doing 1963c, Freysen 1967a, b, 1970, Kneepkens & Verhoeven 1975, Koch 1974, Landolt 1977, Neuteboom 1974, Pegtel 1974, 1976, Segal 1969, Segal & Westhoff 1959, van der Voo & Westhoff 1961, Westhoff 1947, 1950c, 1951-1966, 1958c, 1959, 1965b, 1968a, b, 1971c, Westhoff et al. 1970-1973, Westhoff & Doing Kraft 1959, Westhoff & Heimans 1949, Westhoff & Ketner 1967, Westhoff & van Leeuwen 1960, Westhoff & Mörzer Bruijns 1956, Westhoff & Passchier 1958, Westhoff & Reinink 1967); (3) biocoenological research, in which one or more animal taxocoenoses have been treated as components of the ecosystem and not merely as ecological factors (e.g. van der Aart 1975, Barkman & den Boer 1961, van Heerdt & Bongers 1967, van Heerdt & Mörzer Bruijns 1960, van Leeuwen 1953, van der Maarel 1965, Mörzer Bruijns 1945, 1947, 1953, 1954, Mörzer Bruijns & Westhoff 1951, Westhoff & Westhoff-de Joncheere 1942); (4) application in nature conservation and conservational management, dealt with by P.A. Bakker in this volume.

The following attempt to render the historical development of plant sociological research in the Netherlands may give the impression that the main impulse of this development is historical indeed, and that phytosociological research tended to decline during the last decades. This impression, however, would be at variance with the real situation; it would be the result of the method used by the editor of this volume. It is a normal process in a developing science that it diverges more and more. This centrifugal trend has brought about that four major topics of phytosociological research, which have evolved during the last 25 years, are dealt with by other authors, viz. competition ecology, symmorphology (vegetation structure research), nature conservation research, and numerical data processing. As a consequence, this review starts with the broad field of vegetation science and gradually narrows to only some of its aspects.

## 2. Historical development from c. 1925-c. 1950

The first precursor of vegetation research in the Netherlands was Franciscus Holkema in his admirable study 'De plantengroei der Nederlandse Noordzee-eilanden' (1870), which means: 'Flora and vegetation of the West-Frisian islands'. It presents a quantitative method of vegetation analysis using cover and abundance estimations, apparently unaware of the first quantitative vegetational approach by von Post (1851). Though Holkema did not deal with homogeneity and minimal area, nor did aim at any classification of vegetation or any synecological approach, there is no doubt that this method presents the starting point of vegetation analysis in the Netherlands.

It took nearly half a century before a coherent vegetation study started to develop. This starting point was marked by Bijhouwer (1926) in his thesis, a geobotanical investigation of the ecocline area between the dune system rich in lime and that poor in lime near Bergen (Alkmaar, North-Holland). This study, based on the methods of Hult (1881) and Sernander (1894, 1898), was followed by the thesis of D.M. de Vries (1929) on the vegetation of the semi-natural moist grasslands (litter fen, *Cirsio-Molinietum* in the present classification) of the Krimpenerwaard. The latter study had a much greater impact than the former. De Vries turned out to be the pioneer of a group of young scientists declaring themselves the 'Nordic approach' ('Noordse School'); the main other adherents were Scheygrond (1932), van Dieren (1934a) and Feekes (1936). Largely based on the work of Du Rietz (1921, 1930a, b) their main diagnostic criteria in vegetation description were dominance and constancy of species. They started from synusiae as the fundamental components of vegetation. Their basic unit, 'Verband', is identical to 'society' in the sense of Trass (1964); as to the next higher units, their 'horizontaal complex-verband' is more or less identical with 'stratocoenose' sensu Balogh (1958), their 'verticaal complex-verband' with 'sociatio' sensu Du Rietz 1930 (a, b) (cf. Trass & Malmer 1973).

Application of the 'Nordic approach' appeared to be most successful in a coarse vegetation pattern characterized by sharp patch boundaries and built up by stands with one or two dominant species. Such communities are to be expected in pioneer vegetation or in other highly dynamic ecosystems summarized as 'convergent systems' in the sense of van Leeuwen (1965). In a country like the Netherlands, sites of both types are well represented: pioneer communities are found in the coastal dunes, the salt marshes, the 'verlanding' — autogenic succession in eutrophic fresh water, — and the reclaimed polders of the former Zuyder Zee; in many other areas, disturbance by human impact is adding dynamics to the ecosystem, bringing about the coarse-grained vegetation pattern which is suitable for the 'Nordic approach'. A good example of a 'Nordic' study of the latter type of ecosystem is the investigation of the reed swamps with *Sphagnum* undergrowth ('*Arundinetum-Sphagnetum*') by Scheygrond (1932).

The last major achievement of the joint workers of the 'Nordic approach' has been their monograph of the small Wadden Sea island of Griend (Brouwer et al. 1950), precursed by preliminary studies by van Dieren (1934b), Feekes (1940) and Westhoff (1940).

Partly because of the untimely death of J.W. van Dieren († 1935), who was the talented leader of the 'Nordic' group, this 'school' lost much of its former impetus. It found, however, a most important continuation in the grassland research by D.M. de Vries and his collaborators, which will be dealt with in section 3.1.5.

Some years later than the 'Nordic approach', about 1930, the floristic-sociological approach of Braun-Blanquet (1928), at that time designated as 'French-Swiss school', started to find adherents in the Netherlands. The method was introduced mainly by five scientists: W.C. de Leeuw, Th. Weevers, J. Heimans, G. Kruseman and J. Jeswiet.

W.C. de Leeuw, then a retired chemist and a personal friend and disciple of Braun-Blanquet, was (1) a good organizer, (2) an eloquent and gifted teacher of theory and methods, and (3) a research worker, mainly on dry grasslands (e.g. de Leeuw 1938). His publication on the plant communities of the West-Frisian island of Ameland (Braun-Blanquet & de Leeuw 1936) is a classic pioneer study of Western European coastal vegetation and can be considered as a first Braun-Blanquet type counterpart of the 'Nordic' study of the island of Terschelling by van Dieren (1934a) mentioned above. De Leeuw's compilation work 'The Netherlands as an environment for plant life' (1935b) proved to be a useful documentation. He also compiled the first Dutch plant sociological bibliography (de Leeuw 1935a).

Th. Weevers, professor of botany at the University of Amsterdam, was a physiologist as well as a floristic botanist and a phytosociologist. His main interest was focussed on the coastal vegetation (Weevers 1936, 1939, 1940), but he also was among the first scientists studying woodland communities (Weevers 1934a, b. 1938). His main disciples were G. Kruseman and M.J. Adriani, the former studying the vegetation of arable land (Kruseman & Vlieger 1939), the latter studying halophytes and halophyte communities. Adriani was a disciple of Braun-Blanquet too and studied the syntaxonomy and synecology of salt marshes in Montpellier and in the South-Western estuary of the Netherlands (1946).

J. Heimans, after World War II professor of systematic botany and genetics at the University of Amsterdam, stressed the importance of the dispersal factor in vegetation development and introduced the concept of 'accessibility' (Heimans 1933, 1940). He also introduced (Heimans 1939) the term 'kensoort' in Dutch to avoid the germanism 'karaktersoort', then the usual denomination of a character species; later the Dutch term returned to German as 'Kennart' (Tüxen 1950, Westhoff & van der Maarel 1973). In the successive editions of his 'Flora of the Netherlands', Heimans (1942 seq.) introduced a key to the phytosociological alliances occurring in the Netherlands.

J. Jeswiet, professor of plant taxonomy and dendrology at the Agricultural University, Wageningen, was the first to introduce phytosociology in the Netherlands as a discipline taught on university level. As a teacher of forestry students he contacted R. Tüxen, who was at that time already the pre-eminent leader of the Braun-Blanquet approach in Germany. Though Jeswiet himself hardly did carry out any phytosociological investigations his achievement in this field was his training of W.H. Diemont, E. Meyer Drees, G. Sissingh, J. Vlieger and J.F. Wolterson, five Wageningen scholars who have significantly contributed to the development of vegetation science in the Netherlands. Diemont and Sissingh, who carried out their first investigations under the direct supervision of R. Tüxen, became the latter's main Dutch disciples; Sissingh and Vlieger were also direct disciples of Braun-Blanquet, working in Montpellier under his supervision.

E. Meyer Drees (1936) produced a detailed and classic study of the deciduous forest associations of the 'Middle East' of the Netherlands. He was, in this work, the last scholar to defend the monoclimax hypothesis on the base of pedological considerations; this was just before Tüxen & Diemont (1937) left that hypothesis and introduced the concepts 'climax group' and 'climax swarm'. In later years, Meyer Drees (e.g. 1951), working in Indonesia, published some important methodological studies.

J. Vlieger (1937a, 1938; Braun-Blanquet, Sissingh & Vlieger 1939) was largely concerned with the classification framework of higher syntaxa, many of those still carrying his author's or co-author's name. Apart from that he studied many other topics: woodland vegetation (1935, 1936, 1937b), arable land communities (Kruseman & Vlieger 1939), and mire vegetation (1938, 1939). Together with Adriani he was the first to establish long term succession research on permanent plots (Vlieger 1942). After the second World War, he lost his interest in vegetation science since it proved to be impossible to attain a stabilized nomenclature of vegetation units. A review of the phytosociological work in the Netherlands during the years 1940-1945 has been published by Vlieger in 1949.

Apart from his important contributions to nature preservation which should not be referred to in this paper, Sissingh (1950, 1952) published a comprehensive account of the plant communities of arable land and ruderal sites. A remarkable point in this study is the ethological synecology worked out by the use of life form spectra. His earlier study of the *Nanocyperion* in the Netherlands (1940, together with Diemont and Westhoff), is a classical account of these communities which nowadays, in the Netherlands as well as in the surrounding countries, have been nearly completely destroyed by the human impact on the landscape (see als Sissingh 1957). In later years, Sissingh contributed to the study of roadside communities (1969), grasslands (1942, 1974) and, above all, to the development of vegetation mapping (e.g. Sissingh & Tideman 1960). See also 3.1.4 and 5.1.

The next younger generation of vegetation scientists consisted mainly of J.J. Barkman (Leyden), whose main achievement is his classic comprehensive monograph on the ecology and phytosociology of cryptogamic epiphytes (1958a), J. Meltzer (autodidact) and V. Westhoff (Utrecht, the latter studying the plant communities of the West-Frisian islands (1943, 1947, 1950b, 1952, 1959, 1970; see also van der Maarel 1966a). Both Barkman and Westhoff are direct disciples of Braun-Blanquet, though the latter was educated also by W.C. de Leeuw and E. Meyer Drees. Meltzer (1940) was the first to describe the dune scrub association of the calcareous dunes as *Hippophaeto-Ligustretum*. Meltzer & Westhoff (1942) published the first Dutch textbook on phytosociology and Westhoff et al. (1942, 1946), the first surveys of plant communities of the Netherlands. To the same generation may be reckoned A.C. Boer (1942), working on eutrophic reed swamp communities.

Special mentioning deserve the activities and achievements of the Plant sociological Working Group of the Dutch Youth League for the Study of Nature (N.J.N.). This group, formed in 1938 and still active forty years later, consisted of autodidacts and students in biology and agronomy up to 23 years of age. The results of their investigations have been published in the hard-to-trace periodical 'Kruipnieuws'; a selection of these papers has been republished by Smittenberg (1973). The main impact of the group has been (1) an important contribution to the education of young vegetation scientists, and (2) the investigation of many interesting plant communities which now have been destroyed by agricultural intensification.

The next generation of vegetation scientists consisted mainly of disciples of J. Heimans at the University of Amsterdam (C. den Hartog, G. Londo, E. van der Maarel, W. Meyer, P.J. Schroevers, S. Segal, R.J. de Wit), those of J. Lanjouw at the University of Utrecht (J. van Donselaar, W.A.E. van Donselaar-ten Bokkel Huinink, J. Lindeman, J.Th. de Smidt, A.L. Stoffers), as well as those of D.M. de Vries, H.J. Venema and V. Westhoff at the Agricultural University, Wageningen (A. Bakker, W.G. Beeftink, Th. de Boer, J.H.A. Boerboom, A.W.H. Damman, H. Doing, J.G.P. Dirven, L.G. Kop, F.M. Maas, E. Stapelveld and I.S. Zonneveld), all mostly during the years 1950-1960.

It may be more appropriate, however, to deal with this later development in separate chapters and according to the subdisciplines involved.

In conclusion, it is stressed that the contemporaneous development of the 'Nordic' method and the Braun-Blanquet method in the Netherlands — an exceptional feature in Europe — has essentially contributed to a comprehensive, unbiassed and many-sided approach to vegetation science (see also de Vries 1939). It has proved to be fortunate that the adherents of both 'schools', though in the first years in highly controversial dispute, were discussing together in the 'Commission for Phytosociology and Peat research' of the Royal Botanical Society of the Netherlands. Such discussions were reflected, for instance, in the classic study of the coastal dunes of the West-Frisian island of Terschelling by J.W. van Dieren (1934a); this author designating the Braun-Blanquet approach by the sobriquet 'sociofloristics' (contrary to the true 'sociology' considered by him to be an achievement of the 'Nordic approach' only). There has been, of course, no question of 'elimination' of one of these approaches by the other, but a development of mutual understanding and a certain level of integration. Examples of 'Nordic' influence in the Braun-Blanquet approach are the use of sociations within the Braun-Blanquet framework (Westhoff 1949a, after the proposal by Nordhagen 1936), as well as the increasing stress on structural characters in vegetation classification (see Beeftink 1962, Tüxen 1952, 1962, van der Maarel 1966b, Westhoff 1967, Westhoff & van der Maarel 1973, 1978). It has not been stressed before that a characteristic element of the 'Dutch approach' within the general Braun-Blanquet method has likewise been a consequence of the 'Nordic approach': this is the use of all available sample plot analyses (relevés) for classification purposes, on the condition that those relevés are analytically correct. This procedure is in variance with the usual Braun-Blanquet approach, in which only the most 'typical' relevés are maintained in the final vegetation tables. It is not correct to discriminate the latter method as 'subjective', as has often been stated: it is very well possible to develop objective criteria by which a given relevé is finally accepted or rejected. The difference can be better formulated by stressing that the 'Dutch approach' tries to describe, within a given area, the vegetation pattern as a whole, whereas the usual Braun-Blanquet approach contents itself by describing the preponderant units, 'eating the raisins from the cake'. This difference is also due to the position of the Dutch vegetation workers, who, in large quantity, are studying a very small country very intensively, thus 'chewing and ruminating' every patch of vegetation, whereas other vegetation scientists – particularly outside Europe – have to deal with vast extents of vegetation cover as yet hardly studied.

### 3. Description and classification of vegetation from c. 1950 onwards: symmorphology and syntaxonomy

It would go too far to discuss the vast amount of vegetation descriptions published in the last 30 years; the reader is referred to the bibliographies by Westhoff (1961b) and Schenk (in press) and furthermore to the general account by Westhoff & den Held (1969, 1975). We will first give a short survey on the base of some major landscape units; then we will enter on some methodological contributions by Dutch authors.

### 3.1. *Survey based on landscape units*

#### 3.1.1. Coastal dunes, salt marshes, marine tidal belt

After the studies on both dunes and salt marshes by Braun-Blanquet & de Leeuw (1936), Westhoff (1947) and Brouwer et al. (1950), and those of salt marshes by Adriani (1945), D.M. de Vries (1935) and D.M. de Vries et al. (1940), the dunes of the Wadden District (poor in lime) have been studied incidentally only (e.g. den Hartog 1951, 1953, 1973, Schroevers 1951, Westhoff 1950, 1952, 1959, 1970). The interesting narrow ecotone between Wadden District and Dune District, which was the object of the first proper vegetation study in the Netherlands (Bijhouwer 1926), has been studied according to the Braun-Blanquet approach by Hoffman & Westhoff (1951). A general survey of dune woodland, dune scrub and dune heath communities of both Wadden District and Dune District has been published by Westhoff (1952). Main emphasis has been given to the dunes of the Dune District (rich in lime). Boerboom (1957) studied the dry dune grasslands of the Dune District and presented a monographic treatment of the dune system near the Hague (Boerboom 1960; see also Boerboom & Westhoff 1974). Adriani & van der Maarel (1962, 1968), van der Maarel (1961a, b, 1963, 1966a, 1975a, b, 1978a, b), van der Laan (1974), van der Maarel & Westhoff (1964), and Sloet (1976) carried out basic research on vegetation structure and classification, vegetation dynamics and synecology of the dunes of Voorne. This area stands out in two respects: (1) it still presents the best-developed wet dune slacks within the Dune District; (2) it bears the optimal and most varied dune shrub communities (*Berberidion*) in the Netherlands. In 1966, van der Maarel presented a comprehensive account of the vegetation research carried out in the Dutch coastal dune area, integrated with important fundamental considerations on the methodology of vegetation science. In later years, Doing (1958, 1960, 1966b, 1969, 1974) has largely contributed to the knowledge of the Dutch dune vegetation by studying the dune area in its totality and, by exploring and mapping vegetation complexes, introducing a generalizing approach, and in fact presenting the first achievement of landscape ecology in the Netherlands and an early version of the modern trend of describing 'sigmassociations' (see e.g. Tüxen 1973, 1977, Géhu 1974, 1976, 1977).

Among the minor studies of dune vegetation within the Dune District that of Westhoff et al. (1962) may be mentioned, one of the few presenting also quantitative data about the edaphic synecology.

The more recent studies on the salt marshes of the Netherlands are largely due to Beeftink (1957, 1962, 1965, 1966, 1977), one of the outstanding European specialists on halophyte communities. Among other salt marsh studies may be named those by Fresco (1967a, b), den Hartog (1958, 1973), den Hartog & van der Velde (1970), Joenje et al. (1976), Kortekaas et al. (1976), Mörzer Bruijns et al. (1953), Westhoff (1950b), Westhoff & Beeftink (1950), Westhoff & Mörzer Bruijns (1956).

A recent account of the vegetation of dunes and salt marshes of the Netherlands is found in Westhoff et al. (1970); a more recent comprehensive treatment of European dunes and salt marshes has been presented by Westhoff & Schouten (in press).

The algal communities of the marine tidal belt have been studied by den Hartog (1959, 1960, 1973), Nienhuis (1978) and Simons (1978).

### 3.1.2. Eutrophic wetlands and freshwater tidal delta

One of the first accounts on the vegetation of eutrophic wetlands, which are highly characteristic of the Netherlands, was the monographic treatment of the Naardermeer by van Zinderen Bakker (1936), which was followed by a more general study (1947). The study by Boer (1942) has been mentioned before. In 1949, Westhoff gave the first description of a mesohalinic wetland, viz. the 'polder' area of Botshol near Amsterdam. The Amsterdam school of J. Heimans contributed significantly to the knowledge of wetland ecosystems. From this group W. Meyer (1948, 1949, 1950, 1951, 1953, 1955) should be mentioned first. Meyer strongly opposed to the classification system of Braun-Blanquet. He used, however, the analytical method of the approach, and presented most valuable vegetation studies of both brackish and fresh wetland areas in Holland (s.s.) and Overijsel. De Wit (1951, republished 1973; 1955), van Dijk (1955), Kuiper (1958, republished 1973) and Reijnders (1959) worked in the same field, but they were more inclined to join the usual classification approach. The next important scholar after Meyer was Segal (1964, 1965, 1966, 1968a, b, Segal & Groenhart 1967, Segal & Westhoff 1959), who mainly studied the large swamp and fen area of North-west Overijsel, though he presented a thesis on wall vegetation (1969). Segal did not reject classification, but he attempted to enlarge the criteria of the system by paying more attention to structure, particularly to the dominant life form; in aquatic vegetation his classification is largely based on synusial units (den Hartog & Segal 1964).

More recently, the vegetation of ditches has been treated by de Lange (1972) and that of the oligohalinic wetlands by den Held (1973) and den Held et al. (1970,

1976). A general survey of wetland communities has been presented by Westhoff et al. (1971) and Westhoff (1973b).

A special type of wetland is found in the Fluviatile District, particularly in the former river beds. These areas have been studied by a team of investigators. Van Donselaar (1961, 1972, 1973) and Kop (1961) have published the phytosociological results of this research.

### 3.1.3. Heaths, moorland, pools and bog

The atlantic heathlands, which in the beginning of this century covered vast areas of the pleistocene part of the country, have since 1920 gradually been reduced to some scattered nature reserves (Stoutjesdijk 1953, Westhoff et al. 1972). Though they have been studied regionally or incidentally by a number of authors, e.g. van Andel & Waterbolk (1945), Waterbolk (1948), Damman (1956), Stoutjesdijk (1959), Willems (1969), Barkman & Westhoff (1969), Zonneveld (1959, 1965), it was not before 1975 that a comprehensive study of the heathland communities of the Netherlands was published (de Smidt 1975), though some precursory studies had been published earlier (de Smidt 1962, 1966, 1967). Heathland as a man-made ecosystem, and their proper management have been dealt with by several authors (Sissingh 1968, Stoutjesdijk 1953, Westhoff 1961).

Ombrotrophic bog largely has been reclaimed and dug off before phytosociological research was well developed. Therefore, bog vegetation has been studied only incidentally (e.g. Wassink 1938, 1950, Barkman 1963a, van Leeuwen 1962, Westhoff 1963, Reijnders 1967). Westhoff & van Leeuwen (1964) have attempted to present a scheme of European bogs in the transition from oceanic blanket bog via the Central European raised bog to the continental woodland bog in Eastern Europe, as well as in the thermal gradient from south to north; the Dutch and North-west German bog is considered to be a particular type, 'flat bog' or 'heathland bog' (see also Casparie 1969, 1972).

The vegetation of oligotrophic and mesotrophic ponds and pools has been thoroughly studied by a team of investigators (cf. Schoof-van Pelt 1973), but only part of the results have been published (van der Voo 1962, 1964, 1966, Schoof-van Pelt 1973). The heath pools of the Drenthian District have been dealt with by Barkman & Westhoff (1969). Van Donselaar (1970) stressed that the oligotrophic pools on postglacial drift sand that is very poor in nutrients, constitute a highly remarkable and characteristic feature of the Netherlands which merits conservation from an European viewpoint. As far as it concerns the class *Littorelletea*, Schoof-van Pelt (1973) has published a monograph of the Dutch plant communities together with those of western France, Ireland and Scotland. A minutely detailed study of the vegetation complex and the synecology of oligotrophic moorland pools as well as of the mesotrophic systems arising from it by guanotrophy and agrarian eutrophication has been published by Strijbosch (1976).

An overall treatment of the oligotrophic Dutch ecosystems can be found in Westhoff et al. (1973).

### 3.1.4. Woodland and scrub

Following the initial descriptions of forest communities dealt with in section 2, the main contribution to our knowledge of Dutch woodlands is due to Doing (including Doing Kraft) (1954a, 1962a, b, c, 1963a, c, 1969a, b). Doing is at variance with the usual Braun-Blanquet approach in two respects: (1) he uses sociological groups as differentiating taxa groups; (2) he tries to find a connection between the syntaxonomical class and the physiognomic formation. Sociological groups, presented also by Scamoni & Passarge (1959, 1963, Scamoni et al. 1965), are groups of species that are distributionally related and consequently occur in and characterize particular syntaxa (see also Westhoff & van der Maarel 1973, 1978). The method of Doing has been followed by Bakker (1969) in her study of the Dutch *Quercion robori-petraeae*.

Particular attention has been given to the position of the beech in Dutch, and in general in Western European woodlands, a problem studied by Doing & Westhoff (1959), Sissingh (1963) and Westhoff (1963).

Among the regional studies of special types of woodland, the investigation of the Cretaceous District in South Limburg should be mentioned first (van den Broek & Diemont 1966, Diemont 1968, Barkman 1948, Westhoff 1950d, 1973a). Woodland communities of the Drenthian District have been described by Barkman & Westhoff (1969). The plant communities of the sources and rivulets within or outside woodland have been described by Maas (1959); the periodically flooded woods of the former fresh water tidal delta (Biesbosch) by Zonneveld (1960); woodlands along the rivers upstream of the tidal influence a.o. by Kop (1961) and van Eck (1973); woods of the dunes and the inner dune border by Boerboom (1960), Doing (1962c) and Westhoff (1952). Barkman (1973, p. 470-474; 1976; Barkman & Westhoff 1969; Barkman et al. 1977; see also Barkman in Westhoff & den Held 1969, p. 246, 254-255) has undertaken a careful and detailed study of the juniper scrub communities in Europe, particularly in the Netherlands, with special emphasis on their microstructure, microclimate and their macrofungi. A major part of these investigations still have to be published.

Since a major part of Dutch woodland consists of plantations of exotic or at least probably hardly indigenous trees (*Pinus sylvestris* being an example of the latter category), it is not surprising that phytosociologists have tried to study and to classify the communities developing in such forests and to use such units in vegetation mapping (Westhoff 1958a, 1959, 1962, Sissingh 1970a, b, Zonneveld 1966, 1968, Stapelveld 1956, Doing 1970a).

### 3.1.5. Grassland

In the Netherlands, grassland vegetation has been studied more comprehensively and systematically than any other formation. This is partly due to the extensiveness and the high economic importance of Dutch grasslands, but largely to the work of D.M. de Vries and his school (see also section 2).

Botanical grassland research has developed in the Netherlands in such a way, that since twenty years much is known about the floristic composition of the grasslands and about the ecology of herbage plants and plant communities. In the last decennia the research has focussed on fundamental problems such as indicator value and life strategies of plants. Since these aspects are covered by W.H. van Dobben and P. van den Bergh in this volume, in this paragraph only the study of symmorphology, synecology, dynamics and classification of grassland will be mentioned. Reviews of this work have been published by De Vries (1954, 1957, 1958).

In the Botanical Division of the State Experimental Station for arable crops and grassland at Groningen the foundation was laid for the dry weight analyses of herbage samples in large quantities by K. Zijlstra (1937, 1940). After D.M. de Vries had been appointed at this institute, he studied the methods of analysing herbage (De Vries 1933a, b, 1937, 1938, 1940a, b) as well as the fluctuating equilibrium between the floristic composition of grassland and its habitat under the influence of season and weather conditions (de Vries 1941a, b, 1942, 1943, de Vries & Koopmans 1948). Moreover, the new pastures on the reclaimed Zuyder Zee bottom were investigated (De Vries & Zijlstra 1934, 1944), as well as the changes in botanical composition of unfertilized mown litter fen (*Cirsio-Molinietum*) by drainage (Zijlstra & de Vries 1935). During this period the Danish frequency-method for vegetation analysis, combined with the 'order method' (by which the species sequence according to decreasing cover value was estimated) were introduced; this procedure proved to be most useful if combined which a measurement of biomass of the separate species.

Since 1940, botanical grassland research has been greatly extended at the Central Institute of Agricultural Research (C.I.L.O., later I.B.S. + P.A.W., nowadays C.A.B.O.) at Wageningen. The floristic composition and the environmental conditions of 1600 farm grasslands spread all over the Netherlands have been investigated (de Vries 1948a, b, 1950, 1953). Since 1942, for applied purposes the vegetational assemblage of a sward has been indicated by one value: the grade of quality, composed of the valuation numbers and the dry weight percentages of the species occurring in the stand (de Vries et al. 1942, de Vries et al. 1949, de Vries & de Boer 1949, de Vries et al. 1951).

The use of the combined frequency and order method, however successful it proved to be, also showed some disadvantages. The degree of homogeneity and the vegetation pattern could hardly be detected in this way; moreover, the method did not provide an appropriate basis for any classification. To overcome these problems, de Vries developed an eclectic procedure by using several methods; (a) the frequency and order method; (b) the criterion of cover value by distinction of dominance communities (Ennik & de Vries 1950, Sanders & 't Hart 1951, de Vries 1962) to be compared with sociations; (c) the use of associations in the sense of the Braun-Blanquet approach (Damman & de Vries 1954); (d) a classification on the base of 'grassland types', distinguished by a choice of frequently occurring species selected firstly by their agricultural quality and secondly by their indicator value (de Vries 1948a, b, de Vries & de Boer 1949).

Before the computer era facilitated numerical work, de Vries introduced the study of species correlation as an objective basis for community classification (de Vries 1953, de Vries et al. 1954, Damman & de Vries 1954). He produced a plexus or constellation figure from a matrix of similarity coefficients by placing species with high positive correlation close together and negatively correlated species far apart (see also McIntosh 1973). This scheme was taken over by Ellenberg (1956) and Braun-Blanquet (1964), who reaffirmed de Vries' conclusion 'that the resulting species groups broadly coincide with the associations of the Zürich-Montpellier school'.

It is to be regretted that the opportunity to study the extensive unfertilized fen grasslands (*Cirsio-Molinietum*) of the Netherlands hardly has been used. Nowadays it is too late; these grasslands have been destroyed nearly completely and are preserved only in some small nature reserves. Only provisional or regional investigations have been carried out (Eisma-Donker 1964, Sissingh 1942, van der Kloot 1939, van Dijk 1946, Meyer 1947, Reijnders 1949, Segal & Westhoff 1959, Schimmel 1955, de Wit 1947, Smeets et al. 1979).

Among other grassland studies (apart from dunes and salt marshes: see section 3.1.1.) particularly the investigation of the chalk grasslands in the Cretaceous District in South-Limburg should be mentioned (Diemont & van de Ven 1953, Barkman 1953, Willems 1973, van Haperen 1973, Hilgers et al. 1968, Willems & Blanckenborg 1975).

## 3.2 Methodology; general concepts

Contributions of Dutch authors to the development of methods and concepts in phytosociology have been summarized by van der Maarel (1966) and Westhoff & van der Maarel (1973, 1978); see also Barkman (1958a), Westhoff (1965a) and Westhoff & den Held (1969).

The theoretical base of phytosociology has been critically considered in a paper by Westhoff (1950a), the first to give a logical sequence of definitions, starting with a definition of the general concept vegetation; the definition is presented here in the slightly adapted form published by Westhoff & van der Maarel (1973): 'a system of largely spontaneously growing plant populations, growing in coherence with their sites and forming part of the ecosystem together with the site factors and all other forms of life occurring in these sites'. The distinction between concrete and abstract communities was stressed and expressed in separate definitions of the concrete concept, named 'phytocoenosis' (term of Gams 1918), and the abstract concept, named 'coenon' (Barkman et al. 1958, 1964, Westhoff et al. 1959) and specified into 'phytocoenon' by van der Maarel (1965). The term 'phytocoenosis' is replacing the unfortunate term 'association-individual', which has been severely criticized (most thoroughly by Whittaker 1962) and at present is of historical interest only (Westhoff & van der Maarel 1973, 1978). Other synonyms are 'stand' (English) and 'Bestand' (German), which have no international scientific equivalent however.

The term 'phytocoenon' is meant to replace the terms 'community type' (Whittaker 1956, 1962) and 'nodum', the latter term proposed by Poore (1956, 1962) but used in a more specific meaning by Williams & Lambert (1961).

With this terminology the unclear circular reasonings of Braun-Blanquet can be elucidated, by which the association is defined a.o. as having character species and the character species as characterizing an association. This 'petitio principii', first mentioned by Meltzer & Westhoff (1942), was due to Braun-Blanquet using the term 'association' in two different ways, viz. as the fundamental hierarchical unit with a specific rank as well as in the sense of a general vegetation unit, distinguished before its place in the classification hierarchy has been worked out, thus synonymous with 'phytocoenon' (see discussion in Westhoff & van der Maarel 1973, p. 655).

Phytocoena which are units of any rank in the Braun-Blanquet classification have been termed 'syntaxa' (Barkman et al. 1959, Westhoff et al. 1959, Westhoff & van der Maarel 1973).

Much discussion has been held on the characterization of syntaxa, particularly of the association. Meyer Drees (1951a) was among the first stating that character species in a proper sense were not obligatory (by his definition 'an association is a plant community identified by its characteristic taxa combination, including one or more (local) character taxa or differentiating taxa'). As the number of described associations grew, the role of character taxa gradually diminished while the diagnostic importance of differential taxa increased. In an increasing number of association descriptions, particularly by German authors, the character taxa are not mentioned separately but are given together with the differential taxa ('Kenn- und Trennarten'). In such a presentation the fact is obscured, that in a number of cases no character species is present at all. This situation is at variance, however, with the explicit statement of the Symposia of the International Society for Plant Geography and Ecology, that for the distinction of an association, at least one character taxon is required. It is clear that theory and practice tend to diverge here. To clarify the situation, Dutch authors stressed that associations can also be validly described by a character combination of species (none of the species of this cluster has to be a character species itself (Beeftink 1965, Westhoff & den Held 1969, Westhoff & van der Maarel 1973, p. 628). Unfortunately, the term character combination much resembles the term 'characteristic species combination' (Braun-Blanquet 1928, 1953, 1964) which has a different meaning.

The use of the term 'character taxon' instead of 'character species' does not only relate to the fidelity of subspecific paramorphs, but also to that of higher taxa, e.g. the genus. An example is the genus *Spartina*, character taxon of the class *Spartinetea* (Beeftink 1965, 1968, Westhoff & den Held 1969; see also Westhoff & van der Maarel 1973, 1978).

Knapp (1948) has suggested to give two independent classification systems of syntaxa, one based on edaphic-ecological ('vertical') and another based on historic-geographical ('horizontal') criteria. This suggestion has been amply discussed by Dutch authors (Westhoff 1950, Meyer Drees 1951, Barkman 1958, Beeftink 1965,

Westhoff & van der Maarel 1973, 1978). The present Dutch practice is to recommend that the main axis of the classification should be vertical at all levels of the hierarchy. Given this main direction, secondary to it horizontal vicariant groupings may be recognized on any syntaxonomical level.

Another syntaxonomic proposal has been the replacement of the designation 'typicum' by 'inops' for subassociations which do not present differential taxa and which, on top of that, have a relatively poor characteristic species combination (Westhoff 1965, Westhoff & van der Maarel 1973, 1978).

In the presentation of data, the differential use of groups of species as indicators for syntaxa, though implicit in the Braun-Blanquet approach, has been neglected by the majority of authors. It has been common use to distinguish character species of the association, alliance, order and class under discussion (at the best for each of these categories separately), to add the differential species indicating the variation within the matrix, and to lump all other species as 'companions' or even 'indifferent species'. Segal & Westhoff (1955) and van der Maarel (1969 et seq.) have developed an overall treatment in terms of syntaxonomic groups, so that every species (if possible) is assigned to a syntaxonomical group of which it is characteristic. By this procedure, sociological structure, synecological relationships and dynamic state of the phytocoena are much better elucidated; the community appears to be a pattern of species groups indicating the position of every stand in a spatial environmental gradient as well as in the course of succession.

Among the Dutch contributions to the analytical research phase we may mention the distinction between 'vitality' and 'fertility', brought forward by Barkman et al. (1964), the refinement proposals concerning the combined estimation by Doing (1954) and Barkman et al. (1964), and the studies and considerations on the minimum area by van der Maarel (1966b) and Werger (1972).

Much work has been done on the problem of vegetation boundaries; we may refer to van der Maarel & Leertouwer (1967), Fresco (1972), van der Maarel (1974, 1976), van Leeuwen (1965, 1966, 1970), Westhoff (1971a, b, 1976), Westhoff & van Leeuwen (1966), Westhoff & van der Maarel (1973).

During the last decennia, structural characters, in the first place that of stratification, have gradually become more important. Though the original association concept implied the physiognomic uniformity of the association, there are many cases in which floristically consistent units are structurally heterogeneous. In a number of cases the discrepancy between floristic and structural uniformity has been solved by a refined syntaxonomic treatment. The problem has been discussed by Westhoff (1967, see also Westhoff & van der Maarel 1973, 1978) and is dealt with by J.J. Barkman in this volume.

Apart from attempts to use structural criteria in a floristic-sociological classification, various suggestions for an integration of the latter system with physiognomic systems have been put forward (Doing 1962a, b, 1966a, 1972a, Passarge 1966, 1968, Scamoni et al. 1965, Westhoff 1967, Westhoff & den Held 1969; see also Westhoff & van der Maarel 1973, 1978).

## 4. Synecology and vegetation dynamics

In the Netherlands, synecology has been investigated mainly in three ways:

(1) mesological synecology: qualitative and quantitative investigation of a number of factors within the subsystems climate (mostly microclimate) and soil that were supposed to be master factors for the ecosystem considered, as well as (mostly qualitative) investigation of the biotic factors; the correlation between such factors and the composition, structure and distribution of plant communities has been studied in a number of cases;

(2) ethological synecology (cf. Braun-Blanquet 1964, Westhoff 1954a): investigation of the interrelations between vegetation and habitat factors by studying the adaptive characters of the vegetation itself. Mostly this has been done by studying various aspects of life form spectra. Besides the usual Raunkiaer spectra also life form spectra sensu Iversen (1936) as well as dispersal spectra and synchorological spectra (all introduced in the Netherlands by Westhoff 1947) have been used. Zonneveld (1960) introduced scleromorphy as a character to be used in life form spectra; Rijpert (1977) applied the life form system presented by Hejný (1960) on Dutch wetland communities with *Gratiola officinalis*;

(3) integrated survey of synecology and vegetation dynamics (see below).

Of course, a combined or even integrated treatment of (1) and (2) is the comparatively best way to study the ecosystem as a whole from a phytosociological viewpoint. Examples are studies by Westhoff (1947), Sissingh (1950, 1952), Barkman (1958), Zonneveld (1960) and Eijsink et al. (1977).

A general treatment of ecological factors by Heybroek (1955) has been taken over by Westhoff (1965; see also Westhoff 1955). Climatic and microclimatic factors have been studied a.o. by Westhoff (1947, 1948), Stoutjesdijk (1959), van der Poel & Stoutjesdijk (1959), Barkman (1958), Barkman et al. (1975), Sissingh (1945, 1952), and Boerboom (1964a). Cosmic factors, viz. the tidal influence of sea water (mostly incorrectly assigned to 'edaphic' factors), have been dealt with by many authors, a.o. Adriani (1946), Westhoff (1947), Boerboom (1956), Beeftink (1955, 1965, 1966, 1968), Ketner (1972), and Rozema (1978). The cosmic factor in the freshwater tidal delta has been studied by Zonneveld (1960) in a most outstanding monograph. A general treatment of biotic factors has been presented by Westhoff (1958).

Among the large number of synecological studies dealing with specific syntaxa or landscape units not mentioned above, the grassland publications of D.M. de Vries are outstanding; these have been referred to in section 3.1.5. Long-term investigation of the edaphic factors correlated with the plant communities of calcareous dunes has been carried out by Adriani (Adriani & van der Maarel 1968, Westhoff et al. 1962) and Boerboom (1963). The hypothesis brought forward by van Dieren (1934a) about nutrient deficiency in sea dunes behind the coastal range has been tested by V. de Vries (1947, 1948).

The ecological implications of zinc as an edaphic factor and the specific communities on substrates containing Zn have been discussed by Heimans (1961) and Ernst (1967, 1974a, b).

Vegetation dynamics have been much studied in the Netherlands, as a result of the preponderantly dynamic character of the Dutch ecosystems involved. A large part of this work has been carried out in pioneer ecosystems such as adjoining dune and salt marsh areas (van Dieren 1933, 1934a, Feekes 1935, 1936, 1939, 1940, 1941, Harmsen 1937, Boer 1955, Feekes & Bakker 1954, D. Bakker 1958, Beeftink 1966, 1970, Beeftink et al. 1971, Joenje 1974, 1978a, b, J.P. Bakker 1978, Nienhuis 1978, V. de Vries 1950, 1961) and in areas that have been inundated (Bakker 1947, 1950a, b). Specific problems were dealt with in a study of the infiltration basins in dune areas established for drinking water extraction (Londo 1971) and the investigation of recreational damage on dune areas (van der Werf 1970). The consequences of the destruction of the freshwater tidal delta ecosystem (Biesbosch) by closing the estuaries have been studied by de Boois (1973) and Gottenbosch (1968).

Direct succession study on experimental permanent quadrats has been initiated by Vlieger (1942) in coastal areas of the former Zuyder Zee and Texel, in an attempt to study the influences of desalination and grazing. This study has been pursued by Westhoff (1969) and Westhoff & Sykora (in prep.). An important development has been the careful and very detailed small-scale succession research on dune grassland by van der Maarel (1966, 1975a, b, 1978). For the sake of conservation management in nature reserves, a large number of permanent quadrats has been studied (e.g. Westhoff & van Dijk 1952, Londo 1971, 1974). These studies have resulted in the relation theory of C.G. van Leeuwen (1966, 1970) on the interdependence of pattern and process (see also van der Maarel 1966, van der Maarel & van Leeuwen 1967). In a further development, van Leeuwen has attempted to integrate the different adaptive reaction types of plant species during succession in the general concept of requisite environmental dynamics for each species in relation to the gradient structure of its environment (see e.g. Westhoff et al. 1970-1973).

Whereas these studies are based on repeated analyses of permanent quadrats, another direct method of succession study consists in studying macrofossile or microfossile remnants deposited in the vegetation's substratum. The work of van Donselaar-ten Bokkel Huinink (1956) is one of the few examples of such a study based on macrofossile remnants. Studies based on microfossiles (palynological research) are not covered in this paper.

General considerations on the process of succession have been brought forward by Segal (1967), who himself studied the autogenic succession ('verlanding') in eutrophic swamps. Several other authors also investigated this topic (e.g. van Wirdum 1973). The ecological implications of succession on forest glades (caused by felling or by storms) have been dealt with by van Andel (1977). Van der Maarel & Werger (1979) presented a detailed review of the possibilities for evaluation of succession data.

## 5. Dutch studies abroad

Dutch scholars are increasingly contributing to the study of vegetation in foreign countries, in Europe and further away. In most cases such studies are restricted to general surveys or classification work, but particularly in the Caribbean area and Surinam more sophisticated investigations have been carried out. A world survey of *Zostera* communities has been presented by den Hartog (1976).

### 5.1. *Europe*

A small number of comprehensive studies have dealt with a rather large part of European countries. Examples are the monograph on European epiphyte communities by Barkman (1958a), the studies on salt marshes by Beeftink (1968, 1977), that on dunes and salt marshes by Westhoff & Schouten (in press), that on wall vegetation by Segal (1973), and those on thermophile woodland fringes by van Gils et al. (1975, 1977a, b, 1978). The studies on European juniper scrub (Barkman 1973, 1976, Barkman et al. 1977) are for the greater part still unpublished. In France, several Dutch phytosociologists working at the S.I.G.M.A. under the supervision of Braun-Blanquet, have published studies about the vegetation of Languedoc (e.g. Adriani 1945, Tideman 1947, Barkman 1958b). A team of Utrecht palynologists and vegetation scientists is studying the Vosges. Schoof-van Pelt (1973) studied *Littorellion* communities of Western France, Westhoff (1955) the woodland of Fontainebleau, Willems (1973) limestone grasslands, Sissingh (1974) dune grasslands of the Atlantic coast, and Doing Kraft (1960) the coastal range of Atlantic dunes. In Great Britain, Barkman (1950) has investigated dunes of North Scotland, de Smidt (1965) heathland communities, Schoof-van Pelt (1973) *Littorellion* associations, and Willems (1978) limestone grasslands. On the vegetation of Ireland the publications of de Smidt (1965), Klein (1975), Schoof-van Pelt & Westhoff (1972), Schoof-van Pelt (1973) and Westhoff (1971) may be mentioned. For Germany we refer to the *Fagetalia* woodland study by Diemont (1938) and the description of mountain bogs of the Harz by Barkman (1963). Mountain grasslands in Austria have been studied by Eijsink et al. (1978) and Werger et al. (1978); plant communities of Crete by Gradstein & Smittenberg (1977); Danish beech woods by Sissingh (1963); and Swedish heaths by Damman (1956).

### 5.2. *Outside Europe*

Tundra and epilithic communities of Greenland have been surveyed by Daniëls (1968, 1973). In subarctic and temperate America, the studies of Damman (1964, 1977), Becking (1957) and Janssen (1967) should be mentioned. For historical reasons the Caribbean and Guyana regions of tropical America were preponderant in the attention of Dutch vegetation scientists: for Surinam we may mention Lanjouw (1936), Lindeman (1953), Schulz (1960), Jonker (1961), van Donselaar (1963, 1965, 1969), van Donselaar-ten Bokkel Huinink (1966), Dirven (1965), and

Boerboom (1964b); for a bibliography see Boerboom (1970). Lindeman (1971) worked in Brasil. Stoffers (1956) studied the terrestrial vegetation of the Netherlands Antilles; van Loenhoud & van der Sande (1977) the intertidal biotic communities.

As yet, only a few Dutch vegetation studies have been published about the Indo-Malaysian and the Polynesian subkingdom (Polak 1964, Meyer Drees 1951b, Meyer 1954, 1970). Sprangers & Balasubramaniam (1978) studied some coastal woodlands in India. Doing (1966c, 1970b, 1972b) investigated Australian vegetation.

Phytosociological research in Southern Africa started recently and is now expanding rapidly. Werger (1973, Werger et al. 1978), van Zinderen Bakker (1973), and van der Meulen (1979) studied the South African paleotropic area (Transvaal, Oranje Vrijstaat, Cape Province, Rhodesia), and Werger et al. (1972) presented the first account of plant communities in the Cape Kingdom. A recent account of ecological publications on Southern Africa is presented by Werger (1978). Gremmen (1976) carried out a detailed study of the vegetation of subantarctic Marion Island; the main results still have to be published.

## 6. Evaluation; trends of future development

Probably the greatest merit of the Braun-Blanquet approach in phytosociology is that it results in a reference system for all geobotanical and ecological studies carried out in the various parts of the world. This reference system is above all important for: (1) autecological studies and population studies; (2) various aspects of land use for agricultural, sylvicultural and conservational purposes; (3) all studies of pattern and process in vegetation, as they are dealing with the interrelation between vegetation structure and vegetation dynamics.

Autecological and population ecological studies frequently meet the problem of geographical differentiation in the ecological amplitude of species (see e.g. Westhoff & van der Maarel 1973, 1978, Werger 1974, Werger & van Gils 1976). This problem has two aspects: (1) the law of relative habitat constancy of H. Walter; (2) ecotypic differentiation. Without entering on these problems it may be clear, that such problems even cannot be formulated, leave alone being solved, without a reference framework enabling us to compare geographically different plant communities. It is exactly this reference framework which is furnished by the Braun-Blanquet classification.

On the base of these considerations we may attempt to sketch the possible future development of phytosociology, particularly in the Netherlands. Eleven developmental trends may be discerned.

(1) It is necessary to further develop the classification system itself, putting some emphasis on a more complete integration of knowledge about Dutch vegetation in that of Western Europe. Even some major vegetation units, e.g. the alliance *Agropyro-Rumicion crispi*, are insufficiently known. In this connection it should also be considered whether or not the rapid deterioration and impoverishment of

Western European vegetation will enforce us to accept an adapted and simplified classification system. The relevant ideas of Kopecky & Hejny (1973) have been applied already by Dutch students (e.g. Strijbosch 1976, Braakhekke & Braakhekke-Ilsink 1976).

(2) The above statement is particularly dealing with the taxocoenoses of spermatophytes and pteridophytes. In most vegetation studies, bryophytes and lichens have been insufficiently taken into account, and most often fungi are ignored altogether. The occurrence, the structural and syntaxonomical position, and the ecological niche of cryptogams in the biotic communities have still largely to be investigated.

(3) Though biocoenological research on the base of phytosociological relevés started, in the Netherlands, as early as 1942, much work has to be done in this field. Faunal groups most appropriate in this respect are site-bound animals like ants, snails, spiders and grasshoppers as well as the soil fauna and the fauna of the tidal zone.

(4) As a consequence of the last mentioned topic, research on functional relations between vegetation and fauna has to be developed: zoochory, diaspore predation, zoogamous pollination ecology, the ecology of grazing, the function and distribution of zoocecidia and their relation to vitality and fertility of plants, the function of ant-heeps and mole-hills in the life strategy of grassland species, are only a few examples of research problems in this field.

(5) The investigation of synecological relations is as yet only poorly developed. We do know, in broad outline, which soil types, which phreatic levels, which microclimatological regimes and which major human and animal impact are typical for a number of plant communities; however, we know only very superficially how such an ecosystem is functioning. One of the main problems to be studied is the proper character of the relation between vegetation pattern and environmental stability (relation theory, mentioned above).

(6) Succession research on permanent quadrat plots, though being carried out since decennia, has to be stimulated in two respects: (a) the need of publishing a wealth of data stored in archives; (b) concerning the experimental approach, by burning, cutting sods, fertilizing, mowing, pasturing, etc. Such experimental research is presently carried out in Utrecht (e.g. on chalk grassland by J.H. Willems) and in Nijmegen. An important difficulty hampering this kind of research is the high population pressure of the Netherlands which makes it difficult to keep in order any material arrangement in the field. Another method of studying vegetation dynamics consists of repeated vegetation mapping with intervals of about ten years. Areas whose vegetation has been mapped painstakingly, like the dunes of Voorne (van der Maarel & Westhoff 1964) are most suitable for this type of investigation.

(7) Vegetation mapping. Many detailed maps of parts of Dutch vegetation have been published or are otherwise available (for a survey see Doing & van der Werf 1962, and Westhoff 1954b), but an overall systematic vegetation mapping of the country has not been achieved nor attempted, contrary to the trend in neighbouring countries. One of the reasons of these arrears is the circumstance, that major

parts of the Dutch landscape have been impoverished so extremely by recent human impact, that overall phytosociological mapping hardly would be possible and to a large extent even pointless. Therefore, it seems to be more useful to carry out small-scale mapping not of the real vegetation, but of the potential one. Such a vegetation map of the Netherlands recently has been published by Kalkhoven et al. (1976); see also van der Maarel & Stumpel (1975), Stumpel & Kalkhoven (1978). On the other hand, detailed large-scale vegetation mapping (1:100 until 1:10.000), if possible combined with soil mapping and compared with historical land use maps, can yield important results and should be encouraged, for instance for the sake of succession research (see 6).

(8) Perhaps the most important use of vegetation research in the Netherlands is its application in the conservation of nature, as discussed in this volume by P.A. Bakker. This study is (1) concerned with synecology and vegetation dynamics, particularly with the interrelation of pattern and process in vegetation, (b) with vegetation mapping. It can even be said that the development of fundamental vegetation research and that of nature conservation study in the Netherlands are interdependent and have influenced each other (Westhoff 1978). As yet, vegetation research has been mainly applied for the selection as well as for the outward and inward management of nature reserves. The study of the relation between pattern and process resulting from it (relation theory, van Leeuwen 1966) has brought about the theory of requisite environmental variety for each species in relation to the gradient structure of its environment; this requisite variety is supposed to be a major autecological character of each species (see section 4). The results of nature conservation studies are broadening now into (1) overall considerations of environmental management and planning (Van der Maarel 1975c, van der Maarel & Vellema 1975). (2) environmental impact statements, (3) research on the life strategies of species (see 10, below). The integration of several aspects of nature conservation research, mainly based on phytosociological investigation, has recently produced an important and comprehensive 'general ecological model' for land use in the Netherlands (van der Maarel & Dauvellier 1978).

(9) The further development of vegetation research is hardly thinkable without a major impact of numerical data processing, dealt with in this volume by E. van der Maarel. This trend is closely related with the development of ordination.

(10) Insight in the causal complex of vegetation structure and vegetation dynamics can be obtained only by studying autecological aspects, particularly the adaptive strategies ('epharmony') of the composing species, and the effects of their interference. In the Braun-Blanquet approach, a plant community has always been considered as a pattern of evolving and declining populations (e.g. Braun-Blanquet 1964, Duvigneaud 1946, Westhoff 1965a, Knapp 1967), but in this respect much work still has to be done (Whittaker 1970). One of the besic problems is that of community structure and diversity; the co-existence of many species in an apparently uniform habitat, inducing the investigation of niche differentiation (Grubb 1977, Harper 1978). It is not always clearly understood that knowledge of the composition structure, distribution and dynamics of plant communities forms an

indispensable reference system for any population ecological research.

(11) Finally, it should be stressed that Western and Central European specialists try to solve increasingly sophisticated problems about European vegetation,
'knowing more and more about less and less', whereas the plant communities of the
major part of the world have insufficiently or even hardly been studied. Particularly in the countries of the Third World, surveys of vegetation types as well as studies
on synecological relations (concerning climatic, edaphic and biotic factors) and
vegetation dynamics are indispensable for a more appropriate land use (see also
Werger 1977). In this respect, the arid and semiarid zones are as important as the
perhumid ones; erosion progresses in both. For this type of research the Netherlands
have a well-developed tradition, starting with the botanical, pedological and agricultural departments of the universities (including Wageningen) and nowadays largely
concentrated in the Unesco International Training Centre (for Aerial Survey and
Earth Science) of Enschede.

## References

Aalderen, B. van. 1953. (republished in J.C. Smittenberg 1973). Het Ulmetum suberosae langs
   de IJsel. Kruipnieuws 15 (1): 4-9.
Aart, P.J.M. van der. 1975. De verspreiding van wolfspinnen in een duingebied, geanalyseerd
   met behulp van multivariate methoden. Thesis, Leiden. Brill, Leiden.
Adriani, M.J. 1945. Sur la phytosociologie, la synecologie et le bilan d'eau de halophytes de
   la région néerlandaise méridionale ainsi que de la Méditerranée française. Thesis, Amster-
   dam.
Adriani, M.J. & E. van der Maarel. 1962. De duin- en slikgebieden van Voorne en het Brielse
   Gat. Natuurwetenschappelijke betekenis; consequenties van de afdamming van het Brielse
   Gat. Natuur en Landschap 16: 177-197.
Adriani, M.J. & E. van der Maarel. 1968. Voorne in de branding. Amsterdam.
Adriani, M.J. & E. van der Maarel. 1978. Plant species and plant communities: An introduction.
   In: E. van der Maarel & M.J.A. Werger (eds.), Plant species and plant communities. Proc. Int.
   Symp. Nijmegen, 1976. pp. 3-6. Junk, Den Haag.
Andel, M. & T. van & H.T. Waterbolk. 1945. (republished in J.C. Smittenberg 1973). Bodem en
   plantengroei in het dal van het Anderse Diep. Kruipnieuws (extra edition) 1-48.
Althuis, M. van, H. van Gils & E. Keysers. 1978. Groupement de lisière et stades évolutifs du
   Brachypodio-Geranion dans la série septentrionale du chêne pubescent des Alpes nord-oc-
   cidentales et du Jura méridional (partie française). Bull. Soc. Roy. Bot. Belg. (in press).
Bakker, D. 1947. De flora en fauna van Schouwen en Duiveland tijdens de inundatie en na het
   droogvallen. Natura 44: 107-110.
Bakker, D. 1950a. De inundaties gedurende 1944-1945 en hun gevolgen voor de landbouw. 5:
   de flora en fauna van Walcheren en andere inundatiegebieden tijdens en na de inundatie.
   Versl. Landbouwk. Onderz. 56: 1-40.
Bakker, D. 1950b. De flora en fauna van Walcheren tijdens en na de inundatie. Ned. Kruidk.
   Arch. 57: 87-88.
Bakker, D. 1058. Van Zuiderzee tot IJselmeer, botanisch beschouwd. Natura 55: 91-94.
Bakker, J.C. 1969. Vegetatiekundig en oecologisch-geografisch onderzoek van het Quercion
   robori-petraeae in de Nederlandse zandgebieden ten zuiden van de Waal. Meded. Landb.
   Hogesch. 69 (19): 1-44.
Bakker, J.P. 1978. Changes in a salt-marsh vegetation as a result of grazing and mowing. A
   five-year study of permanent plots. Vegetatio 38: 77-88.
Balogh, J. 1958. Lebensgemeinschaften der Landtiere. Akademie Verlag GmbH, Berlin-Buda-
   pest.
Barkman, J.J. 1948. Bryologische zwerftochten door Nederland II. Zuid-Limburg. Publ. Nat.
   hist. Genootsch. Limburg 1: 5-25.

Barkman, J.J. 1950. Duinvegetaties van Noord-Schotland. Ned. Kruidk. Arch. 57: 145-148.
Barkman, J.J. 1953. De kalkgraslanden van Zuid-Limburg. II. De cryptogamen.-Publ. Nat. hist. genootsch. Limburg 6: 21-30.
Barkman, J.J. 1958a. Phytosociology and ecology of cryptogamic epiphytes, including a taxonomic survey and description of their vegetation units in Europe. Van Gorcum, Assen.
Barkman, J.J. 1958b. La structure du Rosmarino-Lithospermetum helianthemetosum en Bas-Languedoc. Blumea suppl. 6: 113-136.
Barkman, J.J. 1963a. Een nieuwe associatie op hoogveenturf. Jaarb. 1963 K.N.B.V.: 37.
Barkman, J.J. 1963b. Enige indrukken van hoogvenen in de Harz. De Levende Natuur 66: 102-114.
Barkman, J.J. 1973. Synusial approaches to classification. In: R.H. Whittaker (ed.), Ordination and classification of vegetation. Handb. Veg. Sci. 5: 437-491. Junk, den Haag.
Barkman, J.J. 1976. Terrestrische fungi in jeneverbesstruwelen. Coolia 19: 94-110.
Barkman, J.J. & P.J. den Boer. 1961. Biosociologisch onderzoek in Wijster. Jaarb. 1961 K.N.B.V.: 51-52.
Barkman, J.J., H. Doing, C.G. van Leeuwen & V. Westhoff. 1958. Enige opmerkingen over de terminologie in de vegetatiekunde. Corr. Bl. Rijksherbarium 8: 87-93.
Barkman, J.J., H. Doing & S. Segal. 1964. Kritische Bemerkungen und Vorschläge zur quantitativen Vegetationsanalyse. Acta. Bot. Neerl. 13: 394-419.
Barkman, J.J., A.K. Masselink & B.W.L. de Vries. 1977. Über das Mikroklima in Wacholderfluren. In: H. Dierschke (ed.), Vegetation und Klima. Ber. Int. Symp. Vegetationskunde Rinteln 1975. pp. 35-81. Cramer, Vaduz.
Barkman, J.J. & V. Westhoff. 1969. Botanical evaluation of the Drenthian district. Vegetatio 19: 330-388.
Becking, R.W. 1957. The Zürich-Montpellier School of phytosociology. Bot. Rev. 23: 412-488.
Beeftink, W.G. 1962. Conspectus of the phanerogamic salt plant communities in the Netherlands. Biol. Jaarb. Dodonaea 30: 325-362.
Beeftink, W.G. 1965. De zoutvegetatie van ZW-Nederland beschouwd in Europees verband. Thesis, Wageningen; Meded., Landb. hogeschool 65 (1): 1-167.
Beeftink, W.G. 1966. Vegetation and habitat of the salt marshes and beachplains in the South-Western part of the Netherlands. Wentia 15: 83-108.
Beeftink, W.G. 1968. Die Systematik der europäischen Salzpflanzengesellschaften. In: R. Tüxen (ed.), Pflanzensoziologische Systematik. Ber. Int. Symp. Stolzenau/Weser 1964. pp. 239-263. Junk, Den Haag.
Beeftink, W.G. 1977. The coastal salt marshes of western and northern Europe: an ecological and phytosociological approach. In: V.J. Chapman (ed.), Wet coastal ecosystems. Ecosystems of the World 1: 109-155. Elsevier, Amsterdam.
Beeftink, W.G., M.C. Daane & W. de Munck. 1970. Tien jaar botanisch-oecologische verkenningen langs het Veerse Meer. Natuur en Landschap 25: 50-63.
Blom, C.W.P.M. 1974. The influence of soil moisture and trampling on germination and development of the seedlings of four Plantago species at various degrees of soil compactness. In: Progress report 1973 of the Inst. Ecol. Res. Verh. Kon. Ned. Ak. Wetensch. 11, 63: 98-105.
Boer, A.C. 1942. Plantensociologische beschrijving van de orde der Phragmitetalia. Ned. Kruidk. Arch. 52: 237-302.
Boer, A.C. 1955. Plant succession on former tidal lands in the Northeastern Polder. Acta Bot. Neerl. 4: 161-166.
Boerboom, J.H.A. 1957. Les pelouses sèches des dunes de la côte néerlandaise. Acta Bot. Neerl. 6: 642-680.
Boerboom, J.H.A. 1960. De plantengemeenschappen van de Wassenaarse Duinen. Thesis, Wageningen. Meded. Landb. hogesch. 60: 1-135.
Boerboom, J.H.A. 1963. Het verband tussen bodem en vegetatie in de Wassenaarse duinen. Boor en Spade 13: 120-155.
Boerboom, J.H.A. 1964a. Microklimatologische waarnemingen in de Wassenaarse duinen. Meded. Landb. hogesch. 64: 1-28.
Boerboom, J.H.A. 1964b. De natuurlijke regeneratie van het Surinaamse mesofytische bos na uitkap. Med. Landb. hogesch.: 1-150.
Boerboom, J.H.A. 1970. Bibliography of the vegetation of Guiana, Surinam and French Guiana (S.A.). Exc. Bot. Sect. B. 10: 269-272.

Boerboom, J.H.A. & V. Westhoff. 1974. Planten van het duin – duinen maken de plant; Samen-levingen van planten in het duin. In: K. Bakker et al. (ed.), Meyendel, duin-water-leven. pp. 59-83. W. van Hoeve, den Haag-Baarn.
Boois, H. de. 1973. Patronen en processen in de Biesbosch voor en na de afsluiting. Contactbl. v. Oecol. 9: 95-99.
Braakhekke, W.G. & E.J. Braakhekke-Ilsink. 1976. Nitrophile Saumgesellschaften im Südosten der Niederlande. Vegetatio 32: 55-60.
Braun-Blanquet, J. 1928, 1951, 1964. Pflanzensoziologie. Springer, Berlin-Vienna: 1st Ed. Berlin, 1928; 2nd. Ed. Vienna, 1951; 3rd. Ed. Vienna, 1964.
Braun-Blanquet, J. & W.C. de Leeuw. 1936. Vegetationsskizze von Ameland. Ned. Kruidk. Arch. 46: 359-393.
Leeuw, W.C. de. 1938. Anthyllis maritima-Silene otites-Ass., Avena pubescens-Medicago falcata-Ass. In: J. Braun-Blanquet & M. Moor, Prodrome des Groupements végétaux 5: Bromion erecti: 51-54. Comité International du Prodrome Phytosociologique (Montpellier).
Braun-Blanquet, J., G. Sissingh & J. Vlieger. 1939. Prodromus der Pflanzenges. 6: Klasse der Vaccinio-Piceetea. Comité International du Prodome Phytosociologique (Montpellier).
Bremekamp, C.E.B. 1962. The various aspects of biology. Verhand. Kon. Ned. Ak. v. Wet. afd. Nat. 2, 54, 2, 199 p. Amsterdam.
Broek, J.M.M. van den & W.H. Diemont. 1966. Het Savelsbos. Bosgezelschappen en Bodem. Centrum voor landbouwpublikaties en landbouwdocumentatie, Wageningen.
Brouwer, G.A., J.W. van Dieren, W. Feekes, G.W. Harmsen, J.G. ten Houten, W.J. Kabos, J.P. Mazure, A. Scheygrond, P. Tesch & van der Werff. 1950. Griend, het vogeleiland in de Waddenzee, historisch-geografisch, hydrografisch en biologisch beschreven. M. Nijhoff, Den Haag.
Bijhouwer, J.T.P. 1926. Geobotanische studie van de Berger duinen. Thesis, Wageningen.
Casparie, W.A. 1969. Bult- und Schlenkenbildung in Hochmoortorf. Vegetatio 19: 146-180.
Casparie, W.A. 1972. Bog development in Southeastern Drente (The Netherlands). Vegetatio 25: 1-271.
Damman, A.W.H. 1956. Een nieuwe indeling van de heidegezelschappen. Meded. Kon. Ned. Bot. Ver. 1955: 28-29.
Damman, A.W.H. 1957. The South-Swedish Calluna-heath and its relation to the Calluneto-Genistetum. Bot. Notiser 110 (3): 363-398.
Damman, A.W.H. 1964. Some forest types of central Newfoundland and their relation to environmental factors. Forest Sci. Monogr. 8: 1-62.
Damman, A.W.H. 1977. Geographical changes in the vegetation pattern of raised bogs in the Bay of Fundy region of Maine and New Brunswick. Vegetatio 35: 137-151.
Damman, A.W.H. & D.M. de Vries. 1954. Testing of grassland associations by combinations of species. Biol. Jaarb. Dodonaea 21: 35-46.
Daniëls, F.J.A. 1968. Shrub heath communities in South-eastern Greenland. Acta Bot. Neerl. 18: 483-484.
Daniëls, F.J.A. 1973. Opmerkingen over en indrukken van licheenvegetaties op steen in arcti-sche en alpiene gebieden. Jaarb. 1973 K.N.B.V.: 34-35.
Daubenmire, R. 1968. Plant communities, a textbook of plant synecology. Harper & Row, New York-London.
Diemont, W.H. 1938. Zur Soziologie und Synökologie der Buchen- und Buchenmischwälder der nordwestdeutschen Mittelgebirge. Mitt. flor.-Soz. Arbeitsgem. Niedersachsen 4: 1-182.
Diemont, W.H. 1968. Systematische Einteilung der Eichenhainbuchen- und Traubeneichen- Bir-kenwälder im südöstlichen Teil der Niederlande. In: R. Tüxen (ed.), Pflanzensoziologische Systematik, Ber. Int. Symp. Stolzenau 1964, pp. 333-337. Junk, Den Haag.
Diemont, W.H., G. Sissingh & V. Westhoff. 1940. Het Dwergbiezenverbond (Nanocyperion flavescentis) in Nederland. Ned. Kruidk. Arch. 50: 215-284.
Diemont, W.H. & A.J.H.M. van de Ven. 1953. De Kalkgraslanden van Zuid-Limburg. I. De Phanerogamen. Publ. Nat. hist. genootsch. Limburg 6: 1-20.
Dieren, J.W. van. 1934a. Organogene Dünenbildung, eine geomorphologische Analyse der west-friesischen Insel Terschelling mit pflanzensoziologischen Methoden. Thesis, Amsterdam. M. Nijhoff, Den Haag.
Dieren, J.W. van. 1934b De vegetatie van het eiland Griend en haar verandering onder invloed van de afsluiting van de Zuiderzee. Ned. Kruidk. Arch. 45: 217.

Dirven, J.G.P. 1965. Some important grassland types in Surinam. Neth. J. Agric. Sci. 13: 102-113.

Doing, H. 1962a. Systematische Ordnung und floristische Zusammensetzung niederländischer Wald- und Gebüschgesellschaften. Wentia 8: 1-85.

Doing, H. 1962b. Nederlandse bossen en struwelen. Jaarboek 1962 K.N.B.V.: 45-46.

Doing, H. 1962c. De buitenplaatsen en bossen langs de binnenduinrand van Noord- en Zuid-Holland. Natuur en Landschap 16: 261-281.

Doing, H. 1963a. Ubersicht der floristischen Zusammensetzung, der Struktur und der dynamischen Beziehungen niederländischer Wald- und Gebüschgesellschaften. Meded. Landb. hogeschool 63: 1-60.

Doing, H. 1963b. Eine Landschaftskartierung auf vegetationskundlicher Grundlage im Masstab 1:25 000 in den Dünen bei Haarlem. In: R. Tüxen (ed.). Vegetationskartierung. Ber. Int. Symp. Stolzenau/Weser 1959. pp. 297-312. Cramer, Weinheim.

Doing, H. 1963c. Over de oecologie der inheemse berken en de systematische indeling der berkenbossen. Jaarb. Ned. Dendr. Ver. 1959-1961, 22: 97-124.

Doing, H. 1966a. Enkele opmerkingen over het begrip hoofdformatie. Gorteria 3: 5-11.

Doing, H. 1966b. Beschrijving van de vegetatie der duinen tussen IJmuiden en Camperduin. Meded. Landb.hogeschool 66: 1-65.

Doing, H. 1966c. Het landschap rondom Australië's hoofdstad. De Levende Natuur 69: 171-239.

Doing, H. 1969a. Sociological species groups. Acta Bot. Neerl. 18: 398-400.

Doing, H. 1969b. De vegetatie van populierenbossen in Nederland. Populier 6 (2): 22-25.

Doing, H. 1970. The vegetation of artificial forests. Acta Bot. Neerl. 19: 454-455.

Doing, H. 1970b. Botanical geography and chorology in Australia. Meded. Landb. hogesch. 13: 81-98.

Doing, H. 1972a. Proposals for an objectivation of phytosociological methods. In: E. van der Maarel & R. Tüxen (eds.), Grundfragen und Methoden in der Pflanzensoziologie, Ber. Int. Symp. Rinteln 1970. pp. 59-74. Junk, Den Haag.

Doing, H. 1972b. Botanical composition of pastures and weed communities in the Southern Tablelands Region, S.E.Australia. Techn. Papers C.S.I.R.O., Div. of Plant Industry, Canberra, 30: 1-40.

Doing, H. 1974. Landschapsecologie van de duinstreek tussen Wassenaar en IJmuiden. Meded. Landb. hogeschool 74: 1-111.

Doing Kraft, H. 1954a. Een poging tot hernieuwing van de bosassociaties van het systeem der Frans-Zwitserse school. Jaarboek 1954 K.N.B.V.: 9-10.

Doing Kraft, H. 1954b. L'analyse des carrés permanents. Acta Bot. Neerl. 3: 421-425.

Doing Kraft, H. 1958. Zonering in landschap en plantengroei van de duinen bij Bloemendaal en Velsen. De Levende Natuur 61: 219-227.

Doing Kraft, H. 1960. Begroeiingen in de zeereep der Westfranse duinen bezien in het licht van het 'retractie-fenomeen'. Jaarboek 1960 K.N.B.V.: 35-36.

Doing, H. & S. van der Werf. 1962. Overzicht der Nederlandse vegetatiekaarten. Belmontia II, Meded. Landb. hogeschool 62: 1-30.

Doing, H. & V. Westhoff. 1959. De plaats van de beuk (Fagus sylvatica) in het Midden- en West-Europese bos. Jaarb. Ned. Dendr. Ver. 21: 226-254.

Donselaar, J. van. 1961. On the vegetation of former river beds in the Netherlands. Wentia 5: 1-85.

Donselaar, J. van. 1965. An ecological and phytogeographic study of Northern Suriname Savannas. Wentia 14: 1-163.

Donselaar, J. van. 1969. Observations on Savanna vegetation types in the Guianas. Vegetatio 17: 271-312.

Donselaar, J. van. 1970. De Nederlandse natuurbescherming gezien in internationaal verband. In: J.C. van de Kamer (ed.), Het verstoorde evenwicht. pp. 231-244. Oosthoek, Utrecht.

Donselaar, J. van. 1972-1973. Phragmitetalia-gemeenschappen in de uiterwaarden. 1: Gorteria 6: 61-67, 1972; 2: Gorteria 6: 109-117, 1973.

Donselaar-ten Bokkel Huinink, W.A. van. 1961. An ecological study of the vegetation in three former river beds. Wentia 5: 112-162.

Donselaar-ten Bokkel Huinink, W.A.E. van. 1966. Structure, root systems and periodicity of savanna plants and vegetations in western Suriname. Wentia 17: 1-162.

Du Rietz, G.E. 1921. Zur methodologischen Grundlage der modernen Pflanzensoziologie. Akad. Abhandl., Uppsala.

Du Rietz, G.E. 1930a. Vegetationsforschung auf soziationsanalytischer Grundlage. Handb. Biol. Arbmeth. 11 (5): 293-480.

Du Rietz, G.E. 1930b. Classification and nomenclature of vegetation. Svensk Bot. Tidskr. 24: 489-503.

During, H.J. 1973. Some bryological aspects of pioneer vegetation in moist dune valleys in Denmark, the Netherlands and France. Lindbergia 1: 99-104.

Duvigneaud, P. 1946. La variabilité des associations végétales. Bull. Soc. Roy. Bot. Belg. 78: 107-134.

Dijk, J. van. 1955. Bosvegetaties en bosvorming in het Kortenhoefse Veengebied. In: W. Meijer & R.J. de Wit (eds.), Kortenhoef. pp. 60-66. Stichting Commissie voor de Vecht en het O. en W. Plassengebied, Amsterdam.

Dijk, J. van. 1946. (republished in J.C. Smittenberg 1973). Bedreigde blauwgraslanden uit het Nieuwkoopse veengebied. Kruipnieuws 8 (3): 5-8.

Eck, W. van. 1950. (republished in J.C. Smittenberg 1973). Rivierdalbossen. Kruipnieuws 12 (4): 9-11.

Egler, F. 1942. Vegetation as an object of study. Phil. of Sci. 9: 245-260.

Egler, F. 1951. A commentary on American plant ecology, based on the textbooks of 1947-1949. Ecology 32: 673-695.

Eisma-Donker, E.M. 1964. Een analyse van het Cirsio-Molinietum. Jaarboek 1964 K.N.B.V.: 39.

Eijsink, J., G. Ellenbroek, W. Holzner & M.J.A. Werger. 1978. Dry and semidry grasslands in the Weinviertel, Lower Austria. Vegetatio 36: 129-148.

Ellenberg, H. 1956. Grundlagen der Vegetationsgliederung. I: Aufgaben und Methoden der Vegetationskunde. In: H. Walter (ed.), Einführung in die Phytologie 4, 1. Ulmer, Stuttgart.

Ennik, G.C. 1965. The influence of management and nitrogen application on the botanical composition of grassland. Neth. J. Agric. Sc. 13: 222-237.

Ennik, G.C. & D.M. de Vries. 1950. Waardering en oecologie van dominantie-gezelschappen. Verslag C.I.L.O. 1949: 15-17.

Ernst, W. 1967. Bibliographie der Arbeiten über Pflanzengesellschaften auf schwermetallhaltigen Böden mit Ausnahme Serpentins. Exc. Bot. B. 8: 50-61.

Ernst, W. 1974a. Schwermetallvegetation der Erde. Fischer, Stuttgart.

Ernst, W. 1974b. Mechanismen der Schwermetallresistenz. Verhandl. Ges. Ökologie 1974: 189-197.

Feekes, W. 1935. Snelle veranderingen in vitaliteit van pionierplantengezelschappen op maagdelijke bodem en de oorzaken daarvan. Ned. Kruidk. Arch. 45: 211-213.

Feekes, W. 1936. De ontwikkeling van de natuurlijke vegetatie in de Wieringermeerpolder. Thesis. Wageningen. Ned. Kruidk. Arch. 46: 1-295.

Feekes, W. 1939. Botanische Untersuchung in Bezug auf den Nord-Ost-Koog. Ned. Kruidk. Arch. 49: 70-73.

Feekes, W. 1940a. Plantengroei in verband met den bouw van het eiland. Ned. Kruidk. Arch. 51: 53-56.

Feekes, W. 1940b. Een nieuwland-begroeiing. In: Botanische landschapstudies in Nederland. pp. 101-150. Noordhoff, Groningen.

Feekes, W. 1941. Buitenlanden langs Oost- en Westkust van de Zuiderzee, voor en na de afsluiting. Ned. Kruidk. Arch. 51: 63-67.

Feekes, W. & D. Bakker. 1954. De ontwikkeling van de natuurlijke vegetatie in de Noordoostpolder. Van Zee tot Land 6: 1-92.

Fresco, L.F.M. 1967a. De vegetatie van de Groningse en Friese waddenkust. Waddenbulletin 2: 11-14.

Fresco, L.F.M. 1967b. De vegetatie van de Dollard en de Dollard-plannen. De Levende Natuur 70: 230-236.

Fresco, L.F.M. 1969. Factor analysis as a method in synecological research. Acta Bot. Neerl. 18: 477-482.

Fresco, L.F.M. 1972. A direct quantitative analysis of vegetational boundaries and gradients. In: E. van der Maarel & R. Tüxen (eds.), Grundfragen und Methoden in der Pflanzensoziologie. Ber. Symp. Int. Vegetationskunde Rinteln 1970. pp. 99-111. Junk, Den Haag.

Freysen, A.H.J. 1967a. A field study on the ecology of Centaurium vulgare Rafn. Thesis, Utrecht.

Freysen, A.H.J. 1967b. Some observations on the transition zone between the xerosere and the halosere on the Boschplaat (Terschelling) with special attention to Centaurium vulgare Rafn. Acta Bot. Neerl. 15: 668-682.

Freysen, A.H.J. 1970. Die Pflanzengesellschaften von Centaurium vulgare. In: R. Tüxen (ed.), Gesellschaftmorphologie. Ber. Int. Symp. Rinteln 1966. pp. 142-156. Junk, Den Haag.

Gams, H. 1918. Prinzipienfragen der Vegetationsforschung. Vierteljahrschr. Naturf. Ges. Zürich 63: 293-493.

Géhu, J.M. 1974. Sur l'emploi de la méthode phytosociologique sigmatiste dans l'analyse, la définition et la cartographie des paysages. C.R. Acad. Sc. 279: 1167-1170.

Géhu, J.M. 1976. Sur les paysages végétaux ou sigmassociations des praires salées du Nord-Ouest de la France. Doc. Phytosoc. 15-18: 57-62.

Géhu, J.M. 1977. Le concept de sigmassociation et son application a l'étude du paysage végétal des falaises atlantiques francaises. Vegetatio 34: 117-125.

Gils, H.A.M.J. van. 1978. Forb fringes and forb stages between semi-dry grasslands and deciduous forests loess and hard bedrock in Europe; distribution, classification and site. Thesis, Nijmegen.

Gils, H. van & E. Keysers. 1978. Staudengesellschaften mit Geranium sanguineum und Trifolium medium in der (Sub) montanen Stufe des Walliser Rhônetals. Folia Geobot. Phytotax. 13: in press.

Gils, H. van, E. Keysers & W. Launspach. 1975. Saumgesellschaften im Klimazonalen Bereich des Ostryo-Carpinion orientalis. Vegetatio 31: 175-186.

Gils, H. van & A.J. Kovács. 1977. Geranion sanguinei communities in Transsylvania. Vegetatio 33: 175-186.

Gils, H. van & A.B. Kozlowska. 1977. Xerothermic forb fringes and forb meadows in the Lublin and Little Poland highlands. Proc. Kon. Ned. Akad. v. Wetenschappen C 80: 281-296.

Gottenbos, A.J. 1968. Successie-onderzoek in de Biesbosch. Jaarboek 1967/1968 K.N.B.V.: 33-34.

Gradstein, S.R. & J.H. Smittenberg. 1977. The hydrophilous vegetation of western Crete. Vegetatio 34: 65-86.

Gremmen, N.J.M. 1976. De subantarctische eilanden Marion en Prince Edward. Vakblad voor Biologen 20: 310-315.

Grubb, P.J. 1977. The maintenance of species-richness in plant communities: the importance of the regeneration niche. Biol. Rev. 52: 107-145.

Haperen, A. van. 1973. De vegetatie van het Schiepersbergcomplex (Z.-L.), met name de kalkgraslanden. Contactbl. v. Oecol. 9: 91-92.

Harmsen, G.W. 1937. De onder invloed van de afsluiting der Zuiderzee ingetreden veranderingen in het plantendek van het eilandje Griend. Ned. Kruidk. Arch. 47: 92-93.

Harper, J. 1978. Population biology of plants. Academic Press, London.

Hartog, C. den. 1958. De vegetatie van het Balgzand en de oeverterreinen van het Balgkanaal. Wetensch. Med. K.N.N.V. 27: 1-28.

Hartog, C. den. 1959. The epilithical algal communities occurring along the coast of the Netherlands. Wentia 2: 1-241.

Hartog, C. den. 1960. Wetmatigheden in het littorale zoneringssysteem van rotskusten. Med. K.N.B.V. over het jaar 1959: 36-37.

Hartog, C. den. 1970. The sea-grasses of the world. Verh. K. Ned. Akad. Wet. Afd. Nat. 59 (1): 1-275.

Hartog, C. den. 1971. The border environment between the sea and the fresh water, with special reference to the estuary. Vie et Milieu Suppl. 22: 739-751.

Hartog, C. den. 1972. Klassifikatie van zeegrasgezelschappen. Jaarboek 1972 K.N.B.V. 32-33.

Hartog, C. den. 1973. Preliminary survey of the algal vegetation of salt-marshes, a littoral border environment. Hydrobiol. Bull. 7: 3-14.

Hartog, C. den & S. Segal. 1964. A new classification of the water-plant communities. Acta Bot. Neerl. 13: 367-393.

Hartog, C. den & G. van der Velde. 1970. De flora en de vegetatie van het Balgzand. Wet. Med. K.N.N.V. 86: 20-36.

109

Heerdt, P.F. van & W. Bongers. 1967. A biocenological investigation of salt marshes on the south coast of the isle of Terschelling. Tijdschr. v. Entomol. 110: 107-131.

Heerdt, P.F. van & M.F. Mörzer Bruijns. 1960. A biocenological investigation in the yellow dune region of Terschelling. Tijdschr. v. Entomol. 103: 225-275.

Heimans, J. 1933. De transportfactor in de sociologie. Ned. Kruidk. Arch. 44: 96-98.

Heimans, J. 1939. Plantengeographische elementen in de Nederlandse flora. Ned. Kruidk. Arch. 49: 416-436.

Heimans, J. 1940. Accessibiliteit en plantenverspreiding. Ned. Kruidk. Arch. 50: 74.

Heimans, J. 1961. Taxonomic, phytogeographic and ecological problems around Viola calaminaria, the zinc violet. Publ. Nat. hist. Genootsch. Limburg 12: 55-71.

Heimans, J., H.W. Heinsius & J.P. Thijsse. 1942 seq. Geillustreerde flora van Nederland. 12th. and following ed. Versluys, Amsterdam.

Hejny, S. 1960. Oekologische Charakteristik der Wasser- und Sumpfpflanzen in den Slowakischen Tiefebene. Vydavatel 'stvo Slovenskej akadémie ved., Bratislava.

Held, A.J. den. 1973. A comparative study of the vegetation and flora of recently formed peat areas in the fens of the western part of the Netherlands. Acta Bot. Neerl. 22: 264-265.

Held, J.J. & A.J. den Held (eds.). 1976. Het Nieuwkoopse plassengebied. Thieme, Zutphen.

Held, A.J. den, J.J. den Held & E.X. Maier. 1970. Waterplanten en waterplantenvegetaties in de plassen van De Haak bij Slikkendam (Z.H.). Gorteria 5: 21-35.

Heybroek, H.M. 1955. Standplaatseisen en onderlinge beïnvloedingen van planten. Jaarb. 1954 K.N.B.V.: 25.

Hilgers, J.H.M. 1968. Populatiestudie in de kalkgraslandvegetatie van de Berghofweide (Zuid-Limburg). Jaarboek 1966-1967 K.N.B.V.: 35-38.

Hilgers, J.H.M., W. Colaris & C. van Driel. 1968. Populatiestudie in de kalkgraslandvegetatie van de Berghofweide (Zuid-Limburg). Jaarb. 1966-1967 K.N.B.V.: 3 5-38.

Hoffmann, M.E. & V. Westhoff. 1951. Flora en vegetatie van de Verbrande Pan bij Bergen (N.H.). De Levende Natuur 54: 45-51, 74-79, 92-98.

Holkema, F. 1870. De plantengroei der Nederlandse Noordzee-eilanden. Amsterdam.

Hoogers, B.J. 1967. Time-tables as a method to describe changes in aquatic vegetations during the year. In: W. Holz (ed.), European Weed Research Council. Ergebn. des 2. Int. Wasserpflanzen-Symposium pp. 28-33. Oldenburg.

Hult, R. 1881. Försök till analytisk behandling av växtformationerna. Meddn. Soc. Fauna Flora Fenn. 8: 1-155.

Iversen, J. 1936. Biologische Pflanzentypen als Hilfsmittel bei der Vegetationsforschung. Munksgaard, Kopenhagen.

Janssen, C.R. 1967. A floristic study of forests and bog vegetation, northwestern Minnesota. Ecology 48: 751-755.

Janssen, J.J. 1972. Vegetatiekartering en geschiktheidsclassificatie van grove dennen-opstanden. Ned. Bosb. Tijdschr. 45: 169-176.

Joenje, W. 1974. Production and structure in the early stages of vegetation development in the Lauwerszee-polders. Vegetatio 29: 101-108.

Joenje, W. 1978a. Plant succession and nature conservation of newly embanked tidal flats in the Lauwerszeepolder. In: A.J. Davy & R.L. Jefferies (eds.), Ecological processes in coastal environments. pp. 617-634. Blackwell, Oxford.

Joenje, W. 1978b. Plant colonization and succession on embanked sandflats. Thesis, Groningen.

Joenje, W., V. Westhoff & E. van der Maarel. 1976. De plantengroei. In: J. Abrahamse, W. Joenje & N. v.Leeuwen-Seelt (eds.), Waddenzee. pp. 177-196. Landelijke Ver. tot Beh. v.d. Waddenzee & Ver. tot Beh. v. Natuurmon. in Ned., Harlingen & 's-Graveland.

Jonker, F.P. 1961. De begroeiing van in kreken liggende rotsblokken in de Emmaketen, Suriname, Jaarb. 1961 K.N.B.V.: 56-57.

Ketner, P. 1972. Primary production of salt-marsh communities on the island of Terschelling in the Netherlands. Thesis, Nijmegen.

Klaauw, C.J. van der. 1936. Ökologische Studien und Kritiken III. Zur Aufteilung der Ökologie in Autökologie und Synökologie. Acta Biotheor. 2: 195-241.

Kalkhoven, J.T.R., A.H.P. Stumpel & S.E. Stumpel-Rienks. 1976. Landelijke Milieukartering. Environmental survey of the Netherlands. Staatsuitgeverij, den Haag.

Klein, J. 1975. An Irish landscape. Thesis, Utrecht.

Kloot, W.G. van der. 1939. Blauwgraslanden in Nederland (Molinietum coeruleae). 3 vols. M. Nijhoff, Den Haag.

Knapp, R. 1948. Einführung in die Pflanzensoziologie 1: Arbeitsmethoden der Pflanzensoziologie und Eigenschaften der Pflanzengesellschaften. Ulmer, Stuttgart-Ludwigsburg.

Knapp, R. 1967. Experimentelle Pflanzensoziologie. Ulmer, Stuttgart.

Kneepkens, E.J. & J.T.A. Verhoeven. 1975. Verspreiding en oecologie van de Engelse alant (Inula britannica L.). De Levende Natuur 78: 84-94.

Koch, I. 1974. Over de oecologie van Scirpus planifolius Grimm en S. rufus (Huds.) Schrad. Jaarb. 1973 K.N.B.V.: 33.

Kop, L.G. 1961. Wälder und Waldentwicklung in alten Flussbetten in den Niederlanden. Wentia 5: 86-111.

Kopecky, K. & S. Hejny. 1973. Neue syntaxonomische Auffassung der Gesellschaften ein- bis zweijährige Pflanzen der Galio-Urticetea in Böhmen. Folia Geobot. Phytotax. 8: 49-66.

Kortekaas, W.M., E. van der Maarel & W.G. Beeftink. 1976. A numerical classification of European Spartina communities. Vegetatio 33: 51-60.

Kruseman, G. & J. Vlieger. 1939. Akkerassociaties in Nederland. Ned. Kruidk. Arch. 49: 327-398.

Kuiper, P. & C. Kuiper. 1958. (republished in J.C. Smittenberg 1973). Verlandingsvegetaties in Noordwest-Overijssel. Kruipnieuws 20 (1): 323-326.

Laan, D. van der. 1974. Synecological research of the dune slacks on Voorne; Analysis of vegetational and environmental data. In: Institute of ecological research, Progress report 1973. Also in: Verh. der Kon. Ned. Akad. v. Wet. Afd. Nat., Tweede Reeks, 63: 93-96.

Landolt, E. 1977. The importance of closely related taxa for the delimitation of phytosociological units. Vegetatio 34: 179-189.

Lange, L. de. 1972. An ecological study of ditch vegetation in the Netherlands. Thesis, Amsterdam.

Leeuw, W.C. de. 1935a. Bibliographia phytosociologica, 2: Neerlandia. pp. 61-64. Fischer, Stuttgart.

Leeuw, W.C. de. 1935b. The Netherlands as an environment for plant life. Brill, Leiden.

Leeuwen, C.G. van. 1953. Een biosociologisch onderzoek van een binnenduinpan op Terschelling. Jaarb. 1953 K.N.B.V.: 10-11.

Leeuwen, C.G. van. 1962. De hoogvenen van Twente. Wetensch. Meded. K.N.N.V. 43: 19-36.

Leeuwen, C.G. van. 1965. Het verband tussen natuurlijke en antropogene landschapsvormen, bezien vanuit de betrekkingen in grensmilieu's. Gorteria 2: 93-105.

Leeuwen, C.G. van. 1966. A relation theoretical approach to pattern and process in vegetation. Wentia 15: 25-46.

Leeuwen, C.G. van. 1970. Raum-zeitliche Beziehungen in der Vegetation. In: R. Tüxen (ed.), Gesellschaftsmorphologie, Ber. int. Ver. Vegetationskunde, Rinteln 1966. pp. 63-68. Junk, Den Haag.

Lindeman, J.C. 1953. The vegetation of the coastal region of Suriname. Thesis, Utrecht.

Lindeman, J.C. 1971. Indrukken van de vegetatie van Paraná, Zuid-Brazilië. Jaarb. 1971 K.N.B.V.: 13-15.

Loenhoud, P.J. van & J.C.P.M. van de Sande. 1977. Rocky shore zonation in Aruba and Curaçao (Netherlands Antilles), with the introduction of a new general scheme of zonation. I & II. Proc. Kon. Ned. Akad. Wetensch. series C, 80 (5): 437-474.

Londo, G. 1971. Patroon en proces in duinvallei-vegetaties langs een gegraven meer in de Kennemerduinen. Thesis, Nijmegen.

Londo, G. 1974. Successive mapping of dune slack vegetation. Vegetatio 29: 51-61.

Maarel, E. van der. 1961a. De zonering in landschap en plantengroei van de duinen bij Oostvoorne. De Levende Natuur 64: 223-233.

Maarel, E. van der. 1961b. Een vegetatiekartering van de duinen bij Oostvoorne. Jaarb. 1960 K.N.B.V.: 57-59.

Maarel, E. van der. 1963. Een gedetailleerde vegetatiekartering van de duintjes voor 'Weevers'-duin'. Jaarb. 1962 K.N.B.V.: 40-41.

Maarel, E. van der. 1965. Beziehungen zwischen Pflanzengesellschaften und Molluskenfauna. In: R. Tüxen (ed.), Biosoziologie. Ber. Symp. int. Ver. Vegetationskunde, Stolzenau 1960. pp. 184-197. Junk, Den Haag.

Maarel, E. van der. 1966a. Dutch studies on coastal sand-dune vegetation, especially in the Delta-region. Wentia 15: 47-82.

111

Maarel, E. van der. 1966b. Over vegetatiestructuren, -relaties en -systemen, in het bijzonder in de duingraslanden van Voorne. Thesis, Utrecht. RIVON rapp., Zeist.
Maarel, E. van der. 1969. On the use of ordination models in phytosociology. Vegetatio 19: 21-46.
Maarel, E. van der. 1974. Small-scale vegetational boundaries; on their analysis and typology. In: W.H. Sommer & R. Tüxen (eds.), Tatsachen und Probleme der Grenzen in der Vegetation. Ber. Int. Symp. Vegetationskunde Rinteln 1968. pp. 75-80. Cramer, Lehre.
Maarel, E. van der. 1975a. Observations sur la structure et la dynamique de la végétation des dunes de Voorne. In: J.-M. Géhu (ed.), Colloques Phytosociologiques I: 167-183. Cramer, Vaduz.
Maarel, E. van der. 1975b. Small-scale changes in a dune grassland complex. In: W. Schmidt (ed.), Sukzessionsforschung. Ber. Int. Symp. Vegetationskunde Rinteln 1973. pp. 123-134. Cramer, Vaduz.
Maarel, E. van der. 1975c. Man-made natural ecosystems in environmental management and planning. In: W.H. van Dobben & R.H. Lowe-Mc Connell (eds.), Unifying concepts in ecology. pp. 263-274. Junk Den Haag.
Maarel, E. van der. 1976. On the establishment of plant community boundaries. Ber. Deutsch. Bot. Ges. 89: 415-443.
Maarel, E. van der. 1978a. Experimental succession research in a coastal dune grassland, a preliminary report. Vegetatio 38: 21-28.
Maarel, E. van der. 1978b. Environmental management of coastal dunes in the Netherlands. In: A.J. Davy & R.L. Jefferies (eds.), Ecological processes in coastal environment. pp. 543-580. Blackwell, Oxford.
Maarel, E. van der, & P.J. Dauvellier. 1978. Naar een globaal ecologisch model voor de ruimtelijke ontwikkeling van Nederland. Studierapp. R.P.D. No. 9. Min. V. & R.O., Den Haag.
Maarel, E. van der & C.G. van Leeuwen. 1967. Beziehungen zwischen Struktur und Dynamik in Ökosystemen. In: R. Tüxen (ed.), Syndynamik. Ber. Int. Symp. Rinteln. (still not published).
Maarel, E. van der & J. Leertouwer. 1967. Variation in vegetation and species diversity along a local environmental gradient. Acta Bot. Neerl. 16: 211-221.
Maarel, E. van der & A.H.P. Stumpel. 1975. Landschaftsökologische Kartierung und Bewertung in den Niederlanden. Verh. Ges. Ökologie, Erlangen 1974: 231-240. Junk, Den Haag.
Maarel, E. van der & K. Vellema. 1975. Towards an ecological model for physical planning in the Netherlands. In: Ecological aspects of economic development planning. Report Seminar U.N. Economic Commission for Europe, Rotterdam. pp. 128-143. E.C.A., Genève.
Maarel, E. van der & M.J.A. Werger. 1979. On the treatment of succession data. Phytocoenosis (in press).
Maarel, E. van der & V. Westhoff. 1964. The vegetation of the dunes near Oostvoorne (the Netherlands) with a vegetation map. Wentia 12: 1-61.
Maas, F.M. 1959. Bronnen, bronbeken en bronbossen van Nederland, in het bijzonder die van de Veluwezoom. Thesis, Wageningen. Meded. Landb. hogesch. 59: 1-166.
McIntosh, R.P. 1973. Matrix and plexus techniques. In: R.H. Whittaker (ed.), Handbook Vegetation Science 5: 157-191. Junk, Den Haag.
Major, J. 1958. Plant ecology as a branch of botany. Ecology 39: 352-363.
Major, J. 1961. Use in plant ecology of causation, physiology, and a definition of vegetation. Ecology 42: 167-169.
Meltzer, J. 1941. Die Sanddorn-Liguster-Assoziation. Ned. Kruidk. Arch. 51: 385-395.
Meltzer, J. & V. Westhoff. 1942. Inleiding tot de Plantensociologie. Bruegel, 's-Graveland.
Meyer, W. 1947. (republished in J.C. Smittenberg 1973). Blauwgrasland in Twente, in het bijzonder dat van Lemselermaten en Voltherbroek. Kruipnieuws 9 (2): 1-6.
Meyer, W. 1948. Naar reservaat 'de Oude Kooi' in het Naardermeer. Natura 45: 102-106.
Meyer, W. 1949. Botanische ervaringen in het Naardermeer. In: In het voetspoor van Thijsse. pp. 382-387. Wageningen.
Meyer, W. 1950. Enkele problemen bij vegetatie-studie in West-Nederlandse venen. Ned. Kruidk. Arch. 57: 144-145.
Meyer, W. 1951. Flora en vegetatie van de Kierse Wiede. Ned. Kruidk. Arch. 58: 22-23.
Meyer, W. 1953. Flora en vegetatie van het Zwetgebied bij Wormer en Jisp. Wetensch. Meded. K.N.B.V. 6: 2-11.

112

Meyer, W. 1954. Plantensociologische waarnemingen in de topregionen van Gede en Pangrango. Penggemar Alam 34 (1-2): 9-17.
Meyer, W. 1955. Waterplanten- en oevervegetaties. In: W. Meyer & R.J. de Wit, Kortenhoef. pp. 22-44. Uitg. Stichting Comm. voor de Vecht en het O. en W. Plassengebied, Amsterdam.
Meyer, W. 1970. Regeneration of tropical lowland forest in Sabah, Malaysia, forty years after logging. The Malayan Forester 33: 204-229.
Meyer Drees, E. 1936. De bosvegetatie van de Achterhoek en enkele aangrenzende gebieden. Thesis, Wageningen.
Meyer Drees, E. 1951a. Enkele hoofdstukken uit de moderne plantensociologie en een ontwerp voor nomenclatuurregels voor plantengezelschappen. Report Forest Research Inst. 51: 1-218. Bogor, Indonesia.
Meyer Drees, E. 1951b. Distribution, ecology and silvicultural possibilities of the trees and shrubs from the savanna-forest region in Eastern Sumbawa and Timor. Comm. For. Res. Inst. 33: 1-45.
Meulen, F. van der. 1979. Plant sociology of the Western Transvaal Bushveld, South Africa. Thesis, Nijmegen. Diss. Bot. 49. Cramer, Vaduz.
Mörzer Bruijns, M.F. 1942. Slakken en plantengemeenschappen. Kruipnieuws 4 (5): 15-16.
Mörzer Bruijns, M.F. 1943 (1945). Die Gastropodenfauna einiger niederländischer Wälder. Compt. rend. Soc. Neerl. Zool. 1943: 9-20.
Mörzer Bruijns, M.F. 1947. Over levensgemeenschappen. Thesis, Utrecht.
Mörzer Bruijns, M.F. 1953. Een biosociologisch onderzoek van een vochtige duinvallei in het buitenduin van Terschelling. Jaarb. 1953 K.N.B.V.: 10-11.
Mörzer Bruijns, M.F. 1954. Broedvogeltellingen en natuurlijke bossen. Jaarb. 1954 K.N.B.V.:
Mörzer Bruijns, M.F. & V. Westhoff. 1951. The Netherlands as an environment for insect life. Intern. Congress of Entomol. Amsterdam.
Mörzer Bruijns, M.F., A. Lawalrée, H. Schimmel & F. Demaret. 1953. Vegetatieonderzoek van het Zwin in 1951-1952. Bull. Jard. Bot. Etat 23: 1-123.
Neuteboom, J.H. 1974. Variabiliteit van de grassoort Kweek (Elytrigia repens (L. Desv.) op Nederlandse landbouwgronden. Contactbl. v. Oecol. 10: 15-17.
Nienhuis, P.H. 1978. Dynamics of benthic algal vegetation and environment in Dutch estuarine salt marshes, studied by means of permanent quadrats. Vegetatio 38: 103-112.
Nordhagen, R. 1936. Versuch einer neuen Einteilung der subalpinen-alpinen Vegetation Norwegens. Bergens Mus. Årbok 7: 1-88.
Odum, E.P. 1963. Ecology. Holt, Rinehart & Winston, New York.
Oosterveld, P. 1968. Patroonstudie in duinheidevegetaties van de Berkenvallei, Boschplaat, Terschelling. Jaarboek 1967/1968 K.N.B.V.: 59-60.
Passarge, H. 1966. Die Formationen als höchste Einheiten der soziologischen Vegetationssystematik. Report spec. nov. Regni veg. 73: 226-235.
Passarge, H. 1968. Neue Vorschläge zur Systematik nord- mitteleuropäischer Waldgesellschaften. Report Spec. nov. Regni veg. 77: 75-103.
Pegtel, D.M. 1974. Effect of crop rotation on the distribution of two ecotypes of Sonchus arvensis L. in the Netherlands. Acta Bot. Neerl. 23: 349-350.
Pegtel, D.M. 1976. On the ecology of two varieties of Sonchus arvensis L. Thesis, Groningen.
Poel, A.J. van der, & Ph. Stoutjesdijk. 1959. Some microclimatological differences between an oak wood and a Calluna heath. Meded. Landb. hogesch. 59: 1-8.
Polak, B. 1964. Oligotrofe bosvenen in de tropen. Jaarb. 1964 K.N.B.V.: 13-14.
Poore, M.E.D. 1956. The use of phytosociological methods in ecological investigations. IV. General discussion of phytosociological problems. J. Ecol. 44: 28-50.
Poore, M.E.D. 1962. The method of successive approximation in descriptive ecology. Adv. Ecol. Res. 1: 35-68.
Post, H. von. 1851. Om växtgeografiska skildringer. Bot. Notiser 1851: 110-127, 161-187.
Reijnders, Th. 1959. De Noordhollandse brakwaterveren. Natuur en Landschap 13: 66-81.
Reijnders, T. 1967. De vegetatie van hoogveenrestanten in de Peel. De Levende Natuur 70: 121-131.
Reijnders, W. 1949. (republished in J.C. Smittenberg 1973). Nieuwkoopse paradepaardjes. Kruipnieuws 11 (4): 2-5.
Sanders, M. & M.L. 't Hart. 1951. Droge-stof-opbrengsten van de meest voorkomende dominantie-gezelschappen. Verslag C.I.L.O. 1950: 24-26.

Scamoni, A. & H. Passarge. 1959. Gedanken zu einer natürlichen Ordnung der Waldgesellschaften. Arch. Forstwesen 8: 386-426.

Scamoni, A. & H. Passarge. 1963. Einführung im die praktische Vegetationskunde 2. ed. Fischer, Jena.

Scamoni, A., H. Passarge & G. Hofmann. 1965. Grundlagen zu einer objektiven Systematik der Pflanzengesellschaften. Repert. Spec. nov. Regni veg. Beih. 142: 117-132.

Schenk, W. in press. Bibliographia phytosociologica: Neerlandia 1960-1975. Excerpta botanica sectio B sociologica.

Scheygrond, A. 1932. Het plantendek van de Krimpenerwaard. IV. Sociographie van het hoofdassociatie-complex. Arundinetum-Sphagnetum. Thesis, Utrecht. Ned. Kruidk. Arch. 1932 (1): 1-84.

Schimmel, H.J.W. 1955. De Drentse beken en beekdalen. De Levende Natuur 58: 61, 81, 105, 129.

Schoof-van Pelt, M.M. 1973. Littorelletea. A study of the vegetation of some amphiphytic communities of western Europe. Thesis, Nijmegen.

Schröter, C. & O. Kirchner. 1902. Die vegetation des Bodensees. II. Lindau i.B.

Schulz, J.P. 1960. Ecological studies on rain forest in Northern Surinam. Van Eedenfonds, Amsterdam.

Segal, S. 1964. De vegetaties der moerasgebieden van Noord-West-Overijssel. RIVON report, Zeist.

Segal, S. 1965. Een vegetatie-onderzoek van hogere waterplanten in Nederland. Wetensch. Meded. K.N.B.V. 57: 1-80.

Segal, S. 1966. Ecological studies of the peat-bog vegetation in the north-western part of the province of Overijssel (The Netherlands). Wentia 15: 109-141.

Segal, S. 1968a. Ein Einteilungsversuch der Wasserpflanzengesellschaften. In: R. Tüxen (ed.), Pflanzensoziologische Systematik. Ber. Int. Symp. Stolzenau/Weser 1964. pp. 191-219. Junk, Den Haag.

Segal, S. 1968b. Schwierigkeiten bei der Systematik der Moorgesellschaften. In: R. Tüxen (ed.), Pflanzensoziologische Systematik. Ber. Int. Symp. Stolzenau/Weser 1964. pp. 220-229. Junk, Den Haag.

Segal, S. 1969. Ecological notes on wall vegetation. Thesis, Amsterdam. Junk, Den Haag.

Segal, S. & M.C. Groenhart. 1967. Het Zuideindigerwiede, een uniek verlandingsgebied. Gorteria 3: 165-181.

Segal, S. & V. Westhoff. 1959. Die vegetationskundliche Stellung von Carex buxbaumii Wahlenb. in den Niederlanden. Acta Bot. Neerl. 8: 304-329.

Sernander, R. 1894. Studier öfver den gotländska vegetationens utvecklingshistoria. Thesis, Uppsala.

Sernander, R. 1898. Studier öfver vegetationen i mellersta Skandinaviens fjälltrakter. Öfvers. Förhandl., K. (Svenska) Vetensk.-Akad. 6: 325-367.

Simons, J. 1978. Verbreitung und Dynamik von Vaucheria-Algengesellschaften in den südwestlichen Niederlanden. Vegetatio 38: 119-122.

Sissingh, G. 1942. Graslandtypen om Wageningen. Ned. Kruidk. Arch. 52: 308-309.

Sissingh, G. 1950. Onkruid-associaties in Nederland. Thesis, Wageningen. Versl. Landbouwk. Onderz. 56 (15): 1-224.

Sissingh, G. 1952. Ethologische synoecologie van enkele onkruid-associaties in Nederland. Meded. Landb. hogeschool. 52: 167-206.

Sissingh, G. 1963. De Deense beukenbossen. Jaarboek 1963 K.N.B.V.: 51.

Sissingh, G. 1968. Het ontstaan en vergaan van onze heidevelden. Tijdschr. Kon. Ned. Heidemij. 79 (7-8): 338-345.

Sissingh, G. 1969. Ueber die systematische Gliederung von Trittpflanzen-Gesellschaften. Mitt. flor.-soz. Arb. gem. N.F. 14: 179-192.

Sissingh, G. 1970a. De plantengemeenschappen in onze naaldhoutbossen Ned. Bosb. Tijdschr. 42: 157-162.

Sissingh, G. 1970b. De plantengroei in douglasbossen. Jaarboek 1970 K.N.B.V.: 13-16.

Sissingh, G. 1974. Comparaison du Roso-Ephedretum de Bretagne avec des unités de végétation analogues (contribution à la systematique des associations de dunes grises atlantiques et méditerranéennes). Doc. Phytosoc. 7-8: 95-106.

Sissingh, G. & P. Tideman. 1960. De plantengemeenschappen uit de omgeving van Didam en Zevenaar. Meded. Landb. hogeschool 60: 1-30.

Sloet van Oldruitenborgh, C.J.M. 1976. Duinstruwelen in het Deltagebied. Thesis, Wageningen.

Smeets, P.J.A.M., M.J.A. Werger & H.A.J. Tevonderen. 1979. Vegetation changes in a moist grassland following draining. J. Appl. Ecol. (in press).

Smidt, J.Th. de. 1962. De Twentse heide. Wet. Med. K.N.N.V. 43: 3-18.

Smidt, J.Th. de. 1965. Heidegezelschappen in Bretagne en Ierland. Jaarboek 1965 K.N.B.V.: 60-61.

Smidt, J.Th. de. 1966. The inland heath-communities of the Netherlands. Wentia 15: 142-162.

Smidt, J.Th. de. 1967. Phytogeographical relations in the North West European heath. Acta Bot. Neerl. 15: 630-647.

Smidt, J.Th. de. 1975. Nederlandse heidevegetaties. Thesis, Utrecht.

Smittenberg, J.C. (ed.). 1973. Plantengroei in enkele Nederlandse landschappen. Bondsuitgeverij v.d. Jeugdbond voor Natuurstudie in samenwerking met de Ver. tot Beh. v. Natuurmon., Amsterdam.

Spranger, J.T.C.M. & K. Balasubramaniam. 1978. The tropical dry evergreen forest of Marakkanam, South-eastern India. Trop. Ecol. 17 (in press).

Stapelveld, E. 1956. De bodemvegetaties van lariksbossen in Drente. Biol. Jaarb. Dodonaea 23: 290-305.

Stoffers, A.L. 1956. The vegetation of the Netherlands Antilles. Kemink, Utrecht.

Stoutjesdijk, P. 1953. Vegetatiekundig onderzoek van Veluwse heidevelden. In: Heeft onze heide nog toekomst? pp. 15-46. Uitgave Studiekring v.d. Veluwe, Arnhem.

Stoutjesdijk, P. 1959. Heaths and inland dunes of the Veluwe. Thesis, Utrecht.

Strijbosch, H. 1976. Een vergelijkend syntaxonomische en synoecologische studie in de Overasseltse en Haterse vennen bij Nijmegen. Thesis, Nijmegen.

Stumpel, A.H.P. & J.T.R. Kalkhoven. 1978. A vegetation map of the Netherlands, based on the relationship between ecotopes and types of potential natural vegetation. Vegetatio 37: 163-173.

Tideman, P. 1947. Flora van de garrigues in de Franse Midi. Ned. Kruidk. Arch. 57: 80-82.

Trass, H. 1964. Application and validity of the synusial method in phytocoenology. Izuchenie rastit ostrava Saaremaa. pp. 82-111. Tartu.

Trass, H. & N. Malmer. 1973. North European approaches to classification. In: R.H. Whittaker (ed.), Ordination and classification of communities. Handb. Vegetation Sci. 5: 531-574. Junk, Den Haag.

Tschulok, S. 1910. Das System der Biologie in Forschung und Lehre. Eine historisch-kritische studie. Jena.

Tüxen, R. 1950. Grundriss einer Systematik der nitrophilen Unkrautgesellschaften. Mitt. flor.-soz. Arb. Gem. N.F. 5: 155-176.

Tüxen, R. 1952. Hecken und Gebüsche. Mitt. geogr. Ges. Hamburg 50: 85-117.

Tüxen, R. 1962. Zur systematischen Stellung von Spezialisten-Gesellschaften. Mitt. flor-soz. Arbeitsgem, Stolzenau, N.F. 9: 57-59.

Tüxen, R. 1973. Vorschlag zur Aufnahme von Gesellschaftskomplexen in potentiell natürlichen Vegetationsgebieten. Acta Bot. Acad. Sc. Hung. 19: 379-384.

Tüxen, R. 1977. Zur homogenität von Synassoziationen; ihre Syntaxonomischen Ordnung und ihre Verwendung in der Vegetationskartierung. Doc. Phytosoc. N.S. 1: 321-328.

Tüxen, R. & W.H. Diemont. 1937. Klimaxgruppe und Klimaxschwarm. Jahresber. Naturhist. Ges. Hannover 88-89: 73-87.

Tüxen, R. & V. Westhoff. 1963. Saginetea maritimae, eine Gesellschaftsgruppe im wechselhalinen Grenzbereich der europäischen Meeresküsten. Mitt. flor.-soziol. Arbeitsgem. N.F. 10: 116-129.

Vlieger, J. 1935. Enige waarnemingen omtrent de degradatie van het Querceto-Betuletum. Ned. Kruidk. Arch. 45: 213-214.

Vlieger, J. 1936. Over enkele bosch-gezelschappen van de hooge Veluwe-gronden. Ned. Kruidk. Arch. 46: 401-403.

Vlieger, J. 1937a. Algemene opmerkingen over de hogere plantensociologische eenheden in Nederland. Ned. Kruidk. Arch. 48: 53-54.

Vlieger, J. 1937b. Ueber einige Waldassoziationen der Veluwe. Mitt. flor.-soz. Arbeitsgem. Niedersachsen 3: 193-203.

Vlieger, J. 1938. Over hoog- en laagveenassociaties. Natura 37: 56-60.
Vlieger, J. 1939. Plantengemeenschappen in NW.-Overijssel. Natura 38: 63.
Vlieger, J. 1940. Vegetatie en podsolprofiel. Ned. Kruidk. Arch. 50: 77.
Vlieger, J. 1942. Experimenteel successie-onderzoek. Ned. Kruidk. Arch. 52: 314.
Vlieger, J. 1949. On phytosociological work in the Netherlands during the years 1940-1945. Vegetatio 1: 192-196.
Voo, E.E. van der. 1962. Twentse vennen. Wetensch. Med. K.N.N.V. 43: 39-63.
Voo, E.E. van der. 1964. Over de betekenis en het behoud van de Brabantse Vennen. Brabantia 13 (4): 1-6.
Voo, E.E. van der. 1966. De plantengroei van de leemputten bij de Kleine Meer onder Ossendrecht. De Levende Natuur 69: 253-258.
Voo, E.E. van der & V. Westhoff. 1961. An autecological study of some limnophytes and halophytes in the area of the large rivers. Wentia 5: 163-258.
Vries, D.M. de. 1929. Het plantendek van de Krimpenerwaard. III. Thesis, Utrecht. Ned. Kruidk. Arch. 39: 145-403.
Vries, D.M. de. 1933a. De rangorde-methode. Een schattingsmethode voor plantkundig graslandonderzoek met volgorde-bepaling. Versl. Landbouwk. Onderz. 39: 1-24.
Vries, D.M. de. 1933b. De plantensociologische rangorde-methode. Botan. Jaarb. 24: 37-48.
Vries, D.M. de. 1935. De plantengezelschappen als kenteken van het keukenzoutgehalte van den bodem. Ned. Kruidk. Arch. 45: 97-121.
Vries, D.M. de. 1937. Methods of determining the botanical composition of hayfields and pastures. Rep. 4th Internat. Grassland Congr.: 474-480. Aberystwyth.
Vries, D.M. de. 1938. The plant sociological combined specific frequency and order method. Chronica Botanica 4: 115-117.
Vries, D.M. de. 1939. Zusammenarbeit der nördlichen und südlichen Schule ist zum Heil der gesammten Pflanzensoziologie unbedingt erforderlich. Rec. Trav. Bot. Neerl. 36: 485-493.
Vries, D.M. de. 1940. De drooggewichtsanalytische methode van botanisch graslandonderzoek voor beweid land. Versl. Landbouwk. Onderz. 46: 1-19.
Vries, D.M. de. 1941a. De gevolgen van strenge vorst op grasland. De Nieuwe Veldbode 8, 31: 7-8; 8, 32: 9-10.
Vries, D.M. de. 1941b. Enige gegevens betreffende de periodieke schommeling in gewichtsverhouding tusschen de plantensoorten in grasland. Versl. Landbouwk. Onderz. 47: 61-99.
Vries, D.M. de. 1942. Over den invloed van jaargetijde en weer op de botanische samenstelling van grasland. Ned. Kruidk. Arch. 52: 303-307.
Vries, D.M. de. 1943. Grasland en weergesteldheid. Landbouwk. Tijdschr. 55, nr. 676.
Vries, D.M.de. 1948a. Method and survey of the characterization of Dutch grasslands. Vegetatio 1: 51-57.
Vries, D.M. de. 1948b. De botanische samenstelling van Nederlandse graslanden. I. De typering van graslanden. Versl. Landbouwk. Onderz. 54: 1-12.
Vries, D.M. de. 1950. Graslandtypen en hun oecologie. Ned. Kruidk. Arch. 57: 28-31.
Vries, D.M. de. 1953. Objective combinations of species. Acta Bot. Neerl. 1: 497-499.
Vries, D.M. de. 1954. Die angewandte botanische Grünlandforschung in den Niederlanden. Angew. Pflanzensoz., Festschrift E. Aichinger, 2: 1207-1222. Springer, Wien.
Vries, D.M. de. 1957. Overzicht van de ontwikkeling van het botanische graslandonderzoek in Nederland. Jaarboek van het I.B.S., 1957: 171-181.
Vries, D.M. de. 1958. Grasland. In: Het milieu van onze gewassen. pp. 140-148. Min. v. Landb. en Visserij en Voedselv., Den Haag.
Vries, D.M. de. 1962. Trockenmassenertrag und Bewertung von Dominanzgesellschaften. In: Die Stoffproduktion der Pflanzendecke. pp. 54-60. Ulmer, Stuttgart.
Vries, D.M. de, J.P. Baretta & G. Hamming. 1954. Constellation of frequent herbage plants, based on their correlation in occurrence. Vegetatio 5-6: 105-111.
Vries, D.M. de & P.A. de Boer. 1949. Waardering, typering en kartering van grasland. Maandbl. Landbouwvoorlichtingsdienst 6: 357-368.
Vries, D.M. de, Th.A. de Boer & J.G.P. Dirven. 1949. Waardering van grasland en beoordeling van bodemeigenschappen op grond van de botanische samenstelling. Landbouwk. Tijdschr. 61: 347-356.
Vries, D.M. de, Th.A. de Boer & J.G.P. Dirven. 1951. Evaluation of grassland by botanical research in the Netherlands. Proc. United Nations Scientific Conf. on Conserv. and Utiliz. of Resources 6: 522-524.

Vries, D.M. de, M.L. 't Hart & A.A. Kruyne. 1942. Een waardering van grasland op grond van de plantkundige samenstelling. Landbouwk. Tijdschr. 54: 245-265.
Vries, D,M, de & J. Koopman. 1948. De botanische samenstelling van Nederlandse graslanden. II. De invloed van het jaargetijde op de botanische samenstelling van grasland. Versl. Landbouwk. Onderz. 54: 1-14.
Vries, D.M. de & K. Zijlstra. 1934. Over het plantkundig graslandonderzoek op vroegeren Zuiderzeebodem. Natuurwet. Tijdschr. 16: 165-181.
Vries, D.M. de & K. Zijlstra. 1944. De ontwikkeling van op verschillende tijden na droogmaking aangelegd grasland in den Zuiderzee-Proefpolder. Landbouwk. Tijdschr. 56: 53-69.
Vries, D.M. de, K. Zijlstra, W. Feekes & G.W. Harmsen. 1940. De plantengroei van de aanslibbingen in het Noorden van Nederland. In: Botanische landschapsstudies in Nederland. pp. 47-100. Noordhoff, Groningen.
Vries, V. de. 1947. Over de stikstofvoorziening van de helm in binnen- en buitenduinen. Tijdschr. Kon. Ned. Aardr. Gen. 64: 734-741.
Vries, V. de. 1948. Enkele gegevens over de oecologie van de duindoorn. Jaarb. Ned. Dendr. Ver. 1948: 48-64.
Vries, V. de. 1950. Over de plantengroei der duindalen op Vlieland. De Levende Natuur 53: 29-38.
Vries, V. de. 1961. Vegetatiestudie op de westpunt van Vlieland. Thesis, Univ. van Amsterdam. Noorduijn, Gorinchem.
Wassink, E.C. 1938. De tegenwoordige vegetatie in de Engbertsdijkvenen te Vriezenveen. Ned. Kruidk. Arch. 48: 51-53.
Wassink, E.C. 1950. Enkele sociologische en plantengeografische opmerkingen in verband met de hoogveenvegetatie der Engbertsdijkvenen bij Vriezenveen. Ned. Kruidk. Arch. 57: 21-23.
Waterbolk, H.T. 1948. (republished in J.C. Smittenberg 1973). Landschap en plantengroei van Havelte. Kruipnieuws 10 (1): 3-29.
Weevers, Th. 1934a. Enkele boschassociaties in de Geldersche Vallei. Ned. Kruidk. Arch. 44: 98-99.
Weevers, Th. 1934b. Bosrelikten in de Gelderse Vallei. Ned. Kruidk. Arch. 44: 306-308.
Weevers, Th. 1936. De invloed van het zoutgehalte op de verspreiding van verschillende halophyten. Ned. Kruidk. Arch. 46: 898-912.
Weevers, Th. 1938. De bossen van ons land beschouwd vanuit een sociologisch oogpunt. De Levende Natuur 42: 50-57, 81-87.
Weevers, Th. 1939. The dynamic view of a flora. Ned. Kruidk. Arch. 49: 74-77.
Weevers, Th. 1940. De flora van Goeree en Overflakkee dynamisch beschouwd. Ned. Kruidk. Arch. 50: 285-354.
Werf, S. van der. 1970. Recreatie-invloeden in Meyendel. Meded. Landb. hogeschool 70: 1-24.
Werger, M.J.A. 1972. Species-area relationship and plot size: with some examples from South African vegetation. Bothalia 10: 583-594.
Werger, M.J.A. 1973. Phytosociology of the upper Orange River Valley, South Africa. A syntaxonomical and synecological study. Thesis, Nijmegen. V & R, Pretoria.
Werger, M.J.A. 1974. The place of the Zürich-Montpellier method in vegetation science. Folia Geobot. Phytotax. 9: 99-109.
Werger, M.J.A. 1977. Applicability of Zürich-Montpellier methods in African tropical and subtropical range lands. In: W. Krause (ed.), Application of vegetation science to grassland husbandry. Handb. Veg. Sc. 13: 123-145. Junk, Den Haag.
Werger, M.J.A. (ed.). 1978. Biogeography and ecology of Southern Africa. 2 vols. Junk, Den Haag.
Werger, M.J.A. & H. van Gils. 1976. Phytosociological classification problems in chorological border line areas. J. Biogeogr. 3: 49-54.
Werger, M.J.A., F.J. Kruger & H.C. Taylor. 1972. A phytosociological study of the Cape Fijnbos and other vegetation at Jonkershoek, Stellenbosch. Bothalia 10: 1-19. shorter German version: Vegetatio 24: 71-89.
Werger, M.J.A., P.J.A.M. Smeets, H.P.G. Helsper & V. Westhoff. 1978. Oekologie der subalpinen Vegetation des Lausbachtales, Tirol. Verh. Zool.-Bot. Ges. Wien (in press).
Werger, M.J.A., H. Wild & B.R. Drummond. 1978. Vegetation structure and substrate of the northern part of the Great Dyke, Rhodesia. Environment and plant communities. Gradient analysis and dominance-diversity relationships. Vegetatio 37: 79-89, 151-161.

117

Westhoff, V. 1940. Systematik der Pflanzengesellschaften. Ned. Kruidk. Arch. 51: 57-59.
Westhoff, V. 1943. Plantensociologisch onderzoek, in het bijzonder op de Waddeneilanden. Hand. 29e Ned. Nat. Geneesk. Congr.: 27-41.
Westhoff, V. 1947. The vegetation of dunes and salt marshes on the Dutch islands of Terschelling, Vlieland and Texel. Thesis Utrecht. Summary: 131 pp. C.J. van der Horst, Den Haag.
Westhoff, V. 1948. Invloed van de zomer van 1947 op de vegetatie der rivierduintjes. De Levende Natuur 51: 126-127.
Westhoff, V. 1949a. Landschap, flora en vegetatie van Botshol nabij Abcoude. Stichting Comm. voor de Vecht en het Oostelijk en Westelijk Plassengebied, Baambrugge.
Westhoff, V. 1949b. De betekenis van de phenologie voor het plantensociologisch onderzoek. Ned. Kruidk. Arch. 56: 24-31.
Westhoff, V. 1949c. Beken en beekdalen in Twente. In: A.F.H. Besemer (ed.), In het het voetspoor van Thijsse. pp. 36-64. Veenman, Wageningen.
Westhoff, V. 1950a. An analysis of some concepts and terms in vegetation study or phytocenology. Synthese 8: 194-206.
Westhoff, V. 1950b. De Boschplaat op Terschelling, Nederlands grootste staatsnatuurreservaat. Natuur en Landschap 5: 15-32.
Westhoff, V. 1950c. Drepanocladus revolvens (Sw.) Warnst. en Rhynchostegium rotundifolium (Brid.) B. et S. In: Mosvondsten in Nederland. Ned. Kruidk. Arch. 57: 292-296; 302-303.
Westhoff, V. 1950d. De bossen in Zuid-Limburg. De Wandelaar in Weer en Wind 18: 184-188; 198-203.
Westhoff, V. 1951. Standplaatsbeschrijving (oecologie) van de soorten der familie Gramineae. In: Flora Neerlandica 1, 2, (in totaal 274 p.)
Westhoff, V. 1952. Gezelschappen met houtige gewassen in de duinen en langs de binnenduinrand. Jaarb. Nederl. Dendrolog. Ver. 1952: 9-49.
Westhoff, V. 1954a. Some remarks on synecology. Vegetatio 5-6: 120-128.
Westhoff, V. 1954b. Die Vegetationskartierung in den Niederlanden. Angewandte Pflanzensoziologie, Festschr. Aichinger. pp. 1223-1231. Springer, Wien.
Westhoff, V. 1954c. Standplaatsbeschrijving (oecologie) van de soorten van het genus Carex. In: Flora Neerlandica 1, 3 (in totaal 133 p.)
Westhoff, V. 1955. Samenhang tussen vegetatie en milieu, in het bijzonder de bodem. T.N.O. nieuws 106: 9-16.
Westhoff, V. 1956. Standplaatsbeschrijving (oecologie) van de soorten der familie Cyperaceae minus Carex. In: Flora Neerlandica 1,4.
Westhoff, V. 1957. Een gedetailleerde vegetatiekartering van een deel van het bosgebied van Middachten. Landb. Exportbur. Fonds, Wageningen.
Westhoff, V. 1958a. Boden- und Vegetationskartierungen von Wald- und Forstgesellschaften im Quercion robori-petraeae-Gebiet der Veluwe (Niederlande). Ber. über das intern. Symp. Pflanzensoziologie – Bodenkunde in Stolzenau/Weser. Angewandte Pflanzensoziologie 15: 23-30.
Westhoff, V. 1958b. De plantengroei van het Nationale Park Veluwezoom. Wet. Med. K.N.N.V. 26: 1-40.
Westhoff, V. 1958c. Verspreidingsoecologisch onderzoek van zeldzame planten. De Levende Natuur 61: 193-202.
Westhoff, V. 1958d. Standplaatsbeschrijving (oecologie) van de soorten der familie Orchidaceae. In: Flora Neerlandica 1, 5. 127 p.
Westhoff, V. 1959. The vegetation of Scottisch pine woodlands and Dutch artificial coastal pine forest; with some remarks on the ecology of Listera cordata. Acta Bot. Neerl. 8: 422-448.
Westhoff, V. 1961a. Standplaatsbeschrijving/oecologie van de soorten der families Plumbaginaceae, Primulaceae, Pyrolaceae, Empetraceae, Convolvulaceae, Cuscutaceae, Boraginaceae. In: Flora Neerlandica 4, 1. 140 p.
Westhoff, V. 1961b. Bibliographia Phytosociologica neerlandica. Neerlandia 1935-1959, nonnullis operibus annis 1870-1934 addatis. Excerpta botanica sectio B sociologica 3: 81-220.
Westhoff, V. 1961c. Het beheer van Heidereservaten. Natuur en Landschap 14: 97-118.
Westhoff, V. 1962. Het Tonckensbos bij Norg. De Levende Natuur 65: 229-236.
Westhoff, V. 1963. De systematiek van de beukenbossen. Jaarboek 1963 K.N.B.V.: 49-50.
Westhoff, V. 1965a. Plantengemeenschappen. In: Uit de Plantenwereld. pp. 288-349. De Haan, Zeist-Arnhem.

Westhoff, V. 1965b. Het Klavereiland. De Levende Natuur 68: 146-156.
Westhoff, V. 1965c. Beken en beekdalen. In: Twente natuurhistorisch V: Enige Twentse land-schappen en hun flora. Wet. Med. K.N.N.V. 56: 2-14.
Westhoff, V. 1966. The ecological impact of pedestrian, equestrian and vehicular traffic on flora and vegetation. I.U.C.N. Technical Meeting, Luzern.
Westhoff, V. 1967. Problems and use of structure in the classification of vegetation. The diag-nostic evaluation of structure in the Braun-Blanquet system. Acta Bot. Neerl. 15: 495-511.
Westhoff, V. 1968a. Slangelook. De Levende Natuur 71: 1-12.
Westhoff, V. 1968b. Standplaatsen van Corrigiola litoralis L. Gorteria 4: 137-145.
Westhoff, V. 1969. Langjährige Beobachtungen an Aussüssungs-Dauerprobeflächen beweideter und unbeweideter Vegetation an der ehemaligen Zuiderzee. In: R. Tüxen (ed.), Experimen-telle Pflanzensoziologie. pp. 246-253. Junk, Den Haag.
Westhoff, V. 1970. Vegetation study as a branch of biological science. Miscell. Papers 5, Land-bouwhogeschool; Meded. Botanische Tuinen en Belmonte Arboretum, Landbouwhoge-school, Wageningen, 12: 11-30.
Westhoff, V. 1971a. The dynamic structure of plant communities in relation to the object of vegetation. In: E. Duffey & A.S. Watt (eds.). The scientific management of animal and plant communities for conservation. pp. 3-14. Blackwell, Oxford.
Westhoff, V. 1971b. Choice and management of nature reserves in the Netherlands. Bull. Jard. Bot. Nat. Belg. 41: 231-245.
Westhoff, V. 1971c. Enkele gegevens over de standplaats van Hypericum canadense L. Gorteria 5: 239-248.
Westhoff, V. 1972. Die Stellung der Pflanzensoziologie im Rahmen der biologischen Wissen-schaften. In: E. van der Maarel & R. Tüxen (eds.), Grundfragen und Methoden der Pflanzen-soziologie. Ber. Int. Symp. Rinteln 1970. pp. 1-15. Junk, Den Haag.
Westhoff, V. 1973a. Vegetatie en bodem op de beekdalhellingen van het Krijtdistrict. Nat. Hist. Maandbl. 10: 124-132.
Westhoff, V. 1973b. L'évolution de la végétation dans les lacs eutrophes et les bas-marais des Pays Bas. Les Natural. Belg. 54: 1-28.
Westhoff, V. 1974. Stufen und Formen von Vegetationsgrenzen und ihre methodische Annähe-rung. In: W.H. Sommer & R. Tüxen (eds.), Tatsachen und Probleme der Grenzen in der Vegetation. Ber. Int. Symp. Vegetationskunde Rinteln 1968. pp. 45-68. Cramer, Lehre.
Westhoff, V. 1978. Een halve eeuw wisselwerking tussen wetenschap en natuurbehoud. In: We-tenschap in dienst van het natuurbehoud, pp. 13-25. Nat.wet. Com.
Westhoff, V., P.A. Bakker, C.G. van Leeuwen & E.E. van der Voo (& I.S. Zonneveld). 1970-1973. Wilde planten, flora en vegetatie in onze natuurgebieden. 3 Vols. Ver. Beh. Nat. Mon., Amsterdam.
Westhoff, V., J.J. Barkman, H. Doing & C.G. van Leeuwen. 1959. Enige opmerkingen over de terminologie in de vegetatiekunde. Jaarb. 1958 K.N.B.V.: 44-46.
Westhoff, V. & W.G. Beeftink. 1950. De vegetatie van duinen, slikken en schorren op de Kaloot en in het Noordsloe. De Levende Natuur 53: 124-133, 225-233.
Westhoff, V. & J. van Dijk. 1952. Experimenteel successie-onderzoek in natuurreservaten in het bijzonder in het Korenburgerveen bij Winterswijk. De Levende Natuur 55: 5-16.
Westhoff, V. & H. Doing Kraft. 1959. De plaats van de beuk in het West- en Midden-Europese bos. 21e Jaarb. Ned. Dendr. Ver. 1956-1958: 216-254.
Westhoff, V. & J. Heimans. 1939. Hogere planten. In: V. Westhoff (ed.), Landschap, flora en vegetatie van Botshol nabij Abcoude, pp. 14-35. Baambrugge.
Westhoff, V. & C.G. van Leeuwen. 1960. Is het Waterlepeltje (Ludwigia palustris) een oorspron-kelijk inheemse soort? De Levende Natuur 63: 8-16.
Westhoff, V., J.W. Dijk & H. Passchier. 1942. Overzicht der Plantengemeenschappen in Neder-land. Breughel, 's-Graveland.
Westhoff, V., J.W. Dijk, H. Passchier & G. Sissingh. 1946. Overzicht der plantengemeenschap-pen in Nederland. 2nd. ed. Breughel, 's-Graveland.
Westhoff, V. & A.J. den Held. 1969. Plantengemeenschappen in Nederland. Thieme, Zutphen. (2nd. ed. 1975).
Westhoff, V. & P. Ketner. 1967. Milieu en vegetatie van Carex hartmanii Caj. op Terschelling, in het kader van een oecologische vergelijking tussen deze soort en Carex buxaumii Wahlenb. Gorteria 3: 119-126.

Westhoff, V. & C.G. van Leeuwen. 1964. Geografische differentiatie in de hoogvenen van Europa. Jaarboek 1964 K.N.B.V.: 32-33.

Westhoff, V. & C.G. van Leeuwen. 1966. Oekologische und systematische Beziehungen zwischen natürlicher und anthropogener Vegetation. In: R. Tüxen (ed.), Anthropogene Vegetation. Ber. Symp. int. Ver. Vegetationskunde, Stolzenau 1961. pp. 156-172. Junk, Den Haag.

Westhoff, V., C.G. van Leeuwen, M.J. Adriani & E.E. van der Voo. 1962. Enkele aspecten van vegetatie en bodem der duinen van Goeree, in het bijzonder de contactgordels tussen zout en zoet milieu. Jaarb. 1961. Wetensch. Genootschap voor Goeree-Overflakkee: 47-92.

Westhoff, V. & E. van der Maarel. 1973. The Braun-Blanquet approach. In: R.H. Whittaker (ed.), Ordination and classification of vegetation. Handb. Veg. Sci. 5: 617-726. Junk, Den Haag.

Westhoff, V. & E.,van der Maarel. 1978. The Braun-Blanquet approach. 2nd. ed. In: R.H. Whittaker (ed.), Classification of plant communities: 287-399. Junk, Den Haag.

Westhoff, V. & M.F. Mörzer Bruijns. 1956. De groeiplaats van Scirpus americanus Pers. op het Groene Strand bij West-Terschelling. Acta Bot. Neerl. 5: 344-354.

Westhoff, V. & H. Passchier. 1958. Verspreiding en oecologie van Scheuchzeria palustris in Nederland, in het bijzonder in het Besthmerven bij Ommen. De Levende Natuur 61: 193-202.

Westhoff, V. & K. Reinink. 1967. Osmunda regalis L. op Terschelling, oecologisch en vegetatiekundig bezien. Gorteria 3: 204-209.

Westhoff, V. & M.M. Schoof-van Pelt. 1969. Strandlingsgesellschaften seichter Gewässer in Irland (Littorelletea). Mitt. flor.-soz. Arb. Gem. N.F. 14: 211-223.

Westhoff, V. & M.G.C. Schouten. 1978. The structure and diversity of European coastal ecosystems. In: A.J. Davy & R.L. Jefferies (eds.), Ecological processes in coastal environments, in press. Blackwell, Oxford.

Westhoff, V. & E.E. van der Voo. 1966. Standplaatsbeschrijving van de soorten der families Solanaceae. In: Flora Neerlandica 6 (2).

Westhoff, V. & J.N. Westhoff-de Joncheere. 1942. Verspreiding en nestoecologie van de mieren in de Nederlandsche bosschen. Tijdschr. over Plantenziekten 48: 138-212.

Whittaker, R.H. 1956. Vegetation of the Great Smoky Mountains. Ecol. Monogr. 26: 1-80.

Whittaker, R.H. 1962. Classification of natural communities. Bot. Rev. 28: 1-239.

Whittaker, R.H. 1970. The population structure of vegetation. In: R. Tüxen (ed.), Gesellschaftsmorphologie. Ber. Int. Symp. Vegetationskunde 1966. pp. 39-59. Junk, Den Haag.

Willems, J.H. 1969. Heath communities with Sarothamnus scoparius and Erica cinerea in the eastern part of the Belgian Kempen and the Dutch province of Limburg. Acta Bot. Neerl. 18: 485-486.

Willems, J.H. 1973. Experimenteel botanisch synoecologisch onderzoek in het Gerendal (Z.-Limburg). Contactbl. v. Oecol. 9: 25-27.

Willems, J.H. 1978. Observations on North-West European limestone grassland communities: phytosociological and ecological notes on chalk grasslands of southern England. Vegetatio 37: 141-150.

Willems, J.H. & F.G. Blanckenborg. 1975. Kalkgraslandvegetaties van de St. Pietersberg ten zuiden van Maastricht. Publ. Nat. Hist. Gen. in Limburg 25: 1-24.

Williams, W.T. & J.M. Lamberts. 1961. Multivariate analysis in plant ecology. III. Inverse association analysis. J. Ecol. 49: 717-729.

Wirdum, G. van. 1973. Het verband tussen de successie en enige veranderingen in de eigenschappen van het water in de Weerribben. Jaarb. 1973 K.N.B.V.: 49-50.

Wit, R.J. de. 1947. (republished in J.C. Smittenberg 1973). De Lemselermaten. Kruipnieuws 9: 17-20.

Wit, R. de. 1951. (republished in J.C. Smittenberg 1973). De draadzeggegemeenschap in Noordwest-Overijsel. Kruipnieuws 13 (2): 3-6.

Wit, R. de. 1955. Het Caricetum lasiocarpae by Kortenhoef, in verband met de verbreiding en de systematische plaats van de associatie. In: W. Meyer & R.J. de Wit (eds.), Kortenhoef. Stichting Comm. voor de Vecht en het O. en W. Plassengebied, Amsterdam.

Zinderen Bakker, E.M. van. 1936. Het botanisch onderzoek van het Naardermeer. Jaarboek 1929-1935 Ver. Behoud van Natuurmon. in Nederl.: 159-176.

Zinderen Bakker, E.M. van. 1947. De West-Nederlandse Veenplassen. Biologisch-Natuurhistorisch. L.J. Veen, Amsterdam.

Zinderen Bakker Jr., E.M. van. 1973. Ecological investigation of forest communities in the Eastern Orange Free State and the adjacent Natal Drakensberg. Vegetatio 28: 299-334.

Zonneveld, I.S. 1959. Enkele resultaten van een onderzoek naar bodem en vegetatie op de Kalmthoutse heide. Jaarb. 1959 K.N.B.V.: 45-46.

Zonneveld, I.S. 1960. De Brabantse Biesbosch. Een studie van bodem en vegetatie van een zoetwatergetijdendelta. Thesis, Wageningen. Part A: 1-210; part B: 1-396.

Zonneveld, I.S. 1965. Studies van landschap, bodem en vegetatie in het westelijk deel van de Kalmthoutse heide. Boor en Spade 14: 216-238.

Zonneveld, I.S. 1966. Zusammenhänge zwischen Forstgesellschaften Boden-Hydrologie und Baumwuchs in einiger niederländischen Pinus-Forsten auf Flugsand auf Podsolen. In: R. Tüxen (ed.), Anthropogene Vegetation. Ber. Int. Symp. Stolzenau/Weser 1961. pp. 312-335. Junk, Den Haag.

Zonneveld, I.S. 1968. Een vegetatie-indeling voor akkers en naaldbossen. Versl. en Meded. K.N.B.V. 1966-1967: 87-89.

Zijlstra, K. 1937. Het bepalen van de botanische samenstelling van het grasland en de betekenis daarvan voor de practijk. Landbouw. Tijdschr. 49: 1-15.

Zijlstra, K. 1940. Over de botanische analyse van grasland en de bepaling van de gewichtspercentages der plantensoorten. Versl. Landbouw. Onderz. 46: 343-377.

Zijlstra, K. & D.M. de Vries. 1935. De invloed van de behandelingswijze van grasland op de plantkundige samenstelling der grasmat. Versl. Landbouw. Onderz. 41: 635-654.

# 5. THE INVESTIGATION OF VEGETATION TEXTURE AND STRUCTURE

## J.J. BARKMAN

## 5. THE INVESTIGATION OF VEGETATION TEXTURE AND STRUCTURE

### J.J. BARKMAN*

### 1. Some definitions

All scientific research has to start with a description of its objects. In vegetation science this means the analysis of all visible characters of the vegetation. These characters are often grouped into two different categories: floristic composition and structure. Structure then means: all morphological characters of the vegetation except the qualities and quantities of the plant taxa composing it. This is a practical, but not a very logical point of view. A concrete vegetation stand or phytocoenosis does not consist of a number of species, but consists of a number of plant individuals showing a particular horizontal, vertical and temporal arrangement ('pattern'). These individuals can be classified into abstract groups according to different criteria, yielding different categories, of which taxa are only one. Other groups concern: life forms, growth forms, height classes, leaf type and size classes, pollination and dissemination types, etc. Symmorphology comprises the study of all of them. We can also group the taxa (abstractions of the first order) into abstract types of a second order: ecological groups, plant geographical elements, etc. The study of these, however, does not belong to symmorphology, and therefore not to the structural analysis of the vegetation because the parameters used do not form part of the vegetation stand as such.

Following Doing (unpubl.), and in analogy with soil science, it seems more logical to distinguish between vegetation structure and texture. Texture is then defined as the qualitative and quantitative composition of the vegetation as to different morphological elements (in the widest possible sense, see above) regardless of their arrangement, whereas structure is concerned with the spatial (horizontal and vertical) arrangement (the architecture) of these elements. Some authors, including myself, also regard the temporal arrangement of these elements as part of the vegetation structure, but only inasfar as it shows a certain pattern inherent to the phytocoenosis in question. This includes daily and seasonal periodicity, and not succession to other phytocoenoses. A growth form spectrum, therefore, is part of the texture of a phytocoenosis, but the distribution of these growth forms over different layers or over elements of the horizontal pattern is structure, as are also the stratification and the pattern as such.

The term pattern, by the way, is used in at least three different senses: 1. as horizontal pattern only; 2. as synonymous with structure (in the spatial sense or in

* Communication no. 199 of the Biological Station Wijster. Communication no. 37 of the Department of Plant Ecology of the Agricultural University, Wageningen.

the spatial-temporal sense); 3. in a very wide sense including, for instance, the frequency distribution of the importance values or abundance values of the composing species. It is recommended to use the term pattern in the first sense only, since in the second sense it is superfluous, and in the third sense it is a mixtum compositum including aspects of texture (Whittaker (1970) speaks of 'the population structure of communities', however, when he refers to the geographical or ecological distribution of plant species over whole ranges of different communities, and Rejmánek (1977) speaks of the 'population structure of a community', when he refers to the importance value distribution of the species within a community). In modern ecology the term pattern is often understood as being opposed to process (the slogan 'pattern and process' can be heard everywhere). But pattern just is part of structure and vertical structure is related to processes, too.[1]

I should like to go even further and to include functional elements in texture and structure. Functional texture then is concerned with the presence of functional relationships within a biocoenosis, such as competition, allelopathy, parasitism, saprophytism, mycorrhiza and other forms of symbiosis, herbivory, carnivory, etc., as well as with their relative importances. Functional structure is concerned with the arrangement of these relationships ('the web of life'), for instance the question as to whether the food chains are simple, ramified or anastomosing, and to which degree. Functional texture and functional structure together are indicated as 'interactive structure' (Rejmánek 1977). Functional and morphological texture and structure constitute the fundamental characters of an (eco)system but they are not identical with the system, as was suggested by Antomonov (1969), if correctly quoted by Rejmánek (1977).

In this paper I shall restrict myself to morphological texture and morphological structure. For convenience sake, for historical reasons (traditional use), and in view of the fact that species often are defined on characters that play only a minor role in the texture and structure of plant communities, species composition will be regarded here as a notion separate from texture s.s. The term physiognomy will hardly be used here, and certainly not as distinct from structure, as is often suggested (e.g. Fosberg 1967). It is not different from structure in the sense most authors use this concept (i.e. including texture), for it is only a visual expression of part of the texture and structure (mainly the upper vegetation layer) and of no other aspects of the vegetation. As such this concept does not give any extra information and its application can only be regarded as a (useful) first, rough approximation to both texture and structure.

1. Londo (1971) used the term pattern in a fourth sense, namely as the spatial arrangement of both 'structural' (i.e. morphological) and non-structural (f.i. process) elements. With this definition the antithesis pattern-process is not logical either.

## 2. Methods of research

In the study of vegetation texture the following phases can be distinguished:
1. (methodology) the establishment of useful parameters, f.i. the question as to whether we should measure the area or the width of leaves and whether it should be the area per leaf, per plant individual or per soil surface unit (leaf area index).
2. (first abstraction) the establishment of a meaningful subdivision of the range of values of each parameter into classes, f.i. leptophyllous, nanophyllous, etc. for leaf sizes. Even if we measure in cm or mm, we use classes, since we make no distinction between, e.g., 1.3 and 1.4 mm!
3. (practical analysis) the actual measurement of parameters in vegetation and the attribution of plants to the established classes.

From here on we can follow two ways (either 4-6 or 7-9):
4. (first synthesis) calculation of the relative importance (share) of each category, which leads to textural spectra (f.i. leaf size class spectra) of the stand as a whole.
5. (second synthesis) the development of a vegetation typology on the basis of clusters of stands with similar spectra for one parameter.
6. (third synthesis) the comparison of these clusters for different parameters and the establishment of composite clusters on the basis of all parameters.
7. (second abstraction) the delimitation of growth forms on the basis of combinations of all textural characters.
8. (first synthesis) the calculation of growth form spectra for whole stands.
9. (second synthesis) the distinction of vegetation types on the basis of clusters with similar growth form spectra.

Thus we have arrived, along two different lines, at vegetation types based on many textural characters. From here (6 and 9) we proceed to either 10 or 11:
10. (fourth synthesis) the ordination of the vegetation types along one or more axes.
11. (fourth synthesis) the classification of the vegetation types into a hierarchical system.

Now 10 and 11 may lead to:
12. a comparison with floristic ordinations, respectively
13. a comparison with floristic classifications in order to find out where they agree and where they disagree and why.

And finally 13 may lead to:
14. (fifth synthesis) creation of a hierarchical system of vegetation units on the basis of both floristic composition and (other) textural characters ('combined system').

Structure can be investigated directly or via texture analysis. Vertical structure can be studied directly by making vertical sections through the vegetation or the soil (root structure) and by recording the picture thus obtained. There are a number

128

Fig. 1.  Textural and structural analysis of vegetation

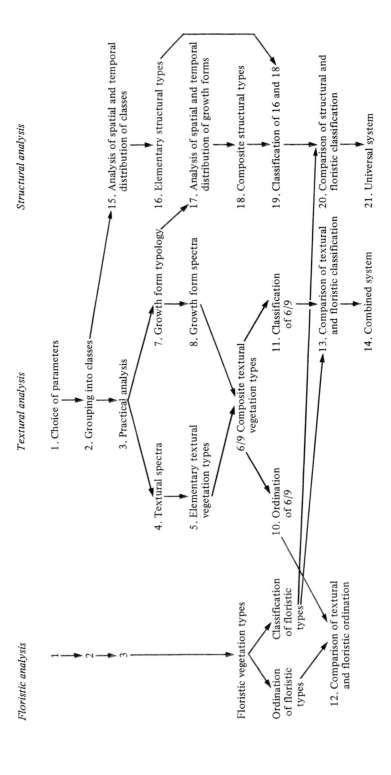

of ways of recording, from very accurate to very schematical, i.e. from stereo colour photographs through exact drawings of all plants and parts of plants (like those of the Russian steppe investigators), through semi-schematical drawings (only the contours of the crowns, for instance in most studies on tropical rain forests), through schematical drawings (using symbols) like those by Dansereau c.s., to, finally, block diagrams, only representing the different layers, their height and their total cover (various publications of the Zürich-Montpellier school). In this sequence more and more information is lost, but the possibility to quantify and to generalize the results is increased. The same applies to analysis or horizontal structure, if we proceed from photographs and exact chartings, through schematical chartings, large scale mapping and, finally, line transect studies.

For indirect methods we have to make a texture analysis first. Then we analyse the spatial and temporal arrangement of the textural elements, (15.) the simple elements, like leaf consistency classes, as well as (17.) the composite ones, like growth forms. We analyse, for instance, the spectra of separate layers. (16/18) We can then group our stands according to similar structures into types, (19) classify these types into higher categories, (20) compare them with floristic types (syntaxa) and, (21) combine them into universal floristic-structural types. The pathways to be followed are summarized in Fig. 1.

All these steps referred to pure description. It is obvious that at each step from phase 3 onwards comparisons can be made with environmental factors from which probably fruitful working hypotheses for experimental work will be generated.

## 3. Historical survey of the investigation of vegetation texture and structure

### 3.1. *General remarks*

So far a great number of publications have been written on what is called 'vegetation structure' but many of them are only concerned with texture, whilst others deal with neither of them but with such items as growth forms of single plant species, biomass, primary production, minimum area, species diversity, floristic composition of synusiae, vegetation mapping, etc. The 1966 international symposium in Rinteln (West Germany) on 'Gesellschaftsmorphologie (Strukturforschung)' (Tüxen 1970) contained 29 lectures, of which only 7 dealt with vegetation structure. In the following survey our main attention is devoted to research in the Netherlands and by Dutch botanists, since we are celebrating with this volume the 100th meeting on vegetation science in this country.

First of all, we must distinguish between (a) the analysis of textural and structural parameters (1, 2, 7; see Fig. 1); (b) the description of vegetation texture on this basis (3, 4, 8); (c) the description of vegetation structure on a direct basis and on an indirect (textural) basis (15, 17); and (d) the typology and classification of vegetation on textural and structural characters (5, 6/9, 11 and 16, 18, 19). The research fields (a) and (to a somewhat lesser extent) (b) are fairly well developed now; much less developed is (c). Most publications deal with (d), yet this field is strongly underdeveloped, as the types are based on either (1) poorly defined characters, or (2) only some aspects of texture and structure, or (3) only some taxonomic plant groups or vegetation layers, or (4) mixed with ecological or geographical characters.

It is very interesting to observe that description of vegetation on a floristic base started rather late, around 1880 in Fennoscandia, around 1900 in Switzerland and France although the flora of these regions was fairly well known long before. On the other hand, description of structural types started much earlier (we may say with Von Humboldt in 1807 and especially Grisebach in 1838), although it was not until the beginning of this century that the parameters for the description of such types were accurately defined: Raunkiaer's life forms (1907), Raunkiaer's leaf size classes (1916), Gams' life forms and growth forms (1918), Iversen's hydrotypes (1936), Molinier & Müller's dissemination types (1938), Luther's (1949) and Segal's (1965) detailed growth forms of water plants, Dansereau's (1959 and later) leaf form and fruit morphology types, and most recently Hallé, Oldeman & Tomlinson's (1978) very detailed architectural types of tropical trees. Only recently (C.T. de Wit 1965) a first effort has been made to classify plants according to the position (inclination) of the leaves (4 categories). As to phenology, there exist very detailed analyses of vegetation types, but detailed classifications in phenological types do not exist (Ellenberg 1974, for instance, distinguished only 4 types of vegetative periodicity, neglecting flowering periodicity). Yet, much is known now of many relevant parameters. However, even recent structural vegetation systems seem to ignore many of these characters. Our present scientific tools for studying the texture and structure of plant communities would allow a much more detailed typology and classification than those found in the literature.

I (unpubl. mscrpt.) recently have tried to give (for the Netherlands) a detailed system of growth forms of individual plants (46 types and a number of subtypes), a detailed system of phenological plant types (8 types with 25 subtypes) and a system of 11 leaf inclination types. On this basis and on the basis of other parameters like hydrotypes, leaf consistency, leaf size, height, density, layering, etc. a system of purely structural vegetation types of the Netherlands has been created, comprising 72 community types with 144 subtypes, united into 35 main types and 6 formation classes.

## 3.2 *Research in the different schools of phytosociology*

Of old most attention has been devoted to structure by Russian and tropical vegetation ecologists. Contrary to other schools structure has always been a fundamental element in the definition of plant communities in both the Leningrad (Sukačov) and Moscow (Aljochin) schools of phytosociology. Sukačov (1928): 'a plant community is a complex of plants which are *interacting*. It is characterised both by a definite composition and *structure* and by a certain interaction with its environment'. Aljochin (1935): 'a plant community is a combination of plant species, caused by historical development, habitat factors and *interaction of plants*. It has a certain *structure*, capacity of regeneration and capacity of modifying its habitat'. In Central Europe only Schmid and Däniker (Zürich) adopted this viewpoint.

Structure was also taken into account by the Russians when they introduced the concept of 'edificator' plants. The interest in structure, however, was more theoretical than practical. In practice the actual community types were based only on floristic composition. In the Leningrad school even no research was devoted to structure analysis. The Moscow school of steppe investigators, on the other hand, paid much attention to the vertical structure of grassland communities by drawing detailed profile diagrams.

In the Anglo-American schools of phytosociology (both the older schools of Clements and Tansley and the recent ordination schools of Bray and Curtis, and Whittaker) hardly any research was done on vegetation structure until recently (but structure was implicitly involved in such Clementsian units as layer societies and aspectal societies). An important exception was formed by tropical vegetationists, mostly in the British Commonwealth, like Richards, Watt, Tansley, Beard (cf. Beard 1973). They were mainly concerned with physiognomy as a basis for classification, particularly with growth-forms, vertical structure, and mosaic structure of primary and secondary forest.

In the last decennia, however, much research has been carried out in Britain on horizontal structure (charting) in relation to (supposed or observed) cyclic succession (Kershaw, Greig-Smith, Gimingham, a.o.). Very recently several research workers in the U.S.A. have turned to rather detailed texture analyses, mostly in chaparral and (semi-)desert vegetation (for instance Parsons & Moldenke 1975, Parsons 1976).

The Scandinavian schools never paid any attention to texture or structure. Their vegetation analyses only comprised species lists with coverage data, disregarding sociability (aggregation), layering, total cover of separate layers, etc. Vertical structure was, however, a very important implicit criterium for the definition of phytocoenoses and coena. We need only mention the 'associations unistrates' of Lippmaa, the socions and unions of Du Rietz, the synusiae of Gams.

In the Braun-Blanquet school, too, floristic composition is and always has been of prime importance. The only structural element included in the definition of an association, adopted not only by this school, but by the International Botanical Congress Brussels 1910, referred to the condition of 'uniform physiognomy'. In 1928 Braun-Blanquet omitted even this faint reference to structure from his association concept[1] and as a result such different units as heathland or garrigue and woodland or forest were sometimes united in the same alliance or even association. Until about 1950 structure found its expression in the hierarchical system of vegetation units of this school only in the sequence of the then highest units (classes) according to 'sociological progression', i.e. the order of increasing complexity of

---

1. However, one of his most outstanding disciples, V. Westhoff, wrote in the first of the theses added to his dissertation (1947): 'The criterion for the acknowledgement of a plant community as such should be the existence of social structure, not the constant (floristic) composition' (translated from the Dutch).

structure (notice that this is actually an ordination superimposed upon a classification!).

In the last 30 years, however, increasing weight is given to structure and texture as diagnostic characters. The incorporation of *Puccinellion maritimae* and *Spartinion* in different classes, the distinction between *Aphyllanthion* and *Rosmarino-Ericion*, between *Violion caninae* and *Calluno-Genistion*, between *Trifolio-Geranietea* and *Festuco-Brometea*, between *Franguletea* and *Alnetea glutinosae*, between *Rhamno-Prunetea* and *Querco-Fagetea*, are ever so many examples of classifications based (or at least inspired) primarily upon structure. In an extreme form this holds true for the units of den Hartog & Segal (*Lemnion minoris, Lemnion trisulcae, Ceratophylletea, Utricularietea*, etc.).

The minor role that texture and structure have played in classification in the Braun-Blanquet school does not mean that these have not been investigated. Some characters are even recorded in each relevé, namely (1) sociability of the species present, which can be regarded as a rough parameter for the horizontal texture (grain size of the pattern), and (2) layering (number of layers, height and density of each layer), being an aspect of vertical structure. In addition, life form spectra according to Raunkiaer are often calculated for tables of relevés, as a contribution to the functional texture of phytocoena. Other aspects of texture and structure are recorded nonsystematically, mostly in specialized investigations. Curiously enough, these deal more often with root structure than with structure above the ground, in spite of the difficulties involved in the investigation of the former. Another curious fact about the Braun-Blanquet school is that attention is paid to vertical layering, not to horizontal pattern within relevés. Yet, different layers are never treated as separate syntaxa (as has been done in Scandinavian schools), whereas elements within the horizontal pattern are more and more often distinguished as separate coena.

Recently much attention is devoted to the analysis of above-ground vegetation texture and structure both by Werger c.s. (South Africa, Netherlands, Austria, India) and by the Austrian MAB programme (cf. Pümpel 1977). The investigations by Werger c.s. will be discussed below. Pümpel investigated phytomass, primary production, horizontal structure (charting), and vertical structure (stratified harvests) of three alpine plant communities. With intervals of only 2 cm the following parameters were measured: phytomass, leaf area index, and leaf inclination.

### 3.3 *Investigations in the Netherlands and by Dutch botanists*

Like all vegetation research in the Netherlands that on texture and structure is mirrored by the lectures presented at the meetings of the Commission for the Study of Vegetation of the Royal Botanical Society of the Netherlands (for an enumeration see elsewhere in this issue). The list shows that texture and structure were not the prime interests of Dutch phytosociologists, but it also shows an increasing interest in this topic. In the past 99 meetings 505 lectures were held, of which only 26 dealt with texture and/or structure in the widest sense (as far as is personally known to the author or can be deduced from the titles). In the first 11 years (1933-1944) not a single lecture was delivered on this topic and of the 26 lectures 18 were given in the last 50 meetings (i.e. from 1962 onwards).

### 3.3.1 Above-ground texture of the vegetation

In 1940 an excellent paper was published by members of the NJN (Netherlands Youth League for the study of nature), dealing with the dry grassland communities of our country. Much attention was devoted to the Raunkiaer life form spectra and their relations to the habitat factors of the various associations (Stafleu 1940). In his monumental and classical thesis on the vegetation of the Wadden Islands Westhoff (1947) not only analysed life form spectra, but also spectra of hydrotypes sensu Iversen, and dissemination spectra of all associations.

Barkman (1958a) elaborated a system of 21 growth forms for epiphytic algae, lichens and bryophytes. He also published two systems of life forms for these plants, one of which was based on height above (or in) the substratum (7 classes), whereas the other was based on their ecophysiological behaviour with regard to the factors water and water vapour in the environment (3 classes). Spectra were made for these growth and life forms for each of the 41 epiphytic associations of the Netherlands.

Den Hartog (1959) published 14 growth forms of epilithic algae. Segal (1965) constructed a system of 11 growth forms for aquatic vascular plants. Van der Maarel (1966), dealing with a stand of dune grassland, proposed a system of 12 growth forms for dune plants (4 bryophytes and lichens, 8 herbs and grasses). He distinguished 13 vegetation types, based on both floristics and on some characters of texture and structure, and calculated their growth form and life form spectra.

Segal (1969) made a thorough study of wall vegetation in Western Europe. He developed a system of 12 growth forms of flowering plants (4 woody plants, 8 herbs and grasses). Growth form spectra, life form spectra, hydrotype spectra and dissemination spectra were calculated for all associations and they were related to geography (macroclimate), but not to the habitat of the communities involved.

Londo (1971), dealing with hydrosere communities in calcareous dunes, also developed his own system of growth forms, based on Warming (1884). His growth forms are 'genotypical' instead of 'phenotypical' (potential instead of actual) and are, for instance, based on the ability to form long runners, irrespective of the question as to whether they are present or not in a particular stand or coenon. He also distinguished 6 height classes of plants (3 for terrestrial herbs and grasses).

Eijsink et al. (1978) calculated life form spectra of various dry grassland communities in Lower Austria, compared them to allied communities from other regions with the same and different climates and discussed their variation in relation to site factors. The latter appeared to be far more important for life form texture than climate. The differences in the shares of hemicryptophytes and chamaephytes were ascribed primarily to their differences in root morphology and to the relation of the latter to soil depth and microclimate.

Werger (1978) analysed the life form spectra of plant communities in the dry Southern Kalahari (South Africa) and showed that codominance of the highly competitive, niche sharing hemicryptophytes and therophytes is confined to very fine soils, where competition for water is eliminated by periodical moisture excess.

Finally, Werger & Ellenbroek (1978) made an interesting study on riverine forests in South Africa along a transect of 2200 km, including four major climatic zones. The composition of the forest communities as to leaf size classes and leaf consistency types was calculated and related to this macroclimatic gradient.

### 3.3.2  Root texture

Westhoff (1947) was the first investigator in the Netherlands who examined the root depth of a great number of plants in natural plant communities.

Zwillenberg & de Wit (1952) distinguished three different root types in *Schoenus nigricans* in the garrigue near Montpellier and emphasized their importance for the ecology of this species. Barkman (1958b) worked in the same area and plant association. He analysed the root depth and its variation and the levels of rootlets and root hairs in 43 species and compared root depths with the height of the plants. As to the mode of branching, 8 different root types could be distinguished, showing a clear relation with the resistance of plants to drought and erosion. Boterenbrood, van Donselaar-ten Bokkel Huinink & van Donselaar (1955) analysed root types of plants in coastal dunes near Montpellier and so did van Donselaar-ten Bokkel Huinink (1961) for aquatic and marsh communities in eutrophic, dead river branches of the Netherlands. This important research contributed substantially to the understanding of the vegetation ecology and succession in eutrophic hydroseres. In 1966 she published a thorough study of root texture of savannah plants in Suriname, showing interesting relations with the extreme habitat conditions (fire, drought, strongly fluctuating ground water level) of this vegetation type.

### 3.3.3  Vegetation structure

Zwillenberg & de Wit (1952) described the horizontal structure of the *Rosmarino-Lithospermetum schoenetosum* near Montpellier, in particular the development of tufts of *Schoenus nigricans* into rings and arcs and its bearing on the microdistribution of other plant species. Barkman (1958b) studied the horizontal and vertical structure in the *Rosmarino-Lithospermetum helianthemetosum* (same region), being a mosaic of dwarf shrubs of *Rosmarinus officinalis* and open places with low herbs. Other species appeared either to prefer the shrub centres, or the shrub edges, or the open spaces in between, or to be indifferent. This phenomenon was studied in relation to the height of the plants and their vegetative parts inside and outside the shrubs. It was possible to explain the horizontal and vertical pattern as a result of grazing and treading by sheep (avoiding *Rosmarinus*), shade, litter and allelopathic effects of the rosemary, and soil erosion. Like *Schoenus*, *Rosmarinus* forms tufts which develop into rings and arcs and eventually die. Consequently the whole horizontal and vertical structure of the community is subject to a cyclic succession.

Barkman (1965) investigated floristic composition and structure of different vegetation types in Drenthe, applying the method of Dansereau (1951, 1958). This method was refined (1) by recording actual heights of plants instead of height classes; (2) by recording variation in height (not only variation in height of the canopy but also variation in height of the crown base of each tree and shrub layer); (3) by recording the horizontal pattern of each layer in relation to other layers; and (4) by recording the actual coverage of each growth form, also if more than 100% (overlapping or interlacing crowns). Van der Maarel (1966) analysed the vertical distribution of biomass in dune grassland over logarithmic height classes. He used sociability values as a basis for a so-called 'exclusion index', which he considered to be a parameter of horizontal structure. This index is the sum of the products of the sociability and the (converted) coverage values of each species.

In 1966 van Leeuwen published the first of a series of papers in which he developed his 'relation theory', a theory that would prove to be of the utmost importance in the understanding of vegetation structure and dynamics and in the management of nature reserves. Van Leeuwen pointed out that connection (com- munication) leads to equality (homogeneity, monotony), and vice versa, and that separation (isolation) leads to inequality (heterogeneity, diversity), and vice versa. The former set of characters is also linked with disorder, noise and instability, the latter with order, information and stability. Every relation of equality has a coun- terpart consisting of a relation of inequality. This applies to spatial as well as to temporal structure of vegetation. Homogeneous, monotonous horizontal pattern is generally linked with sharp discontinuities vertically (disjunct layers) and instability in time (strong fluctuations or rapid succession). Heterogeneous, diverse patterns are usually connected with vertical homogeneity (no distinct layers) and homo- geneity in time (i.e. stability). The former situation is linked with coarse-grained patterns (no internal boundaries, sharp external boundaries: limes convergens, eco- tone), the latter with fine-grained patterns (many internal boundaries, vague exter- nal boundaries: limes divergens, ecocline). In ecotones species diversity is lowest in the centre, in ecoclines it is highest in the centre, as was formulated already by Odum and demonstrated in Dutch vegetation, f.i. by van der Maarel (1966) and van der Maarel & Leertouwer (1967). Dynamic, convergent situations tend to dominate over static, divergent situations. Stability and high diversity of species, life forms, structure, etc. are therefore realized best, where the environment is not homo- geneous but shows gentle gradients in many factors, with non-dynamic (dry, nutrient-poor) areas dominating over dynamic (wet, eutrophic) areas, for instance if the former occupy higher parts of a slope than the latter or occupy larger surfaces. Later on van Leeuwen refined his ideas by introducing variants such as the stable limes convergens and the instable limes divergens. Thanks to his theory we have learned to understand the ecology of a species or a community which occurs over a range of widely different habitats, yet does not occur everywhere. The common denominator is here either stability, or a certain amount of instability of the habitat. The theory also has clarified why plants and communities with a structure of ground rosettes of strongly appressed leaves occur not only in heavily trodden localities, but also in other, highly dynamic habitats.

The relation theory was based on 20 years of observations of permanent sample plots. Unfortunately the original data have not been published. But other workers have tested the theory and published their data, which often confirmed the theory, for instance Barkman (1968a), Londo (1971), and Strijbosch (1976). The first author stated that in oligotrophic fens wet stages of the hydrosere are poor in species, show little spatial variation in pH, and strong pH fluctuations in time (within a year). Drier stages are richer in species with a much greater spatial and smaller temporal variation in pH. Londo, studying succession during 10 years in a number of permanent plots in recently inundated dune valleys, stated that changes in vegetation were greatest in plots with the smallest numbers of species. Strijbosch studied vegetation types as well as many chemical factors of the substratum in oligotrophic, metatrophic and guanotrophic fens during the course of the year. Fens with strong fluctuations of these factors appeared to be much poorer in species (the species-poor communities being more strongly represented) and to have a more coarsely grained vegetation pattern.

Barkman (1968b) introduced a new method for the determination of minimum areas in vegetation. This enabled him to demonstrate that in each of a set of communities differing widely in floristic composition and structure, a number of minimum areas instead of just one could be found. This seems to indicate that even so-called homogeneous stands or phytocoenoses are actually mosaics consisting of finer mosaics, etc. This compound mosaic structure of vegetation seems to be of a general nature. A number of examples were given and the consequences for the typology and classification of vegetation is discussed by the author. Later on attention was focussed on the mosaic structure of juniper scrub, in relation to the microclimatic pattern and the distribution pattern of bryophytes and higher fungi within the scrub. So far results hereof have only briefly been published (Barkman 1973, 1976, Barkman, Masselink & de Vries 1977).

In a recent paper Werger & van der Maarel (1978) lucidly pointed out the importance of research on vegetation structure, both because structure is a valuable indicator of environmental (particularly macro- and microclimatic and biotic) factors, but also because it may contribute towards a better understanding of the functioning of plant communities.

3.3.4  Structure and classification

The first phytosociological investigations in the Netherlands (in the twenties and early thirties) were made according to the Scandinavian methods of Hult, Du Rietz and Raunkiaer (Bijhouwer, Jeswiet, van Dieren, Scheygrond, Feekes, D.M. de Vries). Structure of vegetation only found its expression in the vegetation typology by such concepts as 'enkelvoudige verbanden' (societies), 'horizontale complexverbanden' (stratocoenoses) and 'verticale complexverbanden' (microcoenoses, sociations) of Scheygrond and de Vries.

In the thirties research was started following the Braun-Blanquet tradition. Structure was of no importance in typology and classification in these studies. An important change was introduced by den Hartog (1959) who founded his epilithic algal communities on floristic criteria (both sociations and associations were distinguished), but united them into formations on the basis of (1) his own growth form system, (2) stratification, and (3) habitat factors. Van der Maarel & Westhoff (1964) established a local vegetation system based primarily on textural and structural characters.

Westhoff (1967) analysed the problems of the floristic classification often being non-consistent with a structural classification. He concluded that the discrepancy is manifest only in extreme habitats. In other cases a satisfactory classification based on both structure and floristics seemed possible to him. For categories above the class (or class group) we even have to rely on structure alone, since it is impossible to find character species for coena of those levels. Westhoff made the first attempt to introduce structural formation types in a Braun-Blanquet system, uniting the 38 classes of Dutch vegetation into 14 'structural units', for which the term 'formation' did not seem quite appropriate to him. Westhoff & den Held (1969) reduced the 14 units to 13, which they now called 'formations', although their delimitation is much narrower than that of the formations of older authors. These formations, however, hardly met with any response among colleagues. They are not structural units in the strict sense, since other criteria are equally involved, namely ecological criteria (f.i. formation I: 'water plants and amphibious species'), syngenetical criteria ('pioneer vegetations'), etc. Besides, the formations are often structurally heterogeneous. We must, however, welcome this first attempt because it may become a starting point for further developments. Werger (1973) incorporated his floristic units (syntaxa) in the structural system of Fosberg (1967), in which he succeeded in spite of his justified criticism of Fosberg's system.

### 4. The rôle of vegetation structure in the biocoenosis

Among the three components of a biocoenosis — producers, consumers and reducers — the first-named (green plants) are not only the most important being the only organisms able to produce living matter from inorganic matter, but also because they constitute the bulk of the biomass, and therefore, have a major influence on microclimate and soil. Having a more or less fixed form and a fixed place they make up the architecture of the biocoenosis. Other organisms just have to fit in this skeleton and to adapt their form and place to it. Thus, the influence of vegetation structure upon other organisms is mainly twofold: (1) direct (mechanical), and (2) indirect (via microclimate). This holds true for the influence upon consumers (most animals, parasitic plants) and reducers (fungi, bacteria, protozoa, etc.) but also for the influence of upper vegetation strata upon lower strata.

Within the scope of this paper it is impossible to discuss all known and possible effects of vegetation structure. I shall confine myself to giving some examples in order to illustrate the biological importance of structure.

### 4.1 Direct influence of structure

The most trivial form of direct influence is exclusion. Plants that form dense carpets by rapid vegetative growth do not allow other plants to enter these carpets, especially if the leaves are horizontally positioned and form a continuous leaf mosaic (e.g. *Hedera*, *Oxalis*). Exclusion can also take the more active form of suffocation: *Cladium mariscus* and *Molinia coerulea* do not grow so densely as not to leave enough room for other plants, but they produce large quantities of litter decomposing very slowly and therefore accumulating on the soil; small herbs and especially bryophytes and lichens are thus suffocated. The phenomenon is also well-known from woods with trees with large, flat leaves, like *Castanea*, *Quercus rubra*, *Aesculus*, *Fagus*, especially in regions and habitats where these trees are not indigenous and decomposition of their litter is non-optimal. It is more difficult for small herbs to break through large, heavy flat leaves than to push aside small leaves or needles. Besides, the latter are more easily blown away by the wind. In oak-woods on undulating soil one often notices that a moss layer is only present (1) near the windward edge, (2) on the hummocks of the soil, and (3) around the tree bases.

The *Dicrano-Quercetum* is a type of oak-wood with hardly any herbs, but with many bryophytes forming extensive carpets. It occurs on sand dunes where the wind blows away the litter. In some cases the establishment of *Empetrum nigrum* in this wood type has been observed. In its colonies bryophytes disappear immediately because *Empetrum* brakes the wind and accumulates the oak leaves.

Active mechanical removal of other plants is seen on trees with flaking bark, where the epiphytes are thrown off with the flakes, and in woodland with a dense sward of grasses: the grass leaves, moved by the wind, 'sweep' the mosses and lichens from the bark of the tree base.

Dense thickets, particularly of thorny shrubs, are impenetrable for larger animals. Thus they may enable plants which cannot stand trampling or grazing to grow up in their shelter. If not too dense, they form a hiding biotope for row-deer, and a nesting biotope for many singing-birds. Thus the *Prunetalia spinosae* are not only different from the *Querco-Fagetea* in their structure, but probably also in their fauna.

What scrub means to larger animals, means a moss carpet to small animals. A moss carpet of *Pleurozium schreberi* is too high and too loose for medium-sized carabid beetles to walk over. They will fall in the carpet, where it is too dense, however, to walk. This may be the reason why *Empetrum* heaths are so very poor in carabid beetles (den Boer 1967). By its dense growth *Empetrum nigrum* has a very peculiar effect on microclimate, favouring the growth and dominance of *Pleurozium*.[1] *Calluna* heath, very rich in carabids, forms loose canopies with a much

---

1. As soon as *Empetrum* establishes itself in *Calluna* heath, the moss layer develops and expands. Thus we see that the same species (*Empetrum nigrum*) may have both a detrimental (oak-wood) and a favourable (heath) effect on bryophytes, depending on circumstances.

drier microclimate and a mosaic of bare litter and low, dense, even moss swards (*Pohlia, Campylopus, Hypnum cupressiforme*), on which beetles probably can walk easily. Only on northerly exposed slopes, at the edge of forests, and in old *Calluna* heath with many invading trees, *Pleurozium* can become dominant and thus affect the carabid fauna.

Structure and thickness of litter also seem to be of prime importance to these insects. This is partly a question of availability of food, since the preys of these carnivores can hide more effectively in thick litter. Removal of litter appeared to improve the food supply considerably, although some prey animals must have been removed with the litter (J. Szyszko, unpubl.). Other factors play a rôle, too, since conifer forests are much less favourable to carabids than deciduous woods. This probably is a question of moisture, for the loose litter of needles dries out much more rapidly than broad-leaved litter with its extensive air cavities (it is well-known that these cavities play an important rôle in the survival of drought sensitive isopods). Even a sporadic admixture of large leaves in the needle litter may improve the situation, and so may the presence of moss cushions, which again suggests that moisture is the important factor. Juniper shrubs which occur on very dry sand dunes, have very loose litter and retain much rain in their crowns (Barkman, Masselink & de Vries 1977). They are extremely poor in carabid beetles (den Boer 1967).

An excellent example of the influence of vegetation structure upon the distribution ecology of animals was given by Klomp (1953). He showed that habitat selection of the lapwing is not determined by the floristic composition but by the structure of the vegetation. Lapwings prefer low grass vegetation. The height reached by the vegetation in the second half of May, in combination with its density, is the decisive factor: in more open grassland the lapwing can tolerate taller grass than in dense grassland, where a height of 4-7 cm is the maximum it can stand. This biotope is chosen by the animal much earlier in the year when all grass is still low. The choice is not made on the basis of height of the grass sward but mainly on account of the colour of the grass. Klomp showed that structure is involved in a purely mechanical way. In tall dense grass the bird cannot walk because its toes are caught in the grass.

## 4.2 *Indirect influence of structure. Structure and microclimate*

Vegetation structure has an enormous influence upon dependent organisms through its modifying influence upon microclimate. Microclimatic differences in plant communities some 100 meters apart, may be larger than macroclimatic differences over a North-South distance of 1000 km. In the preceding paragraph we touched upon this subject when dealing with litter moisture as a result of litter structure (which, in its turn, depends on leaf structure). Other examples are given in Dutch literature by Stoutjesdijk (1959, 1961, 1966, 1974, 1977), Lensink (1963), Barkman (1958a, 1965, 1977) and Barkman, Masselink & de Vries (1977).

Generally speaking vegetation arrests both incoming and outgoing radiation (or rather: outgoing radiation from the soil is partly compensated by counterradiation of the canopy) so that temperatures in lower strata and on the soil are lower at day-time and higher during the night than those on bare ground. This effect increases with vegetation density. On the other hand, bare soil where vegetation has been present (e.g. felling areas) has much larger temperature amplitudes than bare, mineral soil, because humus has a low albedo (dark colour), low heat capacity and low heat conductivity Comparing (E) *Empetrum* heath (dense canopy), (Ca) *Calluna* heath (loose canopy), (SC) *Spergulo-Corynephoretum* (very open vegetation, sand) and (Cl) *Cladonietum mitis* (degeneration stage of *Calluna* heath after a heather beetle epidemy: open vegetation, humus) we may find the following differences in soil surface temperatures on extreme days:

1. Maximum temperatures on a hot summer day (air temp. $34.4°$): E $21°$, Ca $42°$, Sc $59°$, Cl $65°$.
2. Minimum temperatures on a cold winter night (air temp. $-14°$): E $-8.1°$, Ca $-16.5°$, Cl $-19.0°$.

Or, to put it another way: Undergrowth of *Empetrum* heath in Drenthe (Neth.) has regular frosts only during 4 months per year, *Calluna* 8 months, and *Cladonietum* 12 months. This must have an enormous effect on mosses, lichens, terrestrial arthropods and soil organisms.

Another effect of structure is through snow cover. *Empetrum* has a smooth, dense canopy. The snow forms a continuous blanket on the canopy, protecting the undergrowth from cold. *Calluna* has an uneven canopy with a broken snow cover. The very cold, heavy air formed just above the snow surface in winter nights, will sink through the snow gaps to the soil, thus affecting the organisms living there.

Deciduous vegetation hampers in- and outgoing radiation only in summer when incoming radiation is most important, but not in winter when outgoing radiation is most important. Thus, the overall temperature of the year may be considerably lower in dense deciduous woods, like beechwood (annual temperature about $3°$ lower) than in the open field. It is interesting to note that in the lowland of Western Europe beechwoods constitute a refugium for montane and boreal epiphytic lichens and bryophytes.

Larch plantations (*Larix leptolepis*) in Drenthe are particularly rich in boreal moss species like *Rhytidiadelphus loreus* and *Ptilium crista-castrensis*. Some of them are more or less restricted to these larch woods, although in the same area there are plenty of pine and spruce woods, for which these mosses are characteristic in Northern Europe! The explanation for this distribution type might be that the mosses are confined to conifer woods generally, but on the southern edge of their distribution area they avoid pine and spruce woods as these are not cold enough. Larch woods are the only deciduous conifer woods in the Netherlands. Measurements actually showed that average winter minima on the soil are $1.2°$ lower in larch wood than in pine wood. On very cold nights we may even note temperatures like: spruce wood $-4°$, pine wood $-9°$, larch wood $-13°$, oak wood $-14°$.

Vegetation not only absorbs and reflects incoming radiation but it also changes the spectral quality of the passing light. This effect is most marked in dense woods of trees with thin leaves, like *Tilia* and *Fagus*. Thick leaves, like those of *Ilex* and especially conifers, hardly transmit any light and therefore do not change its spectral quality (a change is only brought about by the fraction of light reflected by the leaves towards the ground). Open woods have a larger fraction of unchanged light. The filtered light is not only unfavourable to photosynthesis but also to the germination of many plants as it contains very little light red light (600-700 nm) and much dark red and near infrared (700-800 nm). The former stimulates germination, the latter inhibits germination. Stoutjesdijk experimentally proved this. He demonstrated the effect of dense canopies of thin leaves (*Crataegus* scrub) on the germination of herbs; this was shown to be even inferior to germination in complete darkness.

Stoutjesdijk also drew attention to the 'open shade' climate at the north side of forests, woods and scrub and its significance for higher plants; this climate is cool and humid with long lasting dew and rime, with an often negative radiation balance, and a relatively high fraction of blue to ultraviolet light. Some bryophytes are typical of this biotope and others fructify only here. It is the most favourable biotope for the fructification of macrofungi in dry years and seasons.

Vegetation not only affects the radiation balance and light quality, it also brakes the wind and moistens the air by its evapotranspiration, thus reducing the danger of desiccation of dependent organisms. It therefore has a strong effect on the undergrowth and the fauna, particularly in tall, dense, many-layered, deciduous woods with large, thin leaves. Glades in such forests are extremely favourable for epiphytic lichens, since the latter demand both a fairly high light intensity and a humid atmosphere, a combination that is not often realized. Natural glades are typical of primeval forests and absent from cultivated woods. This is one of the reasons of the richness in species and life forms and the complex structure of the former. Another, of course, is the low degree of disturbance, and the presence of dead tree stumps and dead and hollow standing and fallen tree boles, so important to saprophytic fungi, myxomycetes, epixylic bryophytes, xylophagous insects, bats, woodpeckers, and other hole-brooding animals.

Also heathlands can show the indirect effects of vegetation structure. The heather beetle (*Lochmaea suturalis*) only attacks old *Calluna*. It has been shown that the larvae can develop only in thick carpets of moss which protect them from drying out. Such carpets are formed by *Pleurozium schreberi*, and this species only develops in old heath, owing to sufficient shade and humus, provided both by *Calluna* and by invading trees. Here we see the interesting phenomenon that one species (*Pleurozium schreberi*) is favourable to one beetle species, whereas it was shown to be unfavourable to others (carabids).

Vegetation not only supplies moisture, it can also hamper moisture supply. Dense juniper scrub absorbs about 80% of the rainfall in the crown. The loose, very dry litter underneath is the only place in the Netherlands where *Geastrum floriforme* has been found. This is a continental fungus occurring mainly in open, dry, warm, sunny places in Eastern Europe. In spruce forests of Northern Sweden the undergrowth of both dwarf shrubs and bryophytes and lichens forms a mosaic of

141

two types, the drier type under the tree canopies, the moister type in between the canopies. This is due to *Picea* having a dense, centrifugal crown conducting the rain water to the periphery.

In park landscapes with scattered dense shrubs the North and South side of the shrubs often have a different undergrowth, and so have (in windy climates like sea coasts) the windward side and lee side. The lee side is not only sheltered from strong winds but in cold, windy climates snow may accumulate here in winter, and thus snow cover may last here much longer. This was actually observed on the alvar of Öland with scattered junipers.

Westhoff & Westhoff-de Joncheere (1942) were the first Dutch authors to make a study on the composition of the fauna in relation to plant associations based on floristic methods (Braun-Blanquet), and on the distribution ecology of animal species in relation to these associations. The study dealt with ants in various types of woods throughout the Netherlands. They clearly showed that (1) some species are characteristic of certain associations (f.i. *Querco-Betuletum*) or even subassociations (f.i. *Q.-B. molinietosum*), whereas (2) others are dependent on certain plants (f.i. *Deschampsia* species, on the roots of which aphids are living, which are the main food of the ants in question). A third category of ant species is clearly dependent on vegetation structure, probably through microclimate. Dune birch woods on calcareous sand and inland birch woods on acid sand are widely different in vegetation but have almost the same ant fauna. Floristically the latter are closely allied to inland acid oak woods (same association), but they are very different in ant fauna. The (dune and inland) birch woods are much lighter and sunnier than the oak woods, owing to the canopy structure of the trees. In the same way tall wood and coppice wood are different in ant fauna. A moss layer is favourable, whereas the dense shade of a *Pteridium* layer is unfavourable to ants.

The second Dutch study along these lines was published by Mörzer Bruyns (1947). It dealt with molluscs in a great variety of plant communities in a restricted area. Many species apparently belonged to Westhoff's type (1), none to his type (2). The influence of vegetation structure (3) was only evident in the differences between the mollusc faunas of tall wood and coppice wood belonging to the same association.

The different results of the Westhoff's and Mörzer Bruyns are easy to understand: terrestrial gastropods are mainly herbivorous and polyphagous. Humidity and calcium content of the soil are the most important factors controlling their habitat preference. And these are the very factors that control also the floristic composition of the vegetation, whereas they have a minor influence on structure.

Lensink (1963) studied the distribution ecology of three grasshopper species in the dunes of Voorne. Vegetation structure was shown to play a predominant role in the distribution and migration of these insects during the season. The vegetation was divided by him into 7 structural types (6 moss-herb and 1 dwarf shrub community). This typology was based on the share of the following elements in the vegetation: (1) bryophytes and lichens, (2) rosettes, (3) carpets of herbs, (4) tall herbs, (5) low shrubs. Vertical profiles of all types were drawn and microclimatic

142

measurements were made. During spring and summer the animals migrate from the open, warm and sunny vegetation types to the denser, cooler and moister types, but females return to the former for laying their eggs. In cool summers the animals stay longer in the open types. It was shown that the inclination of leaves of herbs and grasses (horizontal versus vertical) was of importance to one of the species. Highest population densities were found where vegetation structure was such that suitable places for oviposition and hatching bordered upon places suitable for nymphs and adults to live in. Thus the importance of a mosaic structure of vegetation was demonstrated.

In this study it was quite obvious that vegetation structure affected the animals mainly through microclimate. There are, however, cases where a correlation between vegetation structure and dependent organisms is clear-cut but the causes are still obscure. This is especially true of vegetation structure and fungi. Mycorrhiza-fungi are much more abundant (at least their fruit-bodies!) in young than in old pine woods. In general, light woods, wood edges and isolated trees often have more mycorrhizafungi than dense woods. Mixed woods are better than monocultures, probably because the density of each tree species is lower than in a pure stand of that species. Density of trees seems to be important, but we do not know why.

Lange (1923) and Leischner-Siska (1939) stated that in plant communities the richness in fungi is inversely proportional to that of herbs. In the permanent plots of woodland, heath and bogs studied in Drenthe no correlation could be found between the number of fruiting macrofungus species and either the number of herb species or the number of graminoids or the total coverage of the graminoids. However, there is a strong negative correlation ($\tau = -0.481$, p = 0.00736) with the total coverage of all herbs and grasses. It is not clear how this phenomenon should be interpreted.

## 5. Parameters of texture and structure

### 5.1 *Leaf size*

Raunkiaer (1916) proposed a classification of leaves into 6 size classes, viz. leptophyllous (less than 20 sq. mm), nanophyllous (20-200 sq. mm), microphyllous (2-20 sq. cm), mesophyllous (20-100 sq. cm), macrophyllous (100-1640 sq. cm) and megaphyllous (more than 1640 sq. cm). Werger & Ellenbroek (1978) introduced an intermediate class of subnanophyllous leaves, and gave a different size range to all categories, using the same names. This may give rise to confusion. Besides, it makes it difficult to compare their spectra with those published before. I therefore propose to maintain the limits given by Raunkiaer with one exception, namely 500 sq. cm as the lower limit of megaphyllous. In the humid tropics Raunkiaer's limit may well be practical, but in the temperate zone such large leaves hardly ever occur. The 500 sq. cm limit proved to be very useful here. I also propose to add the category 'bryophyllous' for leaves under 4 sq. mm. The range 0-20 is too wide. Besides, not only many bryophytes are bryophyllous, but also such vascular plants as *Lycopodium complanatum, Selaginella, Azolla, Equisetum,*

143

*Juniperus sabina, Thuja, Chamaecyparis, Ephedra, Wolffia, Calluna, Tamarix, Asparagus, Salicornia*, etc. For detailed research we may subdivide the nanophyllous leaves in nanophyllous s.s. (20-60 sq. mm) and subnanophyllous (60-200 sq. mm) and in the same way microphyllous into microphyllous s.s. (2-6 sq. cm) and submicrophyllous (6-20 sq. cm). (Parsons (1976) distinguished microphyllous 226-1125 sq. mm, and notophyllous 1125-2025 sq. mm; this is not a well-balanced division, however, the latter range being much narrower than the former). The classes used here are:

| | | | |
|---|---|---|---|
| bryophyllous | less than 4 sq. mm | | |
| leptophyllous | 4 – 20 sq. mm | | |
| nanophyllous | 20 – 200 sq. mm | nanophyllous s.s. | 20 – 60 sq. mm |
| | | subnanophyllous | 60 – 200 sq. mm |
| microphyllous | 2 – 20 sq. cm | microphyllous s.s. | 2 – 6 sq. cm |
| | | submicrophyllous | 6 – 20 sq. cm |
| mesophyllous | 20 – 100 sq. cm | | |
| macrophyllous | 100 – 500 sq. cm | | |
| megaphyllous | more than 500 sq. cm | | |

Leaf size spectra of plant communities may give insight in macro- and microclimate, as small leaves are indicative of a cold and/or dry climate, and very large leaves are typical of a warm, humid climate. There may thus be a difference between a North slope and a South slope, between communities on wet and dry soil, between the various strata, etc.

Differences may exist even within species. *Limonium vulgare*, for instance, is nanophyllous in the *Plantagini-Limonietum*, microphyllous in the *Armerio-Festucetum* (Landolt 1977).

Leaf size also determines the size of the light spots on the forest floor and the structure of the litter. Both may be important for soil surface dwelling organisms. I have already pointed out the importance of leaf size in litter (sub 4.1). The size of light spots on the forest floor determines the time of exposure of terrestrial organisms to heating by sunlight. It also determines the frequency of the intermittent insolation to which lower leaves in the canopy are subjected. As short intervals of reduced light are more favourable than long ones, leaf size is inversely proportional to the number of vegetation layers that can be accomodated within a given height interval (Horn 1971).

A few examples may illustrate the relation between leaf size and habitat. 1. In oligotrophic bog pools of Drenthe all plants have small leaves, the largest being microphyllous. But even this class is poorly represented and its proportional cover decreases during the succession from the wet *Sphagnum cuspidatum-Carex rostrata* sociation (stage I) to the dry hummocks of the *Erico-Sphagnetum magellanici* (stage IV):

| Stage of succession | I | II | III | IV |
|---|---|---|---|---|
| Microphyllous plants | 7.7% | 5.6% | 1.6% | 0.2% |

2. On the dry limestone rocks of Öland (Sweden) succession of the alvar vegetation leads from a lichen vegetation, via a *Sedum album-S. rupestre-S. acre* vegeta-

tion ('*Sedetum*'), a *Thamnolia-Festuca ovina* community ('*Festucetum*'), and a *Filipendula vulgaris-Orchis sambucina-Helictotrichum pratense* community ('*Avene-tum*') to juniper scrub. Pioneer and climax vegetation differ strongly in microclimate, as was shown during some measurements in June 1978:

|  | '*Sedetum*' | *Juniperus* |
|---|---|---|
| Maximum temperature at 0 cm | 44.5° | 21.2° |
| Minimum temperature at 0 cm | 2.0° | 6.4° |
| Saturation deficit at 0 cm | 55.0 mm | 2.7 mm |

It is therefore interesting to compare the leaf size class spectra (the spectra given here refer to numbers of species per class as a percentage of the total number of species irrespective of coverage):

|  | '*Sedetum*' | *Juniperus* |
|---|---|---|
| Bryophyllous | 50 | 16 |
| Leptophyllous | 16 | 14 |
| Nanophyllous | 24 | 26 |
| Microphyllous | 11 | 35 |
| Mesophyllous | – | 9 |

During succession there is obviously a shift from species with small leaves to those with larger leaves, along with decreasing daily temperature amplitude and increasing air humidity.

## 5.2 *Leaf consistency*

As regards leaf consistency I follow the widely used classification into malacophyllous (thin, filmy, easily withering), orthophyllous (normal), sclerophyllous (thick, dry, leathery) and succulent leaves. Following a suggestion by Werger I add: felty leaves. This type only occurs among orthophyllous leaves and therefore does not cut across the above classification. Parsons (1976) distinguished between (a) glabrous, (b) upper side hairy, (c) lower side hairy, (d) hairy on both sides. If we regard only felty leaves as distinct from the rest (i.e. moderately hairy to glabrous), we have three categories only, (a), (c) and (d), since leaves with a felty upper and a glabrous lower side do not exist. Thus it is sufficient to subdivide our felty leaves into the two categories (c) and (d). Malacophyllous leaves are indicative of very high air humidity, a very calm atmosphere and a not too cold climate. Sclerophyllous leaves indicate either aridity of climate or lack of nitrogen supply, succulent leaves either aridity or high salt concentrations, and felty leaves dry sunny climates, often with a high proportion of ultraviolet light. The ecological significance of leaf consistency, especially regarding the moisture factor, was discussed from an original point of view by Orians & Solbrig (1977).

In the example of the oligotrophic hydrosere given above (sub 5.1) the percentage of sclerophyllous plants increases from wet (I) to dry (IV):

|  | I | II | III | IV |
|---|---|---|---|---|
| Sclerophyllous plants | 7.7 | 39.0 | 43.0 | 80.8 |

In the other example given above, succession leads from extremely xeric to mesic conditions. According to expectation the shares of sclerophyllous and succulent species diminish:

|  | 'Sedetum' | Juniperus |
|---|---|---|
| Orthophyllous | 53 | 85 |
| Sclerophyllous | 26 | 14 |
| Succulent | 21 | 1 |
|  | 100% | 100% |

## 5.3 Leaf orientation

The orientation of leaves in space is an important feature for the ecophysiology of individual plants (de Wit 1965). It is also an important character of the structure of vegetation, as it affects the vertical distribution of light intensity within the vegetation, the optimum leaf area index, the evapotranspiration of the vegetation as a whole, and also its penetrability for animals and suitability for hiding and nesting.

Leaf orientation should not be confounded with leaf arrangement or phyllotaxis (spiral, decussate, verticillate, etc.). De Wit therefore used the term 'leaf distribution', which I would rather reserve, however, for the distribution of leaf biomass along the stem of an individual plant and over the various strata of a phytocoenosis.

Leaf orientation has two components: exposition (compass direction) and inclination, except in the case of horizontal leaves and cylindrical, vertical leaves, where exposition is irrelevant. Apart from a few typical 'compass plants' like *Lactuca serriola*, leaf exposition seems to be highly variable without a marked frequency optimum. I shall therefore ignore this parameter here.

Leaf inclination can be defined in relation to (1) the direction of the stem (leaf angle), (2) the substratum, (3) the horizontal plane, (4) the direction of the sun rays. I shall use it in the third sense as did Pümpel (1977). Given the latitude of the locality and the time of the day and the year, (4) is determined by (3), (2) is only different from (3) on inclined surfaces, and (1) is a purely morphological character.

Apart from data on some crops, hardly anything is known (at least published) concerning leaf inclination in plants. Parsons (1976), in his study on Californian chaparral and Chilean matorral, paid attention to it. He classified his plants according to leaf inclination into four categories: vertical, 45°, horizontal and mixed (taken from de Wit?). But his result are meagre, his only conclusion being that the vertical type is restricted to evergreen shrubs. He did calculate leaf inclination class spectra for various plant communities, but without drawing any conclusion. A most interesting study was made by Pümpel (1977). She measured leaf inclination, using the classes 0°, 1-10°, 11-20°, etc. and determined the proportional share of each class per height interval of 2 cm for each of three alpine plant communities. In all communities inclination increased with height above the ground. Average inclination was lowest (24°) in the most extreme habitat, viz. a snow bed community (*Salicetum herbaceae*). This low value, favourable for photosynthesis, may be con-

146

nected with the short snow-free season available for productivity. De Wit (1965), dealing with crops, distinguished four categories: (1) planophilous (mainly horizontal), (2) erectophilous (mainly vertical), (3) plagiophilous (mainly inclined, 30-60°), and (4) extremophilous (mainly horizontal and vertical leaves, few inclined leaves).

I have studied leaf inclination in a number of wild plants (unpubl.), and I never found type (4), though I did find plants with all inclinations equally distributed between 0° and 90°. I also found it necessary to add categories of hanging leaves. This has three reasons: (1) Hanging leaves result in a structure different from patent leaves; (2) frequency distribution of leaf angles is different in hanging leaves as compared with patent leaves (so we cannot make for instance one single category: 30° to 70°, including plus and minus); (3) for individual plants the sign of the angle may be important ecophysiologically. Of course the sign does not matter in dense stands nor in isolated plants with thin crowns (if the exposition of the leaves is random, the average angle under which they catch the sunrays is the same in leaves of $+x°$ and $-x°$). But in isolated plants with dense crowns, like trees and shrubs (and these are the very plants where hanging leaves are most frequent), only the East, South, and West sides of the crown receive direct sunlight (on the Northern hemisphere). A simple calculation will show that leaves of say +40° and −40° will receive quite different quantities of light (apart from the differences in interference by mutual shading). The same consideration is valid for East-, South-, and West-oriented forest edges.

Leaf inclinations may be measured in angles but this is tiresome work and in view of the great variability, even in a single plant, it is perhaps more effective to assign whole plants to one of a restricted number of classes. It has proved useful to distinguish 11 classes. They follow here, each with some examples:

1. Sphaerical (s). All inclinations from +90° to −90° frequent. *Pinus sylvestris, Quercus robur, Ulex europaeus, Calluna.*
2. Hemisphaerical (hs). All inclinations from +90° to 0°. *Larix, Nardus, Scirpus caespitosus.*
3. Erect (e). +90° to +60°. *Narthecium, Glyceria maxima, Ranunculus lingua.*
4. Erecto-patent (ep). +70° to +30°. *Poa pratensis, Calla, Andromeda.*
5. Patent (pa). +40° to +10°. *Poa annua, Vaccinium myrtillus, Parnassia.*
6. Spreading (sp). Lower half erecto-patent, upper half horizontal. *Eriophorum angustifolium, Carex rostrata, Sparganium angustifolium.*
7. Arcuate (ar). Lower half erecto-patent, upper half hanging (−30° to −60°). *Milium, Carex arenaria, Orchis purpurea, Digitalis.*
8. Recurved (r). Lower half horizontal, upper half pendent (−40° to −90°). *Chamaenerion angustifolium, Cornus sanguinea, Elodea canadensis.*
9. Horizontal (h). +20° to −20°. *Fagus, Nymphaea, Hydrocotyle, Listera.*
10. Decumbent (d). 0° to −50°. *Ulmus, Urtica, Solanum dulcamara.*
11. Pendent (pe). −40° to −90°. *Tilia, Humulus, Calystegia sepium.*

I must add some remarks in order to avoid misunderstandings. (1). The examples given only mean to say that a species *usually* belongs to that class (it can also belong to other classes). (2). The ranges given above only mean that most of the leaves belong to that range of angles. Sporadic leaves with greater or smaller angles can always be found. (3). The overlapping such as between 4 and 5 does not mean that plants with leaf angles between 30° and 40° cannot be classified. Such plants do not exist. The variation always exceeds 10°, so that a plant nearly always belongs to either 4 or 5. (4). The examples refer to adult plants and to full-grown, not too old leaves in non-withered condition. It is obvious that after a long and severe drought or in windy weather other inclinations will be found. Nevertheless inclination may be rather specific and may differ even between allied species. For instance, *Agrostis canina* is erecto-patent, *A. tenuis* patent; *Betula pubescens* is decumbent, *B. verrucosa* pendent.

Apart from what seems to be genetically determined, inclination depends on the following factors:

1. Age of the leaves. All plants with erect stems and/or branches have erect leaves when they have just budded (exceptions: naked buds like those of *Lonicera periclymenum* with patent to arcuate young leaves, and ferns with spirally incurved young leaves). After budding the leaves either remain erect (type 3) or they change inclination (other types). A special category is constituted by those plants (many trees and shrubs), where the erect leaves first turn to a pendent position, and thereupon straighten themselves up again (e.g. *Fagus, Quercus, Acer*). In *Betula* and *Aesculus*, however, they remain hanging, whereas the pendent stage is lacking in *Alnus*.

Old leaves often assume a different orientation. Erect leaves may become arcuate (*Iris*); arcuate leaves may become horizontal (*Galanthus, Narcissus*); horizontal leaves may become recurved (*Polygonatum multiflorum*). Often these leaves are dead (*Narcissus*), or they have reduced functions: in *Drosera intermedia* only the erect leaves are capable of catching insects, not the old, horizontal leaves.

2. Meteorological conditions. Wind may influence inclination effectively. In dry weather leaves may assume a drooping position (*Urtica dioica*). In some species of *Oxalis* and *Anemone* the leaves fold up and become pendent during the night.

3. Habitat often has a great influence. In crops the nutrient status has been stated to affect leaf inclination. A higher nutrient level induces more erect leaves. It is not clear whether or not this is equally true of wild plants. It seems that in stations rich in nitrogen the leaves of *Taraxacum* sect. *Vulgaria* are more erect than in poorer sites, but this may also be due to lack of light, because the rich sites have a denser and taller grass sward (cf. point 4).

Interesting observations were made in six grasslands in Drenthe, three of which were non-fertilized hayfields (a, β, γ), and three were heavily fertilized meadows (a, b, c). a and a were adjacent grasslands on the same soil (wet), and so were β and b (moist), and γ and c (dry). The average leaf inclination of all plants per stand was as follows:

| | | | |
|---|---|---|---|
| α | : 60° | a | : 66° |
| β | : 41° | b | : 58° |
| γ | : 35° | c | : 54° |

So it seems that inclination increases both with nutrient status and with soil humidity. This is not simply a matter of richer and moister soils having a denser vegetation, inducing the leaves to assume more vertical positions (owing to lack of space and/or light). In each stand the total cover percentage of the herb layer (TC) was estimated, as well as the cover per species (C), also in percentages. Their sum ($\Sigma C$) may be compared with TC, the difference being a measure of the degree of interlacing of plants within a vegetation layer. Leaf inclination appeared neither to be correlated with TC nor with $\Sigma C$:

| Stands: | $\gamma$ | $\beta$ | c | b | $\alpha$ | a |
|---|---|---|---|---|---|---|
| Leaf inclination | 35° | 41° | 54° | 58° | 60° | 66° |
| TC (%) | 65 | 95 | 100 | 70 | 65 | 90 |
| $\Sigma$ C (%) | 105 | 102 | 131 | 72 | 65 | 99 |

In the study by Pümpel (1977) there was, however, a positive correlation between the cover of the vegetation (herb layer) and the average inclination of the herb layer (given here for 0-2 cm above ground, where the bulk of the biomass was concentrated):

|  | Cover | Average inclination |
|---|---|---|
| *Caricetum curvulae* | 90% | 45.2° |
| *Deschampsia caespitosa* vegetation | 80% | 38.3° |
| *Salicetum herbaceae* | 60% | 24.3° |

4. Light intensity and angle of light incidence. Wood edges are usually characterized by a veil of pendent leaves not only of those species that use to have pendent leaves (*Humulus, Calystegia sepium, Polygonum convolvulus*, etc.), but also of the trees themselves. Isolated trees often have pendent leaves at the sides of their crown. This is obviously an adaptation to angle of light incidence. On the upper side of the crown the leaves are often patent, but smaller trees of the same species, growing underneath the canopy of their taller brothers, use to have horizontal leaves. This must be an adaptation to light intensity, for it is a well-known fact that in full sunlight a horizontal position is not optimal for photosynthesis (de Wit 1965). In a wet alderwood in Drenthe the average inclination of the tree canopy was 23°, that of the herb layer 5°.

We may observe the opposite behaviour in herbs of grassland. Species with normally horizontal leaves (*Plantago major, Polygonum aviculare*) or erecto-patent leaves (*Taraxacum, Leontodon, Succisa, Valeriana*, etc.) display vertical leaves when growing in a dense grass sward, notwithstanding the fact that light intensity is reduced and vertical leaves are less favourable regarding light incidence. The causal explanation might be lack of space, the final (functional) explanation might be that vertical leaves, reaching higher in the sward, perhaps after all receive more sunlight (at their tips) in such dense swards than inclined to horizontal leaves.

5. Heavy traffic (rolling or pedestrian) allows only flat-lying plants to survive. In this way species with horizontal leaves and with normally either very short or with creeping stems are favoured, for instance *Hieracium pilosella* on dry sand paths, *Illecebrum verticillatum* on moist sand, *Drosera rotundifolia* on paths in bogs and

wet heath, *Hydrocotyle* and *Callitriche* in wet dune valleys and eutrophic mires, *Lysimachia nummularia* on paths in moist woods. Also some species which normally possess patent leaves (e.g. *Poa annua*), erecto-patent leaves (*Taraxacum officinale* agg.), and even erect leaves (*Plantago lanceolata, Narthecium*) may survive, albeit that these species, particularly the erect ones, are far from flourishing under those conditions. In such sites they all develop horizontal leaves. Under more natural conditions (game tracks, drinking places of game) the same phenomenon is seen, f.i. in *Ranunculus flammula* and *Alopecurus aequalis.*

If I have succeeded in convincing the reader of the interest of studying leaf inclination, a final remark about the method is appropriate. For each layer, each stand and each community type one can calculate a spectrum of leaf inclination classes. This may be a purely qualitative spectrum (number of species per class), or it may be a quantitative spectrum. If one uses the cover percentages of the species as a basis for the latter, one certainly makes a systematic error and one may make another error. One possible error is that some species have their leaves in one layer (f.i. *Oxalis, Anemone, Mercurialis*), others in many layers (f.i. *Stellaria holostea, Urtica*). Their cover percentages may be similar, yet their biomass and their numbers of leaves are not. A correction should be made.

A systematic error is made because erect leaves, by their very nature, will be underrepresented if we take projection on a horizontal surface as a standard. If the leaf angle with the horizontal surface is $\alpha$, a correction factor $1/\cos\alpha$ should be applied. This means that the cover percentages should be multiplied with the following factors: sphaerical: 5, hemisphaerical: 5, erect: 10, erecto-patent: 2, patent: 1.2, spreading: 1.2, arcuate: 1.6, recurved: 1.2, horizontal: 1, decumbent: 1.2, pendent: 7. By the way, this is not only important in the calculation of leaf inclination spectra. The corrections should always be applied to Braun-Blanquet cover estimations, if we want a better approximation of standing crop.

## 5.4  Growth forms

In literature growth forms and life forms have often been confounded. In the 'life form system' of Raunkiaer, for instance, the main divisions are life form classes, the minor groups are growth forms. The 'growth forms' by Drude (1890) are mainly life forms, the 'growth forms' by Schmithüsen (1961) mixta composita of growth forms, life forms, and ecological types. On the other hand most growth forms of cryptogams are mentioned in the literature as 'life forms'. Den Hartog (1955), when dealing with pure growth forms of algae, constantly mixed up the terms life form and growth form, apparently considering them to be synonymous (l.c., p. 127-129).

In my opinion life forms are the morphological expression of the adaptation of organisms to their environment (i.e. their 'epharmony'), whereas growth forms are based on those morphological characters that control the general architecture of the organisms. They are types of habit ('gross morphology'). Characters affecting habit that cannot be interpreted as related to major habitat factors, f.i. mode and density of branching, are therefore used for growth forms, not for life forms. Such characters, on the other hand, as height of winter buds above the ground, their nature (naked or scaly), presence or absence of aerenchym, are important for the definition of life forms, not for growth forms.

Among the genuine growth form systems the classification of plants by Gams (1918) into errant, adnate and radicant (Luther (1949), dealing with water plants, obviously prefers Greek to Latin using planophytes, haptophytes, and rhizophytes as synonyms of Gams' types) is, to my mind, the most fundamental classification of growth forms ever made. For adnate algae of the Dutch coast den Hartog (1955) created a system of 14 growth forms, of which the first four, by the way, were given without a definition or description. For adnate aerial algae, bryophytes and lichens several systems have been published (for a survey, cf. Barkman 1958a). Barkman's (1958a) own system, made for epiphytes only, has now been adapted to all bryophytes and lichens (unpubl.) but nomenclature has been changed. Following Segal (1965, 1969) all types are named after representative genera with the ending '-ids'. With regard to errant and radicant water plants Segal (1965) made an excellent and detailed growth form system.

The classification of terrestrial radicants offers the greatest difficulties because of their great form variety and because of the many possible criteria, which, moreover, are often not correlated. As far as I know, six systems have so far been published, that can be regarded as pure growth form systems of radicant terrestrial plants:

1. Gimingham (1951) classified herbs from dune vegetation into 9 growth forms, on the following (implicit) criteria: (a) growth in tussocks or tufts, (b) stems erect or prostrate, (c) size of plants, (d) branched or simple, (e) leaf distribution (only radical rosettes, only stem leaves, or both), (f) leaves sessile/petioled. In the writer's opinion criteria (a), (b), (d) and (e) are very useful.

2. Dansereau & Arros (1959) distinguished five main types, one of which was an ecological group (epiphytes). The other groups were: (1) erect woody plants, (2) lianas, (3) herbs, (4) bryoids (incl. lichens and algae). Each type was subdivided according to 4 independent criteria: (a) 'function' (i.e. seasonal leaf periodicity: deciduous versus evergreen), (b) height (3 classes per type), (c) form and size of leaves, (d) leaf consistency. In this way countless combinations are created. The authors did not mention or name them, however, but they are implicit in the 'Danserograms'. The system has two advantages: (1) no character is given priority; (2) for each character the spectral composition of a stand can be calculated and compared separately with relevant site factors. Nevertheless, I have not adopted their system, because: (a) leaf periodicity is one aspect of general periodicity and is more in place in a separate system of phenological types; (b) size of plants is highly variable and only important in certain growth forms, like shrubs (shrubs and dwarf shrubs) and graminoid plants (distinct height classes); (c) form of leaves is of minor importance for the texture and structure of vegetation except in special cases (needles, scales, grassblades); (d) the system by Dansereau c.s. neglects important characters such as those put forward by Gimingham.

3. Van der Maarel (1966) published a useful system of growth forms of dune plants, 8 of which referred to herbs. The Pulvinata (cushion plants) hardly occur in the Netherlands. I have adopted his groups Tapeta, Decumbentia, Scandentia, Rosulata, Caespitosa compacta, and Caespitosa, with slight modifications, but I have split up his last and rather heterogeneous group of Erecta.

4. Segal (1969) based his growth form types of herbs mainly on sociability. I fail to see the importance of this criterium for growth forms. Besides, it is already indicated in each vegetation analysis. For woody plants Segal distinguished only four types, which was probably sufficient for his special purpose (wall vegetation), but not for general use.

5. Londo (1971) distinguished 8 types of *potential* growth forms in herbs, based on: (a) life span, (b) presence of runners and/or rhizomes (stolons), (c) length (growth rate) of (b). This subdivision is clearly adapted to his purpose, viz. the study of the colonization of bare soil by pioneers. For general purposes it is both too detailed in one sense and too simple in other respects.

6. Parsons (1976) chose seven (not quite equivalent and independent) morphological characters and classified his plants (mediterranean shrubs) on the basis of these characters. The resulting nine clusters were considered growth forms. These forms, based on chaparral and matorral, might also prove useful for European garrigue, maquis and phrygana, but not for cool temperate vegetation.

For the Dutch flora I have made a provisional system of 46 growth forms which will be published later. The main outline and a few worked examples are given here in order to illustrate the general idea and to show the possibility to create a detailed system of growth forms. The main division is in adnate, errant and radicant plants. The latter are subdivided into: trees, shrubs, dwarf shrubs, lianas and herbs. Trees and shrubs are divided according to mode of branching and inclination of stems and branches, with subdivisions on the basis of form and density of the crown, presence of stolons, spines and thorns.

Herbs are classified into three main groups on the basis of the proportion of stems (branches) and leaves:

I. Caulosa. Mainly stems and branches. Leaves reduced or branch-like (cylindrical).

II. Foliosa. Standing crop consists mainly of leaves.

III. Caulo-Foliosa. Both branches and leaves important.

The Caulosa consist of three growth forms: Equisetids, Orobanchids and Cuscutids. The Foliosa consist of Rosulata (Dryopterids and Primulids), Tapeta (Oxalids, Menyanthids and Petasitids), and Scaposa (Graminoids). The Caulo-Foliosa are subdivided into Scandentia (Convolvulids, Bryonids and Aparinids), Reptantia (Illecebrids and Ajugids), Erecta, and Arcuata (Polygonatids). The Erecta are the largest group of non-graminoid herbs. They are subdivided according to leaf distribution in Rosulato-Scaposa (rosettes and stem leaves) and Scaposa (only stem leaves). Independently they can be divided into simple or slightly branched and much-branched forms (this refers to the vegetative parts; in simple forms the inflorescence, if sharply delimited, may be strongly branched). This results in a subdivision of Erecta into four growth forms, viz. Digitalids (Ros-Scap, simple), Epipactids (Scap, simple), Ranunculids (Ros-scap, branched), and Gypsophilids (Scap, branched).

Few observations have so far been made on growth form spectra of associations, but they can be expected to differ. In three poor, unfertilized grasslands of Drenthe the average proportion of Primulids was 36.9%, of Epipactids 16.1%. In strongly fertilized grassland these percentages were 3.8 and 0.7, respectively.

## 6. Classification of vegetation on the basis of texture and structure

In vegetation science the following typologies and classifications have been elaborated: floristical, textural-structural (physiognomic), ecological, syngenetical, and historical-geographical. Besides, systems have been founded on various combinations of these characters. It has been the great merit of Braun-Blanquet to have acknowledged at an early time and long before others did so, that a natural system of vegetation should be based on as many of its own characters as possible and on no other ones. This excludes systems based on site factors. Our knowledge of the distribution areas of syntaxa, their history, and their place in various seres is too fragmentary to be used for a consistent classification. Besides, these characters are synthetic and can only be determined once the syntaxa in question have been defined. This leaves us texture, structure, and floristic composition as characters of the vegetation that can be used for classification.

As early as 1928 Braun-Blanquet pointed out that among these characters floristic composition must have priority. The arguments were very strong. In view of the fact that every plant species belongs to only one or a few growth forms, life forms, dissemination and pollination types, etc., the quantitative floristic composition of a phytocoenosis also must give a fair idea of its texture. Besides, when the various layers, their height, density and floristic composition are being noted, as is the case in the Braun-Blanquet school, it also gives a good picture of the vertical structure. Even seasonal periodicity is given implicitly, as the phenology of almost every species is roughly known. Structure and texture, however, hardly give any information on floristic composition.

A syntaxon is not only an ecological unit, but also a historical one. This aspect is only expressed in its floristic composition.

Another consideration is the following. In idiotaxonomy the ideal is a phylogenetic classification, based on genotypical, not on phenotypical characters. A similar (not identical!) consideration could be applied to syntaxonomy. In this connection the (genomes of the) taxa of a phytocoenosis can in a way be compared to the genes of an individual plant, and those of a syntaxon to the gene capital of a taxon. Every species (taxon) has lived in plant communities during its whole evolution and so it must have been constantly subject to the selection pressure not only of abiotic factors, but also of all the other species of the biocoena in which it lived and lives. These biotic influences may be very specific and directional. In this way taxa must have been constantly adapting themselves to other taxa of the biocoenon and must thus have contributed to the evolution of the biocoenon (syntaxon) itself.

Just like phenotypical characters of species are often more variable than genotypical ones, so are texture and structure of syntaxa as compared to floristic composition. And this makes the former less useful as diagnostic criteria. One only has to be reminded of human interference by which forest can be transformed into coppice wood, a profound change in structure, often without much consequence for its floristic composition. It would have little sense to transfer such coppice woods to scrub formations, which one might be compelled to do in a purely structural system.

But there are also strong arguments in favour of structural and textural systems. In three instances the floristic system fails and we must rely therefore upon structural characters: (1) in regions where the flora is insufficiently known or where it is difficult to identify all plant species at one given moment, such as tropical rain forest (this problem may, however, be resolved in future); (2) in the case of units above the class or class group, since it is impossible to find faithful or differential taxa above this level; and (3) in the case of communities very poor in species, as is the case in extreme habitats. They cannot be classified into higher units for lack of common species. This leads to unorthodox units like Chapman's *Coenospartinetea*, etc., based on genera, structure, and ecology.

Apart from these negative arguments there are positive considerations. That structural community types are not affected by historical and accessibility factors may also be an advantage, namely if we want to study the relations between habitat and community. From a philosophical point of view it may even be argued that it is illogical to unite plant specimens first into taxa on account of morphological characters, including a great many such as flower and fruit morphology that hardly influence the nature of the vegetation as a whole, and thereupon to unite these taxa into vegetation units on the basis of completely different criteria, namely their coexistence. As a rule the only more or less discrete visible elements in nature are organs, individuals, societies, and phytocoenoses, not the populations (its limits are usually vague or indetermined) or species (which are abstract units). So one might proceed directly from individuals to phytocoenoses. Cases are known where the growth form of the individual plants is more typical for a syntaxon than the species to which it belongs: climbing and non-climbing *Hedera helix* are characteristic of different plant associations, and so are winding and non-winding *Solanum dulcamara*.

We may conclude that both floristic and structural plus textural characters are important criteria for the delimitation of plant communities. It would be ideal to incorporate them in a universal system, if possible. But perhaps Fosberg (1967) is right when he considers this idea illusive.

His opinion is based on the following facts: (1) far too little is known as yet about texture and structure of plant communities; (2) from what we know, these parameters do not vary parallel with floristic composition; (3) this means that in a universal system we have to compromise, and since textural differences cannot be adequately weighted against floristic differences, the choice will often be arbitrary. This contrasts sharply with the situation where different classifications are possible on a floristical basis. 'Objective' means are then at hand to decide which classification gives the greatest differences between and the smallest within the classes to be distinguished.

An effort to make a universal system at all levels has never been made, and attempts to incorporate floristic syntaxa of lower rank into structural types of higher rank have not been satisfactory. The best attempt so far has been made by Westhoff & den Held (1969). Their formations do group floristic syntaxa (classes), but the groups are not homogeneous from a textural or structural point of view. There is little structural resemblance between the *Ranunculo-Rumicetum maritimi* (*Bidention*), the *Spergulario-Illecebretum* (*Nanocyperion*), and the *Salicornietum strictae* (*Thero-Salicornion*), all of which belong to their formation III. The only difference between formations VII and VIII is the presence resp. absence of annual plants, but formation VII contains grassland associations without therophytes, too (f.i. *Agrostietum tenuis* and *Mesobromion*) and even contains dwarf shrub communities (*Anthyllido-Silenetum* p.p.). Formation X is equally heterogeneous. It contains f.i. (1) very open vegetation types of helophytes with a continuous ground layer of appressed, filamentous algae (*Lycopodio-Rhynchosporetum*), (2) high hummocks of bryophytes with scattered chamaephytes and nanophanerophytes (*Erico-Sphagnetum magellanici*), (3) closed dwarf shrub communities (*Ericetum, Calluno-Genistion*), and finally (4) closed grasslands of hemicryptophytes (*Violion caninae*). The deciduous scrub formation XII also contains a wood association (*Salicetum albo-fragilis*) and an evergreen conifer scrub (*Squarroso-Juniperetum*). Formation XIII unites broad-leaved deciduous woods, evergreen conifer woods, and a scrub (*Dicrano-Juniperetum*). The undergrowth in this formation is highly variable.

As to the lower ranks of the syntaxonomic system, the modern trend in the Braun-Blanquet school is towards giving more attention to structure. Nevertheless, the present systems are far from universal. A *Stellario-Carpinetum* rich in lianas *(Lonicera, Clematis)* has a structure very different from one without lianas, and equally so a *Fago-Quercetum* with an evergreen scrub layer of *Ilex* as compared with one without such a layer. The same applies to a *Deschampsia*-facies and a *Vaccinium myrtillus*-facies of the *Querco-Betuletum*. Yet, these units are not even given the rank of subassociation. Hundreds of examples could be added. If we want to make a serious attempt to incorporate structure into the Braun-Blanquet system at all levels, this system has to be modified drastically. Time has not yet come to make such an attempt. It is better to try and make a detailed system based on structure first. The universal system to be made afterwards will then be more balanced. Besides, the comparison of the detailed structural system with a purely floristic system may yield interesting discrepancies worth a further causal analysis.

Structural systems date back to the last and the beginning of the present century (Grisebach, Drude, Schimper, Brockmann-Jerosch, Rübel). These systems were deductive, i.e. they started from the big, world-wide formations and subdivided these into smaller units. They were impure, because they included ecological criteria. They were incomplete, because they were based mainly on the structure of the uppermost layer. They were insufficiently detailed, so that even their lowest units were far too widely conceived to be compared with alliances or associations.

The most important recent structural systems are those by Schmithüsen (1961), Fosberg (1961, 1967) and Ellenberg & Müller-Dombois (1967, also M.-D. & E. 1974). Schmithüsen's system comprises 9 'classes', mainly on a structural basis, divided into formations based partly on structure (e.g. woods and scrubs), partly on habitat characters (e.g. the eight grassland formations). His smallest units still contain (elements of) different classes in the sense of Braun-Blanquet.

Fosberg published a detailed system, free from floristic and ecological elements. It is a consistent, hierarchical system with five ranks, each based on a specific textural or structural criterion. Consequently the system is logical and its application universal. The main objection to Fosberg (cf. also Werger 1973) is that his primary subdivision is based on the coverage of the densest above-ground layer, with the three classes: closed, open, and scarce. Between these classes, however, all transitions may be found. Besides, it is hardly the most fundamental division, as the coverage depends strongly on the stage of development of a community as well as on such hazards as more or less recent cutting or felling, grazing, fire, and storm damage. Fosberg ignores the structure of layers under the densest layer. This is one of the reasons why his smallest units are still very large, comprising many classes of the Braun-Blanquet system.

The most detailed system is by Ellenberg & Müller-Dombois. It also pays some attention to lower layers. But it is less consistent than Fosberg's, as it makes also use of ecological criteria. A further refinement is not only desirable but also possible. To illustrate this we give here two examples:

(1) Their formation IB 3a is defined as closed, summergreen broad-leaved forests without admixture of evergreen trees and with an epiphytic vegetation mainly formed by algae and crustaceous lichens. In the Netherlands this formation includes all forest associations of the *Querco-Fagetea* (except the variant of the *Stellario-Carpinetum* rich in *Taxus*), of the *Quercetea robori-petraeae* (except the *Ilex*-variant of the *Fago-Quercetum*), the *Betulion pubescentis*, the *Alnion glutinosae*, and the *Salicetum albo-fragilis*. These wood types differ not only floristically and ecologically, but also structurally.

(2) Their 'unit' VC 1 e (2) (a), being a subdivision of a subformation, comprises all closed, fertilized hayfields of the temperate lowlands that are devoid of trees and shrubs, that are mowed several times a year, and that are rich in grasses and herbs with soft leaves, and poor in geophytes flowering in early spring. Even this narrowly defined unit comprises (in the Netherlands alone!) two alliances, viz. *Calthion palustris* and *Arrhenaterion elatioris*, as well as one association from a third alliance, the *Medicagini-Avenetum pubescentis*.

It is true that, with structure as a criterion, syntaxa can be discerned on a higher rank than is possible on a floristical basis. But the opposite is not true; at least it has not been proved convincingly, that floristic criteria permit a more detailed classification (smaller units) than structural criteria, notwithstanding the impression given by the examples and systems discussed above.

I have tried to elaborate a purely textural-structural system of Dutch vegetation, making use of many parameters already known in the literature as well as of growth forms defined by myself. I will give here one example, just to illustrate the possibility to make a detailed system. The system will be published in extenso later.

Radicant terrestrial vegetation is divided into: 1. herb vegetation (forbs); 2. grassland; 3. dwarf scrub and heath; 4. scrub; 5. woodland and forest. Herb vegetation is divided into: 1. vegetation of Equisetids (e.g. *Salicornietum strictae, Eleocharito-Hippuridetum, Scirpetum lacustris*); 2. carpets (e.g. *Tussilaginetum, Calletum palustris*, sociations of *Pteridium*); 3. prostrate veils (e.g. *Asplenio-Hederetum* on walls); 4. erect herb vegetation. The latter is divided into low, medium and high (tall) herb communities, the last-named ('Hochstaudenfluren') into those of unbranched (or little branched) and those of strongly branched herbs. The unbranched, erect, tall herb communities are divided into moss-rich communities and those with a scarce moss layer. The latter may be divided into: (1) communities up to 1 m high, consisting mainly of Digitalids, with a high proportion of felty and rough-haired plants (*Echio-Verbascetum*); and (2) communities up to 2 m high, consisting of Digitalids mixed with Epipactids, leaves glabrous, climbers present. Type (2) can be subdivided into: (2a) Perennial herbs. The climbers are herbs (Convolvulids, Cuscutids, and Aparinids): *Senecionion fluviatilis, Sonchetum palustris, Angelicion litoralis* p.p., *Valeriano-Filipenduletum*; (2b) Vegetation consists partly of annual and biennial plants, often with narrow leaves. The climbers are half-shrubs (*Rubus*): *Epilobion angustifolii.*

The system thus arrives sometimes at the association or lower level, sometimes at the level of an alliance, but generally at a unit not corresponding with a floristic syntaxon (f.i. comprising (sub)associations from different alliances). This applies to the 144 smallest units of the system. None of the higher units correspond to syntaxa.

## References

Aljochin, V.V. 1935. Main concepts and main units in phytocoenology. (Russian) Sov. Bot. 1935 (5): 21-34.

Barkman, J.J. 1958a. Phytosociology and Ecology of Cryptogamic Epiphytes. 628 pp. Van Gorcum, Assen.

Barkman, J.J. 1958b. La structure du Rosmarineto-Lithospermetum helianthemetosum en Bas-Languedoc. Blumea Suppl. 4: 113-136.

Barkman, J.J. 1965. Die Kryptogamenflora einiger Vegetationstypen in Drenthe und ihr Zusammenhang mit Boden und Mikroklima. In: R. Tüxen (ed.), Biosoziologie. Ber. Int. Symp. Stolzenau/Weser 1960. pp. 157-171. Junk, Den Haag.

Barkman, J.J. 1968a. Botanisch onderzoek op het Biologisch Station, Wijster 1957-1967. Med. Bot. Tuinen Belmonte Arbor. LH Wageningen 9 (1967): 141-160.

Barkman, J.J. 1968b. Das synsystematische Problem der Mikrogesellschaften innerhalb der Biozönosen. In: R. Tüxen (ed.), Ber. Int. Symp. Stolzenau/Weser 1964 pp. 21-53. Junk, Den Haag.

Barkman, J.J. 1973. Synusial Approaches to Classification. In: R. Whittaker (ed.), Handbook of Vegetation Science. Part V: Ordination and Classification of Vegetation pp. 437-491. Junk, Den Haag.

Barkman, J.J. 1976. Terrestrische fungi in jeneverbesstruwelen. Coolia 19: 94-110.

Barkman, J.J. 1977. Die Erforschung des Mikroklimas in der Vegetation. Theoretische und methodische Aspekte. In: H. Dierschke (ed.), Vegetation und Klima. Ber. Int. Symp. Int. Ver. Veg.kunde 1975. pp. 5-20. Cramer, Vaduz.

Barkman, J.J., A.K. Masselink & B.W.L. de Vries. 1977. Ueber das Mikroklima in Wacholderfluren. Ibidem: 35-81.

Beard, J.S. 1973. The Physiognomic Approach. In: R. Whittaker (ed.), Handbook of Vegetation Science. Part V: Ordination and Classification of Vegetation. pp. 355-386. Junk, Den Haag.

Boer, P.J. den. 1967. Zoölogisch Onderzoek op het Biologisch Station, Wijster, 1959-1967. Med. Bot. Tuinen Belmonte Arbor. LH Wageningen 9 (1967): 161-181.

Boterenbrood, A.J., W.A.E. van Donselaar-ten Bokkel Huinink & J. van Donselaar. 1955. Quelques données sur l'écologie de la végétation des dunes et sur la fonction de l'enracinement dans l'édification des dunes à la côte méditerranéenne de la France. I et II. Proc. Kon. Ned. Akad. Wet. Ser. C, 58: 523-534, 535-547.

Braun-Blanquet, J. 1928. Pflanzensoziologie, Biol. Studienbücher 7. 1st. Ed. Springer, Berlin.

Dansereau, P. 1951. Description and recording of vegetation upon a structural basis. Ecology 32: 172-229.

Dansereau, P. 1958. A universal system for recording vegetation. Contr. Inst. Bot. Univ. Montréal 72: 1-58.

Dansereau, P. & J. Arros. 1959. Essais d'application de la dimension structurale en phytosociologie. I. Quelques exemples européens. Vegetatio 9: 48-99.

Dansereau, P. & K. Lems. 1957. The grading of dispersal types in plant communities and their ecological significance. Contr. Inst. Bot. Univ. Montréal 71: 5-52.

Donselaar-ten Bokkel Huinink, W.A.E. van, 1961. An ecological study of the vegetation in three former river beds. Wentia 5: 112-162.

Donselaar-ten Bokkel Huinink, W.A.E. van. 1966. Structure, root systems and periodicity of savanna plants and vegetations in northern Surinam. Wentia 17: 1-162.

Drude, O. 1890. Handbuch der Pflanzengeographie. Engelhorn, Stuttgart.

Eijsink, J., G. Ellenbroek, W. Holzner & M.J.A. Werger. 1978. Dry and semi-dry grasslands in the Weinviertel, Lower Austria. Vegetatio 36: 129-148.

Ellenberg, H. 1974. Zeigerwerte der Gefässpflanzen Mitteleuropas. Scripta Geobot. 9: 5-97.

Ellenberg, H. & D. Mueller-Dombois. 1967. Tentative physiognomic-ecological classification of plant formations of the earth. Ber. Geobot. Inst. ETH, Stift. Rübel 37: 21-55.

Fosberg, F.R. 1961. A classification of vegetation for general purposes. Trop. Ecol. 2: 1-28.

Fosberg, F.R. 1967. A classification of vegetation for general purposes. In: G.F. Peterken (ed.), Guide to the check sheet for IBP areas. IBP Handbook no. 4: 73-120.

Gams, H. 1918. Prinzipienfragen der Vegetationsforschung. Vierteljahresschr. naturf. Ges. Zürich 63: 293-493.

Gimingham, C.H. 1951. The use of life form and growth form in the analysis of community structure, as illustrated by a comparison of two dune communities. J. Ecol. 39: 396-406.

Hallé, F., R.A.A. Oldeman & P.B. Tomlinson. 1978. Tropical Trees and Forests. An Architectural Analysis. 441 pp. Springer Verlag, Berlin-Heidelberg-New York.

Hartog, C. den. 1955. A classification system for the epilithic algal communities of the Netherlands' coast. Acta Bot. Neerl. 4: 126-135.

Hartog, C. den. 1959. The epilithic algal communities occurring along the coast of the Netherlands. Wentia 1: 1-241.

Horn, H.S. 1971. The adaptive geometry of trees. Princeton Univ. Press, Princeton, N.J.

Iversen, J. 1936. Biologische Pflanzentypen als Hilfsmittel in der Vegetationsforschung. 224 pp. Munksgaard, København.

Klomp, H. 1953. De terreinkeus van de kievit, Vanellus vanellus (L). 139 pp. E.J. Brill, Leiden.

Landolt, E. 1977. The importance of closely related taxa for the delimitation of phytosociological units. Vegetatio 34: 179-189.

Lange, J.E. 1923. Studies in the Agarics of Denmark. V. Ecological Notes. Dansk Bot. Arkiv 4: 1-55.

Leeuwen, C.G. van. 1966. A relation theoretical approach to pattern and process in vegetation. Wentia 15: 25-46.

Leeuwen, C.G. van & E. van der Maarel. 1971. Pattern and process in coastal dune vegetations. Acta Bot. Neerl. 20: 191-198.

Leischner-Siska, E. 1939. Zur Soziologie und Ökologie der höheren Pilze. Beih. Bot. Centralbl. 59B: 359-429.

Lensink, B.M. 1963. Distributional ecology of some Acrididae (Orthoptera) in the dunes of Voorne, Netherlands, Tijdschr. Entomologie 106: 357-443.

Londo, G. 1971. Patroon en proces in duinvalleivegetaties langs een gegraven meer in de Kennemerduinen. 279 pp. Derks, Cuyk.

Luther, H. 1949. Vorschlag zu einer ökologischen Grundeinteilung der Hydrophyten. Acta Bot. Fenn. 44: 1-15.

Maarel, E. van der. 1966. Over vegetatiestructuren, -relaties en -systemen in het bijzonder in de duingraslanden van Voorne. 170 pp. Univ. Utrecht.

Maarel, E. van der & J. Leertouwer. 1967. Variation in vegetation and species diversity along a local environmental gradient. Acta Bot. Neerl. 16: 211-221.

Maarel, E. van der & V. Westhoff. 1964. The vegetation of the dunes near Oostvoorne. Wentia 12: 1-61.

Mörzer Bruyns, M.F. 1947. Over levensgemeenschappen. 195 pp. Kluwer, Deventer.

Molinier, R. & P. Müller. 1938. La dissemination des espèces végétales. Rev. gén. Bot. 50: 1-178.

Müller-Dombois, D. & H. Ellenberg. 1974. Aims and Methods of Vegetation Ecology. John Wiley & Sons, New York — London — Sidney — Toronto.

Orians, G.H. & O.T. Solbrig. 1977. A cost-income model of leaves and roots with special reference to arid and semiarid areas. Amer. Naturalist 111: 677-690.

Parsons, D.J. 1976. Vegetation structure in the mediterranean scrub community of California and Chile, J. Ecol. 64: 435-447.

Parsons, D.J. & A.R. Moldenke. 1975. Convergence of vegetation structure along analogous climatic gradients in California and Chile. Ecology 56: 950-957.

Pümpel, B. 1977. Bestandesstruktur, Phytomassevorrat und Produktion verschiedener Pflanzengesellschaften im Glocknergebiet. Veröff. Oest. MAB- Hochgebirgsprogr. Hohe Tauern 1: 83-101.

Raunkiaer, C. 1907. Planterigets Livsformer og deres Betydning for Geografien. Munksgaard, København.

Raunkiaer, C. 1916. Om Bladstorrelsens Anvendelse i den biologiske Plantegeografi. Bot. Tidsskr. 34: 225-237.

Rejmánek, M. 1977. The concept of structure in phytosociology with references to classification of plant communities. Vegetatio 35: 55-61.

Schmithüsen, J. 1961. Allgemeine Vegetationsgeographie. 2 Aufl. 262 pp. De Gruyter & Co., Berlin.

Segal, S. 1965. Een vegetatieonderzoek van de hogere waterplanten in Nederland. Wet. Meded. Kon. Ned. Nat. hist. Ver. 57: 1-80.

Segal, S. 1969. Ecological Notes on Wall Vegetation. 325 pp. Junk, Den Haag.

Stafleu, F. 1940. Het Corynephorion. In: J. Heimans, J. Meltzer, F. Stafleu & V. Westhoff, Onze droge graslanden. pp. 20-26. Uitg. NJN-Soc. groep. Onze binnenlandse Brometen. Ibidem: 33-38.

Stoutjesdijk, Ph. 1959. Heaths and inland dunes of the Veluwe. Wentia 2: 1-96.

Stoutjesdijk, Ph. 1961. Micrometeorological measurements in vegetations of various structures II and III. Proceed. Kon. Ned. Akad. Wet. ser. C, 2: 180-207.

Stoutjesdijk, Ph. 1966. On the measurement of the radiant temperature of vegetation surfaces and leaves. Wentia 15: 191-202.

Stoutjesdijk, Ph. 1974. The open shade an interesting microclimate. Act. Bot. Neerl. 23: 125-130.

Stoutjesdijk, Ph. 1977. On the range of micrometeorological differentiation in the vegetation. In: H. Dierschke (ed.), Vegetation und Klima. Ber. Int. Symp. Int. Ver. Veg.kunde 1975. pp. 21-34. Cramer, Vaduz.

Strijbosch, H. 1976. Een vergelijkend syntaxonomische en synoecologische studie in de Overasseltse en Hatertse vennen bij Nijmegen. 333 pp. Stichting Studentenpers, Nijmegen.

Sukačov, V.N. 1928. Plant Communities (Introduction to Phytosociology). 4 ed. 232 pp. Kniga, Leningrad/Moscow.

Tüxen, R. (ed.) 1970. Gesellschaftsmorphologie (Strukturforschung). Ber. Int. Symp. Int. Ver. Veg.kunde Rinteln 1966. 360 pp. Junk, Den Haag.

Werger, M.J.A. 1973. Phytosociology of the Upper Orange River Valley, South Africa. A syntaxonomical and synecological study. 220 pp. V. & R. Pretoria.

Werger, M.J.A. 1978. Vegetation structure in the Southern Kalahari. J. Ecol. 66: 933-941.

Werger, M.J.A. & G.A. Ellenbroek. 1978. Leaf Size and Leaf Consistence of a Riverine Forest Formation along a Climatic Gradient. Oecologia 34: 297-308.

Werger, M.J.A. & E. van der Maarel. 1978. Plant species and plant communities: some conclusions. In: E. van der Maarel & M.J.A. Werger (eds.), Plant species and plant communities. Proc. Int. Symp. Nijmegen, 1976. pp. 169-175. Junk, The Hague – Boston – London.

Westhoff, V. 1947. The vegetation of dunes and salt marshes on the Dutch Islands of Terschelling, Vlieland and Texel. 131 pp. C.J. van der Horst, Den Haag.

Westhoff, V. 1967. Problems and use of structure in the classification of vegetation. Acta Bot. Neerl. 15: 495-511.

Westhoff, V. & A.J. den Held. 1969. Plantengemeenschappen in Nederland. 324 pp. Thieme & Cie., Zutphen.

Westhoff, V. & J.N. Westhoff-de Joncheere. 1942. Verspreiding en nestoecologie van de mieren in de Nederlandse bosschen. Tijdschr. Plantenziekten sept.-okt. 1942: 1-76.

Whittaker, R.H. 1970. The population structure of vegetation. In: R. Tüxen (ed.), Gesellschaftsmorphologie (Strukturforschung). Ber. Int. Symp. Rinteln 1966 pp. 39-62. Junk, Den Haag.

Wit, C.T. de. 1965. Photosynthesis of leaf canopies. 57 pp. Pudoc, Wageningen.

Zwillenberg, L.O. & R.J. de Wit. 1952. Observations sur le Rosmarineto-Lithospermetum schoenetosum du Bas-Languedoc. Acta Bot. Neerl. 1: 310-323.

# 6. MULTIVARIATE METHODS IN PHYTOSOCIOLOGY, WITH REFERENCE TO THE NETHERLANDS

## E. VAN DER MAAREL

# 6. MULTIVARIATE METHODS IN PHYTOSOCIOLOGY, WITH REFERENCE TO THE NETHERLANDS

## E. VAN DER MAAREL

### 1. Introduction

Vegetation science may be considered as practically synonymous with phytosociology. For both terms one finds definitions similar or equal to 'the study of plant communities' (Westhoff & van der Maarel 1978, Mueller-Dombois & Ellenberg 1974). Still, phytosociology is considered to emphasize the typological plant community approach. Vegetation research in the Netherlands has been based largely upon the plant community concept according to Braun-Blanquet (1964, 1965) and hence it may also be referred to as phytosociology. Interestingly, the Dutch translation of this term, 'plantensociologie', is not very popular and one usually finds the term 'vegetatiekunde' (= vegetation science) used in Dutch studies. Still the term phytosociology will be used in this contribution also when it is referred to the Dutch situation. Anyhow, this review will concentrate on multivariate methods used in the description and ecological interpretation of plant communities. Other aspects like relationships between the distribution of plant species and chorological factors, or palaeo-ecological data will be mentioned only casually.

Multivariate methods in vegetation science may be described as methods for the establishment of numerical relations between phytosociological entities and between those and their environment. A phytosociological entity can be a relevé, i.e. a standardized description of a phytocoenosis, a concrete stand, or a plant community, a phytocoenon (Westhoff & van der Maarel 1978). Such numerical relations are analysed and accordingly interpreted. Hence, we often find the term multivariate analysis for such methods (e.g. Orlóci 1978a), but we should remember that they form part of a research phase in phytosociology which is called the synthetic research phase (Braun-Blanquet 1964, 1965, Westhoff & van der Maarel 1978).

Multivariate methods are usually divided into classification and ordination. Classification in a phytosociological context can be described as the procedure of creating types of vegetation to which phytosociological entities can be assigned. Any vegetation type (= plant community type = phytocoenon) is an abstraction of a group of entities related to each other in all respects which are considered relevant for the classification.

Phytosociological ordination is the procedure of arranging phytosociological entities along axes of variation in a one- or more-dimensional scheme according to the relations between the entities. In both numerical classification and ordination the relations between the entities are expressed in a numerical value and the subsequent procedures usually involve extensive calculus. The relations between the entities may be based upon various attributes (characters) of the entities, notably floristic and structural attributes, but usually they concern the occurences op plant species which are registered as mere presence data or, more usually, measured quantitatively. They are the variables, or variates subjected to multivariate analysis.

The review of these numerical methods can be conveniently based upon the second editions of two well-known textbooks, (Whittaker 1978a, 1978b, Orlóci 1978a). It will be preceded by a discussion of how the development of multivariate methods has influenced Dutch phytosociology in its various periods and by a much shorter discussion of the impact Dutch phytosociology has had so far on numerical methods (see Westhoff, this volume, for a general account and van der Maarel 1975, for a division into periods).

I completed this review for this special occasion, the 100th Meeting organized by the Commission for the Study of Vegetation, with much pleasure, after I subsequently introduced data-processing, classification and ordination, and finally the essentials of multivariate analysis in vegetation systematics at this forum in 1963, 1964 and 1969 respectively (see list of contributions by Vroman, this volume).

## 2. Periods in Dutch phytosociology and the significance of multivariate methods

### 2.1 *Circa 1930-1950*

Phytosociology in the Netherlands started around 1925. Its development was first of all stimulated by Scandinavian approaches (Du Rietz, Nordhagen, Raunkiaer, see Trass & Malmer 1978) and somewhat later also by the Braun-Blanquet approach (Braun-Blanquet, Tüxen, see Westhoff & van der Maarel 1978). By that time these two 'schools' had developed towards a mature stage already, after continuous developments from c. 1880.

The strong numerical tradition in vegetation analysis according to the Scandinavian traditions, resulting in extensive data on the frequency of species in vegetation units and subsequent statistics (e.g. Raunkiaer 1934), was followed in the Netherlands by Scheygrond and particularly by de Vries (1926, 1933, 1938), who will appear as an outstanding numerical ecologist in the next period.

Soon in this period the Braun-Blanquet approach became the dominant one in the Netherlands, chiefly through the personal influence of Braun-Blanquet and Tüxen. The conceptual framework and theory of phytosociology was well established here, mainly in the work of Westhoff (1947, 1951, Meltzer & Westhoff 1942) in which, moreover, the promising elements of the Scandinavian approaches were integrated (as was advocated by de Vries 1939). This strong theoretical interest is met with in later periods of Dutch phytosociology (e.g. Meyer Drees 1951, Becking 1957), but no numerical tradition whatsoever was established. The first attempts to quantify similarities between relevés (notably by Kulczynski 1928) and correlations between species (Tuomikoski 1942) as a basis for classification were not appreciated here, although these outstanding monographs were known all right. A third classical contribution by Ramensky (1930), which is generally considered as the very beginning of ordination, did not influence Dutch phytosociology either.

The only numerical element in classical phytosociology was the use of numerical values for species occurrences in the comparison of associations, i.e. 'Artmächtig-keit' (Schwickerath 1940, 'species magnitude', Braun-Blanquet 1965) and group abundance (Schwickerath 1931) or group value (Tüxen & Ellenberg 1937).

## 2.2 Circa 1950-1970

The main feature of this period is a series of monographs in which most of the main Dutch vegetation types and their environment were described. They were synthe-sized by Westhoff & den Held (1969, see 1975). However, numerical treatment of the often large materials involved did not occur and no serious attempts towards storage and retrieval systems were developed. Only at the end of this period numeri-cal methods started to be applied to plant community studies.

The first real numerical approach in the Netherlands was developed early in this period by de Vries (1953, de Vries, Baretta & Hamming 1954), but this concerned the occurrence of species combinations as ecological indicators for grassland condi-tions. His diagrams found their way in the international literature and stimulated the further development of what is now indicated with the term plexus techniques (see McIntosh 1978). Later, species combinations have been used in the classifica-tion of plant communities, e.g. by Looman (1963), van der Maarel (1966a) and Segal (1969) and most of all by de Lange (1972) who presented elaborate plexuses of aquatic plants.

The numerical classification of vegetation can be considered to have started with Sørensen's (1948) work on Danish grasslands (For similar Polish studies by Motyka and Matuskiewicz of that time, see Whittaker 1978c and Goodall 1978a). Sørensen applied a similarity coefficient for comparison of stand descriptions which became very popular (see below), and constructed a hierarchy of clusters on the basis of between-relevé and between-cluster similarities, the level of which he tried to compare with the levels in the syntaxonomical system of Braun-Blanquet. The first Dutch study including such an approach was by Barkman (1958a).

A second and further reaching stimulus was given by Williams & Lambert and co-workers in a series of papers 1959-1966 (particularly Williams & Lambert 1959, Lambert & Williams 1966, see Goodall 1978a) dealing with various computer-based clustering techniques. They were first applied in the Netherlands by van der Maarel (1966a), but found their full application only in the next period.

A related approach with the special aim of numerically ordering phytosocio-logical tables started with an attempt by Benninghoff & Southworth (1964) and was introduced into Braun-Blanquet circles by Moore (1966, 1973). Such table rearrangement programs developed rapidly from 1970 onwards (see Westhoff & van der Maarel 1978 for a review).

The second main group of multivariate techniques, numerical ordination, is based upon a paper by Goodall (1954) which deals with the use of factor analysis in plant ecology. This paper forms part of a series dealing with objective methods for the classification of vegetation (see Goodall 1978a), which as such was overshadowed by the more pragmatic series of Williams & Lambert.

The research line of factor analysis in vegetation research was further developed by Dagnelie (1960, 1978), who applied factor analysis in a strict sense to a set of relevés. The technique implies the selection of 'fundamental variables', factors upon which the observed variables, in our case the species or other floristic-structural characters, are supposed to depend. This dependence is supposed to be partly the same for all variables, which is measured as common variances or communalities. Ferrari, Pijl & Venekamp (1957) contributed to this development in the Netherlands, but their lucid paper did not become known outside agricultural research. Because of the ecological uncertainties as to underlying common factors determining the distribution of species, this type of multivariate analysis has not been used much in phytosociology (see Fresco 1969 for a Dutch example) and it will not be discussed further in this contribution (see Dagnelie 1978 for a recent review).

A paper with a much stronger impact on vegetation science was that by Bray & Curtis (1957) who devised a simple ordination technique by constructing axes of variation with pairs of maximally dissimilar relevés at the end of axes and the other relevés placed with proportional distances to the end relevés. This technique was based on earlier direct gradient analysis concerned with the distribution of plant populations along environmental gradients as developed by Curtis and his 'Wisconsin school' (Curtis & McIntosh 1951, Curtis 1959, reviewed by Whittaker 1978d who contributed much to the approach himself). The Bray & Curtis technique was introduced under the name ordination (after Goodall 1954 who adopted the German translation Ordnung of Ramensky's approach) and is now known as polar ordination (term by Whittaker, see Cottam, Goff & Whittaker 1978). It was introduced in the Netherlands by van der Maarel (1966a) working with dune grassland data and de Jong (1964) working with coastal tide mark communities as examples.

Despite the international attention to these approaches underlined by the wide use of the 2nd edition of Greig-Smith's (1964) textbook on quantitative plant ecology, in which these methods were reviewed, the Dutch interest remained small.

In addition to the achievements in monographing Dutch types of vegetation, three further developments in Dutch phytosociology can be discerned (see Westhoff, this volume), viz. (1) the development of elaborate analytical scales (notably Barkman, Doing & Segal 1964); (2) the development of phytosociological (syntaxonomical) spectrum analysis as an aid in descriptive species ecology by Westhoff (see Segal & Westhoff 1959 as a first example, and Adriani & van der Maarel 1978 for a survey of Westhoff's work); and (3) the refinement of of vegetation structure analysis (Barkman 1958a, 1958b, van der Maarel 1966a, Westhoff 1967, Segal 1969, 1970). This latter development includes the elaboration of the synusial approach, particularly by Barkman who (1978) also reviewed this approach. Although the use of numerical methods is obvious in relation to all three developments, there is only one Dutch example to be mentioned, viz. a component analysis of structural characters in a dune grassland (van der Maarel 1969). The situation in European symmorphology is similar (see Tüxen 1970 and Rejmánek 1977 for a review of structural approaches).

166

A different, though related numerical approach is pattern analysis, as developed by Greig-Smith (1952, 1964), Goodall (1952, 1961), and Kershaw (1957, 1973). Through variance analysis of performance data for species in quadrats of growing size, or increasing distance, patterns in the growth and distribution of species can be detected. This technique can also be applied to all species at the same time, but such a multiple pattern analysis (Noy-Meir & Anderson 1970, Walker et al. 1972) is computationally very cumbersome.

Obviously this pattern analysis refers to concrete situations in the field. We should distinguish it from multivariate analysis of data, in which, of course, patterns may be detected as well, but for which we should not use the term pattern analysis without specification (see e.g. Hogeweg 1976, and Williams 1976).

Notwithstanding the interest in vegetation structure no single paper on pattern analysis was published so far in the Netherlands. (See Ludwig & Goodall 1978 for recent advances in this field).

Pattern analysis is related to studies of homogeneity and minimum area of phytocoenoses. Although the first studies date from the early nineteentwenties (Arrhenius, Romell, Kylin, see Goodall 1952, 1970, Dahl 1957, Greig-Smith 1964), it was only in the period 1950-1970 that the subject became fully appreciated through the work of Vestal (1946), Hopkins (1955, 1957), Gounot (1956, 1961, 1969, Moravec (1973), and Looman (1976a, b, c). Dutch interest in minimum area problems was shown by van der Maarel (1966a, 1970) who related minimum area to structural characters, and Werger (1972) who sought a relation with C.B. Williams' (1964) considerations on diversity. As was indicated by Gounot & Calléja (1962) the minimum area of phytocoenoses can also be approached by means of multivariate analysis of quadrats of increasing size. Neither pattern analysis nor homogeneity and minimum area studies will be further reviewed in this contribution because they are not yet sufficiently developed in the Netherlands.

## 2.3 Circa 1970-present

In the present period of phytosociology we may distinguish four further developments in which Dutch multivariate analysis is of significance. Besides, some contributions to multivariate ecology have been made in this country. They will be discussed mainly in the next chapter.

### 2.3.1 The use of ecological theories

The main ecological theory used in Dutch vegetation science is the relation theory of van Leeuwen (1966, 1974, 1977, see also van der Maarel 1976, van der Maarel & Dauvellier 1978). This theory is, of course, of general interest for vegetation ecology, but it also affects phytosociological classification and ordination, both in the choice of vegetation types to be studied and in the interpretation of results.

According to van Leeuwen two contrasting types of vegetational boundary zone should be distinguished, i.e. the limes convergens zone and the limes divergens zone, which can be paralleled with ecotone and ecocline (see van der Maarel 1966b, 1976). The first type is characterized by comparatively sharp boundaries between phytocoenoses, coarse grained patterns of species populations and a dynamic environment, with the *Agropyro-Rumicion crispi* as a characteristic phytosociological alliance. The second type is characterized by numerous small-scale boundaries within phytocoenoses, fine-grained patterns of species populations, and gradual transitions between phytocoenoses in comparatively constant environments, with the complex of *Trifolion medii, Mesobromion* and *Carpinion betuli* as a phytosociological example.

The nature of such boundaries can be approached by means of multivariate methods applied to series of relevés along transects across the boundary zones, as was particularly attempted by Dutch workers (van der Maarel 1974, 1976, van der Maarel & Leertouwer (1967), Londo 1971, Fresco 1972). Fig. 1 shows some results from the latter work. In each square m of a transect of 20 x 1 sq. m in a heathland relative height and some soil factors were measured, while the vegetation data per sq. m was interpreted as follows: floristic similarity between adjacent quadrats ($G_{RQ}$), number of species finding a boundary in a quadrat as fraction of the total number of species in that quadrat ($G_G$), and average product-moment correlation value for pairs of all species occurring in a quadrat ($G_{RS}$) were calculated. The results show that the simple height gradient in fact consists of two parts, from 1-8 and from 10-17. The first part reflects the transition from a *Calluna vulgaris – Gentiana pneumonanthe* heath to a grassy heath with *Agrostis stolonifera*, the second part fades into a *Molinia caerulea -Viola palustris* type.

Van Leeuwen's theory of gradient development can also be linked with the multivariate analysis of vegetational composition along gradients. The idea of clinal variation along gradients, now expressed in the term coenocline, was developed independently by Whittaker (1960, 1970, 1978d), and Westhoff (1947, 1954) and van der Maarel (1966a, van der Maarel & Westhoff 1964). Van der Maarel & Leertouwer (1967), Thalen (1971) and van der Maarel (1971) showed results of measurements of coenocline variation with multivariate methods as has been similarly applied by Whittaker (1967, 1978d). Fig. 2 presents information on a transect in the *Drosera* valley on the West-Frisian island of Schiermonnikoog. A subtle change in height is accompanied by a marked change in vegetation. Although the variation is continuous three types could be distinguished with, respectively, *Radiola linoides, Linum catharticum* and *Parnassia palustris* as differentiating species. Species richness also shows a very clear pattern, with maximal values in the middle part of the gradient. Van der Laan (1971, 1974, 1979), Werger (1973), Sloet van Oldruitenborgh (1976, Sloet van Oldruitenborgh & Adriani 1971) and Fresco (1969) are other Dutch ecologists to have used ordination in the detection of gradients, mainly in dunes.

The relation between spatial and temporal variation in vegetation in relation to such variation in the environment, a major element of van Leeuwen's theory, can also be approached successfully by multivariate methods, notably ordination (van der Maarel & van Leeuwen 1967, van der Maarel 1969, 1978, Londo 1971). On the

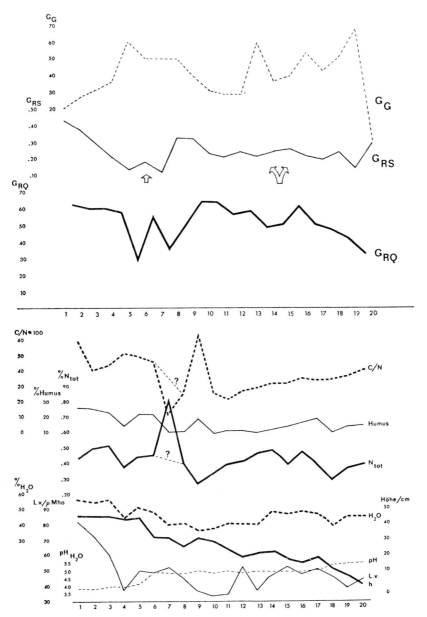

Fig. 1. Trends of vegetational and soil variation in a transect of 20 x 1 sq. m in a heathland (from Fresco 1972).
a. Trends in some soil factors. h = local height in cm above a local minimum; C/N = carbon/ nitrogen quotient; $N_{tot}$ = total nitrogen content (Kjeldahl method); $H_2O$ = g $H_2O$/g wet soil x 100; L.v. = conductivity $\mu$Mho cm$^{-1}$.
b. Trends of three boundary parameters. $G_{RQ}$ = floristic similarity between adjacent quadrats; $G_{RQ}$ = number of species finding a boundary in a quadrat as fraction of the total number of species in that quadrat; $G_{RS}$ = average product-moment correlation value for pairs of all species occurring in a quadrat.

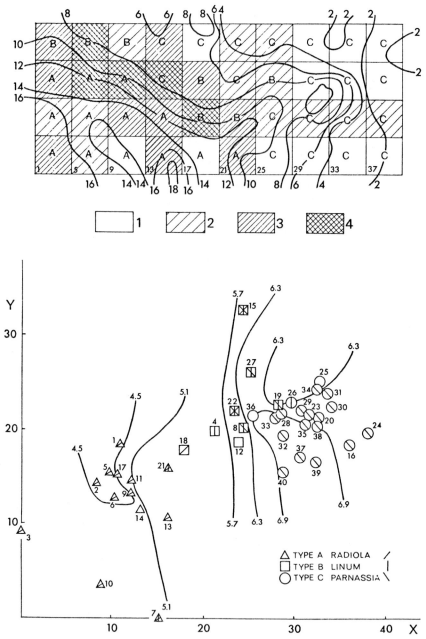

Fig. 2. Trends of variation in a transect in the *Drosera* valley, Schiermonnikoog, from van der Maarel & Leertouwer (1967). a. Distribution of species-diversity, vegetation types and height. Diversity class 1: 17-25 species/ sq. m, class 2: 26-30 sp., 3: 31-35 sp., 4: 36-43 sp. Vegetation type A: *Radiola linoides*, B: *Linum catharticum*, C: *Parnassia palustris*. b. Polar ordination (see section 3.5.2) of 40 quadrats based on Braun-Blanquet estimates of species, with isolines for pH.

basis of results from both rapidly changing and rather stable dune communities van der Maarel & van Leeuwen (1967, see van der Maarel 1978) came to a general hypothesis, which could be tested by further multivariate treatments of data: Along well-developed gradients with a long history and constant management, zones with high diversity and stability will occur in the middle parts, whereas along gradients of a limes convergens type under dynamical environmental conditions low stability may occur in the middle parts as compared to the end parts which are more permanently in a dynamical state. Such conclusions cannot be evaluated without reference to the torrent of publications on diversity and stability which appeared in the last decade (see van Dobben & Lowe-McConnell 1975 and May 1975 and for a recent confirmation of part of our hypothesis Mc Naughton 1978).

The relation theory also has revealed floristic-sociological characteristics of plant communities bound to the various environmental types as characterized by their spatial and temporal variation (Westhoff & van Leeuwen 1966, Beeftink 1965, 1966). Van der Maarel (1966a) put these characteristics into a general diagram (Fig. 3) which suggests a complementary use of classification and ordination in the following way: the coarse-grained vegetation structure of relatively dynamic and homogeneous environments allows a comparatively effective classification of plant communities up to the higher syntaxonomical units fitting into those environments, e.g. *Agropyro-Rumicion crispi, Artemisietea vulgaris, Phragmitetea, Asteretea tripolii*. On the other hand the fine-grained structure of relatively constant and heterogeneous environments allows an efficient ordination of corresponding plant communities, such as *Galio-Koelerion, Junco-Molinion, Violion caninae, Prunetalia spinosae, Trifolio-Geranietea* and *Carpinion betuli*. From the convergent to the divergent environments the classification becomes more difficult, resulting in increasing difficulties in establishing higher syntaxonomical units, while ordination, resulting in reticulate rather than hierarchic systems, becomes more and more obvious. For a pragmatic approach the syntaxonomical system could be used on the alliance level, being a 'gulden mean'.

### 2.3.2 Numerical succession studies

Van der Maarel (1969) introduced ordination as a method of treating data from permanent plots in dune grasslands to reveal successional relationships between plant communities, in a way similar to a Russian 'direct gradient' (Whittaker 1978c) approach to temporal sequences (Utekhin 1969, see Sobolev & Utekhin 1978). This method has been applied successfully to dune slack communities (Londo 1971, 1974, 1978, van der Laan 1979), pioneer communities on desalinating flats after enclosure (van Noordwijk-Puyk, Beeftink & Hogeweg 1979, Hogeweg 1976, based upon a study described earlier by Beeftink, Daane & de Munck 1971). The same approach was followed by Austin (1977) for developments in a lawn. Austin also used numerical classification to detect transitions from one community type into another, as was first attempted by Williams et al. (1969) and also by Hogeweg (1976). Through classification of successional data one can obtain transition ma-

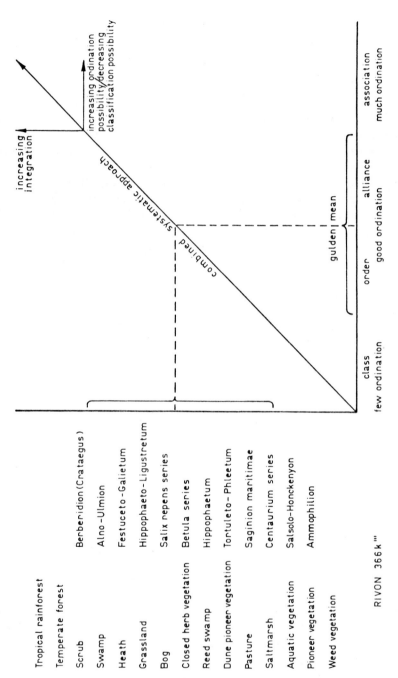

Fig. 3. Relation between the level-of-integration in vegetation and a combined systematic approach based on classification and ordination (after van der Maarel, Westhoff & van Leeuwen in van der Maarel 1966b).

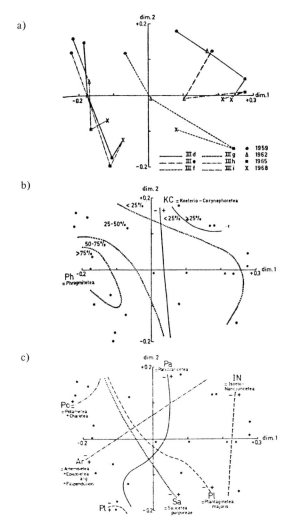

Fig. 4. Succession of six permanent plots in a transect across the zonation of an artificial lake in the dunes near Haarlem expressed by a principal component analysis (ORDINA, see section 3.5.2) on quantitative data. a. Situation of quadrats in four years. b. Isocene pattern for some syntaxonomical groups. c. Ibid. for some other groups (from Londo 1971).

trices (Londo 1971, 1974, Hogeweg 1976) and indicate probabilities of transition (Godron & Lepart 1975). The Dutch van Hulst (1976) is one of the ecologists who has elaborated on this topic (see van der Maarel & Werger 1979 for a review).

Fig. 4 presents an example from Londo (1971) of an ordination applied to succession data from four years from six 4 sq. m quadrats in a transect across the zonation along an artificial dune lake. With the help of isocenes, i.e. lines of equal representation of a group of species characterizing a syntaxon (van der Maarel 1969), the spatial and temporal pattern can be interpreted ecologically. The starting position reflects the position on the transect, with the wet part on the right and the drier part on the left. This sequence remains roughly the same in the course of time, but there are irregularities. The general tendency is a development towards higher vegetation, in the wetter part with forbs, in the drier part with shrubs. It also becomes clear that the total vegetational variation varies. In the dry years 1959 and 1965 the variation is large, while in the wetter years 1962 and 1968 a convergence occurs.

### 2.3.3. Multivariate analysis in landscape ecology

The development of landscape ecology as a discipline of its own (see Troll 1968, Leser 1976) has been stimulated by various Dutch phytosociologists (Zonneveld 1971, Doing 1974, Werkgroep GRAN 1973, Tjallingii 1974, see also Van der Maarel & Dauvellier 1978). However, the use of phytosociological data-processing techniques, including storage, retrieval and subsequent multivariate analyses, for land and land use surveys have hardly been established in the Netherlands, although good examples are available from France (Godron et al. 1968, 1969, Romane et al. 1972), Canada (Anon. 1974), the German Federal Republic (Olschowy 1975) and other countries (see Walker 1974 for a survey), and as far as soil survey is concerned also the Netherlands (e.g. Bie, Lieftinck & van Lynden 1976). The data-processing system developed by the Working Group for Data-Processing in Phytosociology of the International Society for Vegetation Science, which will be discussed in the next section, may be a good starting point (van der Maarel, Orlóci & Pignatti 1976). More regionally oriented data bases are developed at various University centres and research institutes, e.g. the Department of Plant Ecology of the University of Groningen (Fresco) and the Delta-Institute for Hydrobiological Research (Beeftink).

### 2.3.4. Multivariate analysis in syntaxonomy

The tradition of writing phytosociological-syntaxonomical monographs, so characteristic for previous periods in Dutch phytosociology, has been continued after 1970, but neither an updated general syntaxonomical system nor an integrated numerical approach can be reported. Plans for both projects, and at the same time for their integration, are being made now, mainly by the Project Group for Synsystematics and Synchorology within the Foundation for Biological Research in the Netherland, BION.

Numerical syntaxonomy has been promoted particularly by the Working Group for Data-Processing, which started in 1969 (Pignatti, Cristofolini & Lausi 1968). It produced a series of papers in *Vegetatio* (see van der Maarel, Orlóci & Pignatti 1976 for a survey) in which salt marsh communities have been the main object of research (Beeftink 1972). This working group developed a storage system, mainly

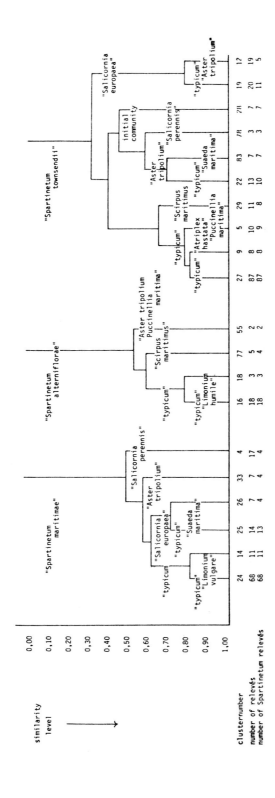

Fig. 5. Average linkage clustering with similarity ratio on quantitative data (see section 3.4.3) applied to *Spartinetum* relevés interpreted in terms of the existing syntaxonomical hierarchy (from Kortekaas, van der Maarel & Beeftink 1976).

175

for relevés already collected in phytosociological tables, whether published or not (Pignatti 1976, van der Maarel, Orlóci & Pignatti 1976). This system now comprises c. 6000 salt marsh relevés and is kept at the Botanical Institutes at Trieste and Nijmegen. In addition the Trieste group has collected some thousands of relevés of alpine vegetation, while the Nijmegen group built up a large collection of dune grassland relevés (dissertation project of W.M. Kortekaas) and is building up a collection of representative Dutch phytosociological tables (project of M. Rijken-Kępczińska).

Dutch contributions to the international project include:

(1)  An elaborated numerical syntaxonomy of European *Spartina* communities by Kortekaas, van der Maarel & Beeftink (see 1976), which was in overall agreement with a classical approach carried out in the same period (Beeftink & Géhu 1973), but revealed some new syntaxa below the association level. Fig. 5 shows the *Spartinetum* hierarchy as obtained with a numerical classification. Clearly the various syntaxonomical levels are linked with a similarity level. From general experience it was concluded (see also Westhoff & van der Maarel 1978) that with the similarity ratio as a measure (see next chapter) associations can be detected on an average similarity level of 0.40-0.60, subassociations on 0.60-0.70 and variants on 0.70-0.80.

(2)  The development of a computer program TABORD for clustering, cluster ordination and subsequent production of a structured phytosociological table (Janssen 1972, van der Maarel, Janssen & Louppen 1978). Fig. 6 presents one example from this study. It deals with the example used by Mueller-Dombois & Ellenberg (1974) to show various multivariate treatments in comparison with the classical structuring of a phytosociological table. The material consists of 25 relevés of the *Arrhenatheretum* in three forms, viz. a *Bromus erectus* type, a *Geum rivale* type and a *Cirsium oleraceum* type (a deviating relevé with *Glyceria fluitans* and *Phalaris arundinacea*, nr. 19, was added). Fig. 6a presents the first half of the initial table from Mueller-Dombois & Ellenberg. TABORD was first applied with the following options: a threshold value of 0.45 was adopted to achieve that relevés with a similarity value of less than 0.45 with any of the clusters were put aside in order to homogenize the clusters; a fusion limit of 0.65 was adopted to allow eventual fusions of homogenized clusters, provided that the similarity between clusters is more than 0.65; a constancy limit of 0.70 was adopted, which results in the construction of the final table so as to obtain species in the top part of the table characterizing one or more clusters, each with a constancy of at least 0.70. Three clusters resulted, but one with 17 relevés. This seemed unsatisfactory and hence a new run was performed with both a higher threshold value and a higher fusion limit (0.55 and 0.80). This results in more and more homogeneous clusters. The final sequence of clusters was determined on the basis of arrangement of the clusters along the first dimension of an ordination. See Fig. 6b.

The TABORD program is available in a small version for 200 relevés, 250 species and 20 clusters, requiring 384 K bytes of computer storage, and now also in a larger version with flexible lengths of arrays for a storage of 512 K bytes, allowing the

176

Fig. 6a.

```
                              0000000000000000000000000000
                              0100102011120221101022131
                              4019534268205233267815479

BROMUS ERECTUS                8677775-----+-------------+-
SCABIOSA CULUMBARIA           211-1---------------3--
THYMUS SERPYLLUM              132--1-------------1--
SALVIA PRATENSIS              -4-3244-----+-------------+-
KOELERIA PYRAMIDATA           3333----------------
FESTUCA OVINA                 -2--33-2----+----------+-
CAMPANULA ROTUNDIFOLIA        2221211-1-2---2----1-12--
VIOLA HIRTA                   -3-3111--1----------
BRIZA MEDIA                   -223---23-----+----------+-
LINUM CATHARTICUM             1-11-1--1----------
GEUM RIVALE                   --------11111--2133421--22
HOLCUS LANATUS                --------2-12323336321
MELANDRIUM DIURNUM            --------2-1--1-113--14221
ALOPECURUS PRATENSIS          --------3--35-4-54562---
LYSIMACHIA NUMMULARIA         --------11---111211--
LYCHNIS FLOS-CUCULI           --------2-----1-1111--1
GLECHOMA HEDERACEA            --------1-11----11---1-
CIRSIUM OLERACEUM             --------11121353633661
DESCHAMPSIA CAESPITOSA        -----------5-734-253
ANGELICA SILVESTRIS           --------21-2---123-
CAREX ACUTIFORMIS             ----------4-23-53
FILIPENDULA ULMARIA           --------1-3-3--1-
PIMPINELLA MAGNA              --------11--2--1-
POLYGONUM BISTORTA            --------1--------4-22
ARRHENATHERUM ELATIUS         351463726656467564546754 6
DACTYLIS GLOMERATA            444346546674555665556654 2
GALIUM MOLLUGO                253335344443553644236643
POA PRATENSIS                 454565685564456343442543 2
PLANTAGO LANCEOLATA           2222221232223541214213441
FESTUCA PRATENSIS             -3-44334455366337355633 35
CHRYSANTHEMUM LEUCANTH        142323412-3312224121-331
RANUNCULUS ACER               -31111111-12231313211123 1
VERONICA CHAMAEDRYS           1211111-2-221131121122-2 1
ACHILLEA MILLFOLIUM           25433331452434141---1162-
DAUCUS CAROTA                 122212-21212113312-41-2-
RUMEX ACETOSA                 1-1-1---2-22121213331232 2
```

Fig. 6. Rearrangement of a phytosociological table of *Arrhenatheretum* relevés from Mueller-Dombois & Ellenberg (1974) with the TABORD program according to van der Maarel, Janssen & Louppen (1976) (see section 3.4.4).
a. Initial table (species which are neither constant nor differentiating have been left out).
b. Structured table with threshold value 0.55 and fusion limit 0.80, with resulting clusters arranged according to a configuration obtained with ORDINA.

Fig. 6b.

```
                              0000000000000000000000000
                              0000112020111122000111221
                              1439054235236802678417159

                              0000000000000000000000000
                              0000000010000000000111110
                              11333332244444445555333337

                              BBBBBBBBGGCGGGGCCCCCCCCC

LINUM CATHARTICUM             1111-------1----------
CAREX FLACCA                  33---2----3------3--3
KOELERIA PYRAMIDATA           33-33-----------------
BROMUS ERECTUS                7877675-----------------
CAMPANULA ROTUNDIFOLIA        2211221-2-2---1-1-1--
VIOLA HIRTA                   --13111----1------
SALVIA PRATENSIS              --43424-------------
SCABIOSA COLUMBARIA           12--11-3------------
FESTUCA OVINA                 --3-23-2------------
SILENE INFLATA                ---1-3------1-----
TRISETUM FLAVESCENS           --43-43345-45456----45-4
CREPIS BIENNIS                --1---11432--2222-51111-
HERACLEUM SPHONDYLIUM         1-2---1-23-7112114--2-1-
HELICTOTRICHON PUBESCENS      2--41-22-624647353--41-
AJUGA REPTANS                 --2---1-121-1111231212 1-
GEUM RIVALE                   -------112111324-31212 2
CIRSIUM OLERACEUM             -------21-1--11566633331
FILIPENDULA ULMARIA           --------1---3-31----
ANGELICA SILVESTRIS           -------2-1---23-2---1-
HOLCUS LANATUS                -------12-2----233333361
LYCHNIS FLOS CUCULI           --------2---1-111111
DESCHAMPSIA CAESPITOSA        --------------53-754-3
CAREX ACUTIFORMIS             -----------5--4-233
GLYCERIA FLUITANS             ---------------------6
PHALARIS ARUNDINACEA          ---------------------7
```

177

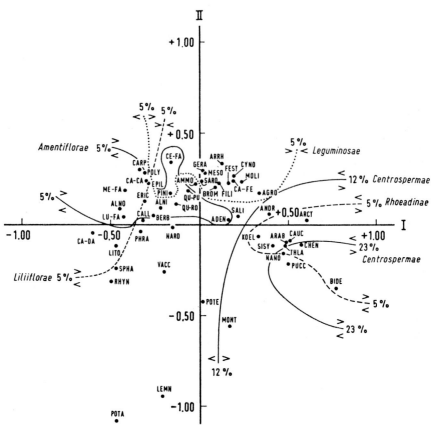

Fig. 7. Principal component analysis (ORDINA see section 3.5.2) of synoptic data from plant communities mainly from South Germany (Oberdorfer 1957) based on plant order representation. a. Dimensions 1 and 2. b. Dimensions 3 and 4.

handling of e.g. 500 relevés with 150 species in 50 clusters. The program has been implemented in 15 centres and will be regularly updated.

(3)  The description and application of a clustering and table ordering, based upon species correlation, as devised by Stockinger & Holzner (1972, Holzner & Stockinger 1973) by Holzner, Werger & Ellenbroek (1978). This program is suited for data sets with relatively many species.

(4)  The development of computer programs for handling very large data sets, in the order of thousands of relevés. Fresco (1971) developed a 'compound analysis' in which a principal components ordination is applied to overlapping series of relevés, whereafter relevés are grouped according to similar positions along the dimensions of the various resulting ordinations. Janssen (1975, Louppen & van der Maarel in prep.) developed CLUSLA, a program which clusters relevés by comparing similarity values between relevés in the sequence they are entered in the program and grouping relevés together if their similarity exceeds a threshold value,

178

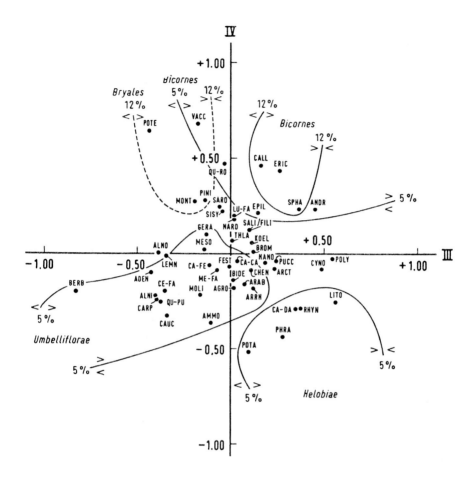

while making a new cluster with any relevé not reaching the threshold similarity value with previous clusters. The resulting clusters are homogenized through re-allocation of relevés, like in TABORD. The latter program can then be applied to one or some related clusters arrived at with CLUSLA.

(5) The introduction of the use of taxonomical units above the species level, notably families and orders as floristic characters in numerical treatments (van der Maarel 1972). This approach has been adopted by others, e.g. Dale & Clifford (1976), Dale (1977) and del Moral & Denton (1977). The general outcome is that such higher taxonomical categories are revealing interpretable ecological results. Dale suggested to proceed with a numerical classification of a comparatively large material by starting with using higher taxa and changing to taxa of lower rank as the material gets split up into more homogeneous groupings.

Van der Maarel (1972) tried to detect major lines of variation in an ordination of a wide spectrum of plant communities (mainly from South Germany and described by Oberdorfer 1957) on the basis of their plant order composition. He found a first line showing the sociological progression from pioneer communities of alliances such as *Bidention, Polygono-Chenopodion* and *Nanocyperion* to mature communities like *Fagion* and *Carpinion*. The orders *Centrospermae* and *Rhoeadales* are characteristic for the first group, the *Amentiflorae* and *Liliiflorae* for the second group of communities. The second line of variation roughly reflects a general moisture gradient, with *Eu-Potamion* and *Lemnion minoris* at one side and dry grasslands at the other side, and *Spadiciflorae* and *Helobiae* versus *Leguminosae* as characterizing orders (Fig. 7a). Between dimensions 3 and 4 (Fig. 7b) two further lines of variation can be discerned. Line 3 runs from *Phragmition* and *Polygonion aviculars* to *Potentillion caulescentis, Cardamino-Montion* and *Vaccinio-Piceion*, with *Helobiae* and *Bryales* as contrasting orders. This clearly suggests a shading gradient. The fourth line of variation runs from alliances characteristic for raw humus soils with a relatively high C/N ratio, such as *Ericion tetralicis* and *Sphagnion fusci*, on one side to alliances like *Caucalion* and *Berberidion* on the other side, with *Bicornes* and *Umbelliflorae* as contrasting orders.

Not only within the Working-Group for Data-Processing numerical-syntaxonomical or similar studies have been performed. Pignatti & Pignatti (1975) and Feoli-Chiapella & Feoli (1977) studied alpine communities in this way, Adam (1977) British salt marshes, Meyer (1977) dry grasslands between Lake Garda and Lake Locarno, and in our country van Gils, Keysers & Launspach (1975) and van Gils & Kovács (1977) *Trifolio-Geranietea* and van Schaik & Hogeweg (1977) *Calthion* communities. Fig. 8 presents a result from the latter study, viz. an ordination of associations derived from the literature and eight clusters referring to *Calthion* communities described in the Netherlands. The first axis can be interpreted in terms of a pH gradient, and the second axis reflects the geographical position from North-West Europe to South Germany. The 'Dutch' clusters include a *Senecio aquaticus* community (cluster 1), a *Senecioni-Brometum* (cl. 3 and 4), a *Scirpetum sylvatici* (cl. 5), a community with *Lychnis flos-cuculi* and *Hypericum tetrapterum* (cl. 7), and a community of *Orchis morio* and *Ophioglossum vulgatum* (cl. 8).

2.3.5 Program packages

For the rapid and efficient application of multivariate analyses various packages of computer programs have been composed. These include: (1) the extensive set at the CSIRO Division of Computing Research at Canberra developed by Williams et al. (2) CLUSTAN, a large set of clustering programs with many similarity and correlation coefficients, developed at the Computing Centre of the University of St. Andrews (Wishart 1969, 1975); (3) a set of ordination and clustering programs developed by Goldstein & Grigal (1972) for the Oak Ridge National laboratory; (4) a set of mainly novel multivariate programs developed by Orlóci (1978a); and (5) ORDI-FLEX, a set of ordination programs developed at the Ecology and Systematics group at Cornell University under R.H. Whittaker (Gauch 1977).

In the Netherlands more or less complete program packages, including simple statistics, classification and ordination have been developed at the Bio-informatica Group of the University of Utrecht (Hogeweg & Hesper 1972), the Plant Ecology Laboratory of the University of Groningen (Fresco), and the Division of Geobotany of the University of Nijmegen (Janssen, Louppen, van der Maarel).

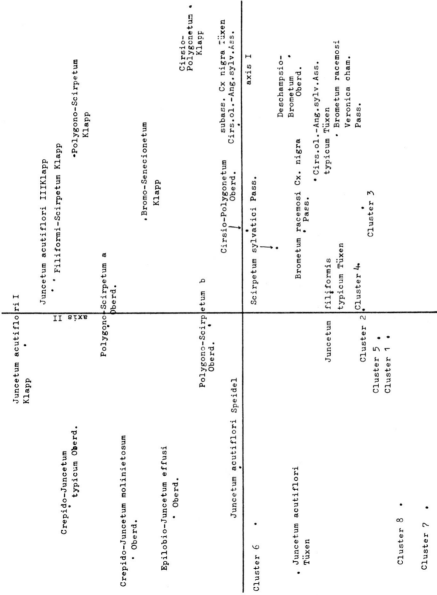

Fig. 8. Principal coordinate analysis (see section 3.5.1) of synoptic data from *Calthion* communities (from van Schaik & Hogeweg 1977).

181

# 3. Classification and ordination: current methods and problems

## 3.1 *Outline*

In the foregoing chapter many multivariate methods have been referred to without specification. Such information should now be presented, certainly since so much difference of opinion exists as to the pros and cons of these methods. The survey of methods must be restricted to the current methods and only in as far as they are useful in phytosociology. For a more general survey the handbooks mentioned in the Introduction (Whittaker 1978a, b, Orlóci 1978a), as well as some other reviews are available (Cormack 1971 and Williams 1971 for classification, Dale 1975 and Noy-Meir & Whittaker 1977 for ordination).

Classification and ordination will be treated separately first and then discussed in their coherence, when two more basic aspects have been introduced, viz. the measurement of numerical relationships between phytosociological entities, and the transformation of data collected with regard to the occurrence of species, or other floristical or structural characters as phytosociological variates. (In the following usually only species will be dealt with).

## 3.2 *Data transformation*

### 3.2.1 Nature and measurement of floristic and structural characters

The result of any classification or ordination will, of course depend on the phytosociological variables we choose to compare phytocoenoses on. The general interpretation of likeness and unlikeness of phytocoenoses will depend on the discriminating power of the variables and the success of the ecological interpretation will increase as the variables better express the underlying environmental differences. In the Braun-Blanquet approach, which is essentially a typological one, we mainly use direct characters such as gross physiognomy and stratification (see Westhoff 1967, Mueller-Dombois & Ellenberg 1974, Rejmánek 1977) and most of all the performance of individual plant species as environmental indicators, but we can also use indirect characters, e.g. life form – or taxonomical spectra (Knight 1965, Knight & Loucks 1969, Whittaker 1972, van der Maarel 1969, 1972, Werger 1978).

The theoretical and practical implications of the selection of phytosociological variables have been discussed by Greig-Smith (1964), Lambert & Dale (1964), Goodall (1970), Smartt, Meacock & Lambert (1974), Pielou (1974) and have been reviewed thoroughly by Orlóci (1978a). Some of these implications are not or less relevant to the Braun-Blanquet approach because of both the particular way of sampling and the choice of the variables. The sampling is what Orlóci (1978a) called preferential, as different from random and systematic sampling. This approach ('not entirely without advantages' Orlóci concludes) has been elucidated by van der Maarel (1966b), Westhoff & van der Maarel (1978) and Werger (1974a, b). The variables are the combined estimation values for cover-abundance of participating (higher) plant species.

182

Following Orlóci (1978a) we may make some general remarks on these estimates. First, they represented discrete variables, at least after an arithmatical transformation. A simple transformation was applied by van der Maarel (Westhoff & van der Maarel 1978, van der Maarel 1979) to the original Braun-Blanquet values to make them liable to calculations. This transformation includes an extension of the Braun-Blanquet scale in splitting up the symbol 2 into 2 m for a high abundance, 2a for a cover of 5-12,5%, and 2b for a cover of 12,5-25%. This extension was adopted by Westhoff (Westhoff & van der Maarel 1978) from the extended scale of Barkman, Doing & Segal (1964). The transformation orders the Braun-Blanquet estimates from 1 to 9 (hence it is called ordinal transform) as follows:

r   +   1   2m   2a   2b   3   4   5
1   2   3   4    5    6    7   8   9

(Note that in this way the abundance region of the scale now includes four grades, r, +, 1 and 2 m, and the cover region six, with scale values r − 2m, 2a, 2b, 3 and 4-5 forming roughly a geometric series like the Hult-Sernander-Du Rietz scale (see Trass & Malmer 1978).

If cover values were taken as such, i.e. as percentages of the total surface, they would represent a continuous variable. What we would need for further calculations is a ratio scale, i.e. a scale with a natural zero point; with a continuous cover scale we would have one. A discrete scale which includes only integers, such as for abundance or density determinations, can also be calculated with. However, the Braun-Blanquet scale is an ordinal scale, as Orlóci (1978a) concludes and as we already implied in the name of the arithmetical transformation given above. Of course one cannot apply arithmetical operations such as addition and multiplication on the basis of an ordinal scale, but this is disregarded in phytosociology.

Apparently the Braun-Blanquet scale is used as a (discrete) scale of some sort of relative importance values. (Note that the importance value of Curtis & McIntosh 1951 is a composite index, based on relative frequency, density and dominance which has a similar aim, but a different basis). This interpreted importance value characteristic of the Braun-Blanquet scale, which is also emphasized in the ordinal transformation, should be given more attention. Any consequences of such an interpretation will be discussed later.

Orlóci (1978a) discussed 8 criteria to consider when choosing phytosociological variables, of which 4 are directly relevant to the Braun-Blanquet combined estimation. First the four other will be mentioned: (a) commensurability, i.e. with a common unit of measure; (b) additivity, i.e. with statistical or theoretical independence. Both criteria apply to the Braun-Blanquet scale; (c) distribution properties. Braun-Blanquet sampling variables are not supposed to have a probability distribution, as e.g. frequency might have; (d) measurability. The cover-abundance scale is typically an interval estimation scale. As such the cover estimates have proven to be sufficiently accurate, at least within the rather broad interval limits. This was shown in a comparative analysis of estimation results from 15 Dutch phytosociologists concerning simulated and real patterns. (This analysis has never been published, but van der Maarel 1966b made some remarks on it).

The four relevant criteria concerning variables are:

(a) Dependence on the sampling unit. The abundance categories in the Braun-Blanquet scale are usually loosely defined and it should be considered to define them as density categories, (see Barkman, Doing & Segal 1964) which would make the entire scale independent of the size of the relevé.

(b) Seasonal variation (and we may add year-to-year fluctuation). Both abundance and cover are subject to such variations. The seasonal variation can, however, be overcome by sampling once or more often at fixed periods to be chosen separately for each community type. The yearly fluctuations should be appreciated as ecologically interpretable differences.

(c) Relativity. It is advantageous if a variable can be transformed into a relative quantity. This transformation would be especially feasible for biomass values, or cover values as approximative values, but it could also be considered for the Braun-Blanquet ordinal transforms, if seen as relative importance values.

(d) Ecological informativeness. This is the most important criterion, but at the same time the most difficult one. There is circumstantial, but conflicting, evidence available on the relative informativeness of quantitative (cover) data versus presence-absence data. If, for instance, a phytosociological data set is relatively homogeneous in species composition while the species number per relevé is high, e.g. in an *Arrhenatheretum*, the presence-absence differences could be rather small and little informative. Or, in relevés of related species-poor dominance type communities, such as a *Spartinetum* and a *Salicornietum*, the species assemblages may be similar whereas the differing quantitative values of the leading species entirely determine the two types as distinct associations (e.g. Kortekaas, van der Maarel & Beeftink 1976). On the other hand a strong quantitative difference in one species within an otherwise fairly homogeneous data set, e.g. in the case of an association with a facies, could easily disturb the coherence.

Beyond these rather simple cases a more complicated pattern exists which will first be approached with some experiences on data transformations and then be placed in the wider framework of the crucial question as to which measures of species performance do reflect which kind of interrelationships between species and their abiotic and biotic environment. Transformation is used here as a general term including weighting and standardization as different though related forms.

3.2.2 Some general aspects of transformation

The various forms of transformation will be described with the help of algebraic expressions which refer to a matrix representation of a phytosociological table (Fig. 9). The occurrence of species $k$ in relevé $i$ is noted as $x_{ki}$, any transformation as $y_{ki}$. Weighting implies as the relative change in quantitative species scores $x_{ki}$ irrespective of their distribution in the rows and columns of the matrix. Standardization involves some sort of change in species scores $x_{ki}$ relative to either the species vectors $x_k$ in which they appear: 'standardization by species', or the relevé vectors $x_i$: 'standardization by relevés', or both: 'double standardization'.

184

| relevé | | 1<br>$X_{.1}$ | 2 ...... $i$ ........ $j$ ........<br>$X_{.2}$ $\quad X_{.i}$ $\quad X_{.j}$ | | | | |
|---|---|---|---|---|---|---|---|
| species | 1 | $X_{1.}$ | $x_{11}$ | $x_{12}$ | $x_{1i}$ | $x_{1j}$ | $\sum\limits_i x_{1i}$ $\quad n_1$ |
| | 2 | $X_{2.}$ | $x_{21}$ | $x_{22}$ | $x_{2i}$ | $x_{2j}$ | $\sum\limits_i x_{2i}$ $\quad n_2$ |
| | $k$ | $X_{k.}$ | $x_{k1}$ | $x_{k2}$ | $x_{ki}$ | $x_{kj}$ | $\sum\limits_i x_{ki}$ $\quad n_k$ |
| | $l$ | $X_{l.}$ | $x_{l1}$ | $x_{l2}$ | $x_{li}$ | $x_{lj}$ | $\sum\limits_i x_{li}$ $\quad n_l$ |
| | | | $\sum\limits_k x_{k1}$<br>$m_1$ | $\sum\limits_k x_{k2}$<br>$m_2$ | $\sum\limits_k x_{ki}$<br>$m_i$ | $\sum\limits_k x_{kj}$<br>$m_j$ | $\sum\limits_k \sum\limits_i x_{ki}$ $\quad n$<br>$m$ |

Fig. 9. Phytosociological table presented as a species – relevé data matrix $[x_{ki}]$ with $m$ species in rows and $n$ relevés in columns. The relevés are presented as vectors $[x_{.1}], [x_{.2}], \ldots$ $[x_{.i}]$, the species as vectors $[x_{1.}], [x_{2.}], \ldots [x_{k.}]$. Species scores for species $k, l$ in relevés $i, j$ are presented as $x_{ki}, x_{li}, x_{kj}, x_{lj}$. Relevé totals are presented as $x_{k1}, x_{k2}, x_{ki}$, species totals as $x_{1i}, x_{2i} \, x_{ki}$, relevé species numbers as $m_1, m_2, \ldots m_i$, species frequencies as $n_1, n_2, \ldots n_k$.

A different scheme (slightly modified here) of (species) transformations, under the heading of weighting, was presented by Hogeweg (1976): (a) a priori weighting on the basis of external knowledge (our weighting in a strict sense); (b) on the basis of univariate variance (our standardization); (c) on the basis of multivariate variance, i.e. by scores on principal components; (d) on the basis of separation between classes of relevés and the distribution of species over the classes. The latter two types could be called a posteriori transformations. (see also Williams 1971)

### 3.2.3 Weighting

A priori weighting has been used most commonly. The type of estimate subjected to weighting is usually a cover estimate, which is either directly determined up to 5 or 10% accuracy, or with an interval scale. This can be an entire cover scale, such as the Hult-Sernander-Du Rietz scale, or a combined cover-abundance scale with the abundance classes converted to arbitrary cover values, such as the Domin scale, and particularly the Braun-Blanquet scale (see van der Maarel 1979 for these and many other scales).

Dagnelie (1960), following Goodall (1954) advocated the angular transformation. If cover percentage values are expressed as fractions of unity, $p = x/100$, then the angular transform follows from:

$$y = 2 \arcsin p^{\frac{1}{2}} \text{ for } p \leqslant 0.5$$
$$y = 3,1416 - 2 \arcsin (1 - p)^{\frac{1}{2}} \text{ for } p > 0.5 \tag{1}$$

| Br. Bl. scale | corresponding cover % value | arc sin transf. | ibid. adapted to Br. Bl. scale |
|---|---|---|---|
| r | ( ) | ( ) | ( ) |
| + | 0.1 | 0.06 | 0.12 |
| 1 | 5.0 | 0.45 | 0.93 |
| 2 | 17.5 | 0.86 | 1.78 |
| 3 | 37.5 | 1.31 | 2.73 |
| 4 | 62.5 | 1.82 | 3.77 |
| 5 | 87.5 | 2.42 | 5.00 |

Fig. 10. Braun-Blanquet cover-abundance values, corresponding cover % values according to Braun-Blanquet (1964), angular transformation values (after Dagnelie 1960) and values multiplied by 2.07 to arrive at values within the original range of the Braun-Blanquet scale.

Dagnelie applied this transformation to cover percentage values taken as class averages for the Braun-Blanquet scale, i.e. the group values of Tüxen & Ellenberg (1937), which he considered as not suited for direct calculations because the class averages are not equidistant. Another argument is that percentage estimates are rarely normally distributed, while angular transforms are more often so.

Dagnelie showed that the angular transforms form a series which runs roughly parallel to the original Braun-Blanquet values and thus he confirmed Schwickerath's (1931, 1940) idea to use the Braun-Blanquet scale values as species magnitude values (with conversions for r and +) (see Fig. 10). A slightly adapted angular transformation with an interval from 1 to 9 was used by van der Maarel (1966a), and Fresco (1969). Angular transformations have also been applied directly to cover estimates (see Smartt, Meacock & Lambert 1976).

Another transformation to cover data is a logarithmic one. Jensén (1978), who reviewed earlier attempts, compared various forms, the effects of which are presented in Fig. 11.

The general practical aim of such transformations is to reduce a possibly too large influence of highly dominant species in multivariate analyses (see Hogeweg 1976 for a Dutch example and a similar argument).

Van der Maarel (1979) compared many of the existing modifications of the Braun-Blanquet scale and transformations of such scales and used a general transformation function to relate different transformations to each other:

$$y = ax^w \qquad (2)$$

where $w$ is a transformation exponent. On the basis of the ordinal transformation from 1-9 (see section 3.2.1) it appeared to be possible to reproduce various scales and transformations with values for $w$ from 0 to 4 (Fig. 12). R.S. Clymo (in van der Maarel 1979) similarly devised a transformation function for cover percentage values (which can, of course, as van der Maarel's one, be used for any other set of values) to present a model for Jensén's (1978) series of transformations:

$$y = (1 - e^{ax})/(1 - e^a) \qquad (3)$$

where $e$ is the base of natural logarithms, $a$ the transformation exponent, and $x$ is supposed to vary from 0 to 1. For near-to-0 values $y$ increases almost linearly with

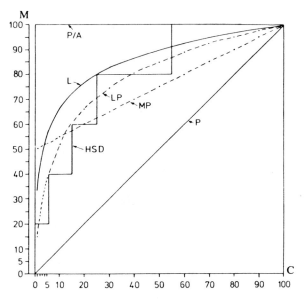

Fig. 11. Relation between cover value (C) and corresponding transformed values in % of maximum M; P = original percentage score, HSD = Hult-Sernander-DuRietz scale values; LP = 100 (log C% + 1)); L = 100 (1 + log C%); P/A = presence-absence MP = (100 + C%) (From Jensén 1978).

$x$, for large values of $a$ a presence-absence situation is approximated. The function generates a series of parabolas with which transformations such as Jensén's ones can be compared.*

Jensén (1978), van der Maarel (1979) and also Campbell (1978) found in many diverging examples that both classification and ordination results obtained with intermediate transformations (in terms of Fig. 12 values of n from 0.5 to 1.5) were more satisfactory as far as resulting data structures and ecological interpretions are concerned.

Weighting on the basis of class separation has been applied by Feoli (1973, discussed in Orlóci 1978a) and Hogeweg (1976). Feoli's measure is:

$$w_k = 1 - \overline{S}_K - / \overline{S}_K + \tag{4}$$

where $\overline{S}_K+$ is the average similarity of relevés with species $k$ and $\overline{S}_K -$ is the average similarity between relevés with species $k$ and those without species $k$., $w$ thus measures the sharpness of isolation of the group with species $k$.

Hogeweg first performed a clustering, defined an optimal number of clusters,

---

* After completion of the manuscript I came across an attempt by Baum (1977) to reduce the dimensionality of a idiotaxonomical dissimilarity matrix (see section 3.5.1.) by applying exactly the same transformation as I used. Moreover, Baum suggested an afternative similar to Clymo's function: $y = (e^{-ax} - 1)$!

187

| Braun-Blanquet scale diff. symbol 2 | Presence-absence transform. p = c⁰ | | Moore transform. | approx. p = 8.67 c$^{0.25}$ | Jensén transform. 1 + log. cover % | approx. p = 0.98 c$^{0.50}$ | combined transformation p = c$^{1.00}$ | | Coetzee & Werger transform. | approx. p = 1,852 c$^{1.50}$ | Schwickerath transform. | approx. p = 0.62 c$^{2.00}$ | Tüxen & Ellenberg average cover %-values | approx. p = 0.01333 c$^{4.00}$ |
|---|---|---|---|---|---|---|---|---|---|---|---|---|---|---|
| r | 1 | 1 | 10 | 8.7 | 1.00 | 0.98 | 1 | 1 | 1 | 2 | ( ) | 0.06 | 0.02 | 0.01 |
| + | 1 | 1 | 10 | 10.3 | 1.00 | 1.39 | 2 | 2 | 5 | 5 | 0.25 | 0.25 | 0.1 | 0.2 |
| 1 | 1 | 1 | 11 | 11.4 | 1.39 | 1.70 | 3 | 3 | 10 | 10 | 1 | 0.56 | 2.5 | 1.1 |
| 2m | 1 | 1 | 11.5 | 12.2 | 1.69 | 1.96 | 4 | 4 | 13 | 15 | 1.5 | 1.0 | 5.0 | 3.4 |
| 2a | 1 | 1 | 12 | 13.0 | 1.90 | 2.24 | 5 | 5 | 17 | 21 | 2 | 1.6 | 8.75 | 8.3 |
| 2b | 1 | 1 | 12.5 | 13.6 | 2.25 | 2.32 | 6 | 6 | 24 | 27 | 2.5 | 2.2 | 18.75 | 17.3 |
| 3 | 1 | 1 | 13 | 14.1 | 2.58 | 2.60 | 7 | 7 | 30 | 34 | 3 | 3.0 | 37.5 | 32.0 |
| 4 | 1 | 1 | 14 | 14.6 | 2.70 | 2.77 | 8 | 8 | 40 | 42 | 4 | 4.0 | 62.5 | 54.6 |
| 5 | 1 | 1 | 15 | 15.0 | 2.94 | 2.94 | 9 | 9 | 50 | 50 | 5 | 5.0 | 87.5 | 87.5 |

Fig. 12. Approximation of current cover-abundance scales through application of an exponential transformation of the form $y = ax^w$ based on ordinal transform values (from van der Maarel 1979).

and calculated a 'character evaluation criterion' by analogy of the Kruskal-Wallis one way analysis of ranks:

$$w_k = (12 / (n(n + 1)) \sum_i r_{ik}^2) / (n_i - 3 (n + 1) / 1-t(n^3 -n)) \qquad (5)$$

where $r_{ik}$ is the sum of ranks of species $k$ in cluster $i$, $n$ the number of relevés and $t$ the number of ties.

A related approach has been developed by Dale & Williams (1978) with the opposite intention of deleting species in subsequent treatments if they are not discriminating between relevés or groups of relevés. Their measure is of the form $y_k = \sum_i |e_{ki} - x_{ki}|$ where $e$ is some expected value, which they consider a measure of the importance of a species. To avoid the latter term they suggest the term 'eident value'. The expected 'eident values' are calculated with the help of the two-parameter method (see Dale & Anderson 1973, Dale & Webb 1975, and also section 3.4.4) which measures a species' discriminating power in cluster formation in relation to the overall cover-abundance (or frequency) of the species. Thus, it is related to Feoli's approach. Similarly an eident value $y_i = \sum_k |e_{ki} - x_{ki}|$ can be calculated for relevés, in which the total richness of the relevé is taken into account.

Finally, a very special form of transformation devised by Swan (1970) is mentioned. Swan calculates the 'degree of absence' of species by comparing the species with those species which are present in a relevé where the species itself is absent over the range of the compared species (in fact with an asymmetric similarity measure to be discussed in section 3.3.2). The reasoning is that an absent species which shows very much the same distribution over the range reflected by the set of relevés (Swan especially thinks of reflected gradients) with species that are present, is not very much absent. Through a simple calculation the 0 value for an absent species is replaced by a positive score which is higher as this similarity with present species is higher.

3.2.4  Standardization

The following forms may be distinguished:

Centering (usually by species) is a standardization to zero vector mean by subtraction with the mean score: $y_{ki} = (\bar{x}_{ki} - \bar{x}_k)$, where $\bar{x}_k = (\underset{i}{\Sigma} x_{ki}) / n$. A full standardization is both to zero mean and unit variance: $y_{ki} = (x_{ki} - \bar{x}_k) / (\underset{i}{\Sigma} (x_{ki} - x_k)^2 / n)^{\frac{1}{2}}$. This is also applied mainly to species. A simpler transformation, called relativization (Gauch & Whittaker 1972) or equalization (Orlóci 1978a) is to unit range, which is particularly applied to relevés: $y_{ki} = x_{ki} / \underset{k}{\Sigma} x_{ki}$. Each species score is thus divided by the relevé total, which may make the scores fractions of something similar to total biomass.

Standardization may be involved in the detection of numerical relationships between relevés or species, or in the subsequent multivariate analysis. Examples are the correlation coefficient were species scores are implicitly standardized to zero mean and unit variance, and the reciprocal averaging ordination method (see section 3.5.2) where, in an iterative process species are standardized by species totals and relevés by relevé totals alternately.

Standardization has been studied more consistently than weighting, both on a priori grounds and in relation to multivariate methods (mainly ordination), and notably by Dagnelie (1960), Greig-Smith and co-workers (e.g. Austin & Orlóci 1966, Austin & Greig-Smith 1968), in the Wisconsin approach (e.g. Bray & Curtis 1957, Goff & Cottam 1967), and later especially by Noy-Meir (e.g. Noy-Meir 1973, Noy-Meir, Walker & Williams 1975, Noy-Meir & Whittaker 1977, see also Noy-Meir & Whittaker 1978).

Some general observations and deductions are:

(1)  Standardization by species implies a reduction in the effect of differences in species frequency and species cover-abundance and consequently the rarer and/or the less abundant species are weighted, while the influence of dominant species is diminished.

(2)  Standardization by relevé implies a reduction in the effect of differences in total cover, or cover-abundance values per relevé and therefore the relevés with low totals will be weighted. This again implies a weighting of species with high cover, or cover-abundance values relative to the other values in the relevé.

(3)  No standardization implies in fact also a weighting, viz. of the species with high frequencies (high cover-abundance values), or species-rich relevés, or both.

### 3.2.5 Some conclusions

It seems justified to pay comparatively much attention to the problems of transformation because: (1) the way species scores are transformed, or taken directly, is of great importance for the results of subsequent multivariate analyses; (2) transformation is often involved in such analyses or in the coefficient used, while the user is not always aware of the consequences; (3) it makes us realize that the parameters we use for the expression of species performance may have different ecological meanings.

Evidently a priori weighting and standardization are related to each other (van der Maarel 1979). The effects of standardization by species and the application of species presence weighting by the use of a low value of the transformation exponent $w$ are similar, and so are the effects of standardization by relevé and application of a high value of the transformation exponent. The conclusion of Gauch, Whittaker & Wentworth (1977) is consistent with ours: application of a log-transformation of cover data prior to (species) standardization 'sometimes gives an advantageous intermediate weight, one that overemphasizes neither the dominants nor the rarities'.

Interestingly Noy-Meir & Whittaker (1977, following Noy-Meir et al. 1975) tried to relate standardization to the principles of the Braun-Blanquet approach by concluding that in proper syntaxonomy the emphasis is on combinations of faithful species which are often rare and of low abundance. Standardization by species would help finding unusual combinations of species (see also Meyer Drees 1949, Goodall 1953b, 1969) and isolating relevé groups on that basis. However, this approach has never been really followed by Braun-Blanquet students!

As to the use of cover-abundance estimates we may first emphasize the idea expressed above that it would be realistic to convert the usual combined estimation values in some sort of importance value (see the ordinal transformation mentioned in section 3.2.1). At the same time such a transformation would be intermediate in the sense of Gauch et al. (1977). This would appeal to an intuitive idea that cover – and maybe also abundance – is a manifestation of exponentially developing processes in plant reproduction (van der Maarel 1969). Finally we may add a suggestion by Austin (in prep.) that species interrelationships such as found in competition experiments may be most effectively expressed in relative biomass estimates. If such an expression would be realistic for the field situation as well, we would tend to apply a standardization by cover-abundance totals per relevé. Both these ideas should be tried out, especially by numerical syntaxonomists in the Braun-Blanquet approach.

### 3.3 *Resemblance functions*

#### 3.3.1 Geometric model

With Orlóci (1972a, 1978a, b) we may adopt the term resemblance function for any measurable relation of pairs of objects, i.e. pairs of relevés or relevé groups, or species. Resemblance, or proximity, can be expressed either as likeness (similarity, for species also: correlation, association) or as unlikeness (dissimilarity, distance, difference).

Resemblance functions should be considered in relation to a geometric model, a 'phytosociological space' (Orlóci 1978a speaks of a sample space) which is either defined by the $m$ species vectors or by the $n$ relevé vectors of a phytosociological data matrix (Fig. 9). A resemblance function $f(i, j)$ regarding two relevés $i$ and $j$ will be based upon a comparison of the relevé vectors $i$ and $j$, whereas a species resemblance function $f(k, l)$ is based upon a comparison of the species vectors $k$ and $l$. In both cases the resemblance function is supposed to measure the spatial positions of points in a multidimensional space. In the case of comparison of relevés the relevé points are in the $m$-dimensional species space, where the position of a relevé $i$ along dimension $k$ is determined by the score of species $k$ (or any transformed value) in relevé $i$.

An important aspect of a resemblance function (however not a necessity!) is whether it is a metric function or not. A metric satisfies four simple 'metric space' axioma's as described by Orlóci (1978a); for two relevés (species) $A$ and $B$ in relation to a relevé (species) $C$ and their distance $d$ in the phytosociological space it follows:

(1)    if $A = B$ then $d(A,B) = 0$                                                                  (6)
(2)    if $A \neq B$ then $d(A,B) > 0$
(3)    $d(A,B) = d(B,A)$
(4)    $d(A,B) \leqslant d(A,C) + d(B,C)$

### 3.3.2  Current measures

Some frequently used relevé resemblance functions will now be discussed briefly.

The Euclidean distance is the most straightforward metric resemblance function:

$$ED = [\sum_k (x_{ki} - x_{kj})]^{\frac{1}{2}} \tag{7}$$

The well-known similarity measures named after Jaccard and Sørensen are, in the form of a dissimilarity measure, examples of a semi-metric, i.e. they satisfy only axioma's 1, 2 and 3:

$$DJ_{ij} = 1 - SJ_{ij} = m_{i \cap j} / (m_i + m_j - m_{i \cap j}) \tag{8}$$

$$DS_{ij} = 1 - SS_{ij} = 1 - 2\, m_{i \cap j} / m_i + m_j \tag{9}$$

where $m_{i \cap j}$ is the number of species common to relevés $i$ and $j$. Note that only presence or absence is taken into account here.

Another often used measure is the precentage dissimilarity according to Bray & Curtis (1957), which goes back to a formula of Czekanowski (1909, see Goodall 1978b):

$$PD_{ij} = 1 - 2 \sum_k \min (x_{kj}, x_{kj}) / \sum_k (x_{ki} + x_{kj}) \tag{10}$$

191

where min $(x_{ki}, x_{kj})$ is the lesser value of the values for species $k$ in relevés $i$ and $j$.

A satisfactory dissimilarity measure in the group of semimetrics is the dissimilarity ratio derived from the similarity ratio as introduced in the CLUSTAN package (Wishart 1973) and as often used in the TABORD program:

$$DR_{ij} = 1 - SR_{ij} = 1 - \sum_k x_{ki} x_{kj} \Big/ (\sum_k x_{ki}^2 + \sum_k x_{kj}^2 - \sum_k x_{ki} x_{kj}) \tag{11}$$

This formula can be efficiently used for both quantitative and presence-absence data. In the presence-absence form it reduces to the Jaccard formula.

A completely different measure is the so-called information measure (Lambert & Williams 1966), which should better be called a heterogeneity measure. $I_i$ is derived from $I_n$ for a set of $n$ relevés:

$$I_n = mn \log n - \sum_k (n_k \log n_k - (n - n_k) \log (n - n_k) \tag{12}$$

Taking $\log_2$ values and $n = 2$ and then taking into account that $n_k$ can only be 0, 1 or 2, where species which occur in both relevés or in neither contribute 2 and species occurring in 1 relevé contribute 0 to the substraction term, one may see that $I_i$ reduces to:

$$I_{ij} = 2 (m_i + m_j - 2 m_{i \cap j}) \tag{13}$$

or in words two times the number of non-common species. It should be noted that for presence-absence data the Euclidean distance becomes related to the heterogeneity measure as follows:

$$ED = \sqrt{\tfrac{1}{2} I} \tag{14}$$

An example of a non-metric resemblance function, i.e. not satisfying axioma 3, which could be realistic in particular cases is a so-called asymmetric similarity measure, which compares a relevé vector $i$ with a relevé vector $j$ only over the range where species occur in vector $i$ and vice versa. Such an approach would be useful when comparing given phytocoenoses with apparently poorly developed ones of the same type. These could be facies or fragments from classical syntaxonomy, but also the basal and derivate (unsaturated) community types distinguished by Kopecký & Hejný (1974) in their new approach to the classification of anthropogenic plant communities. Jensén (1978) used the asymmetric (= reciprocal) index for comparing different classifications of the same relevés.

Austin (pers. comm., in prep.) develops a measure for species (dis)simmilarity which is based upon a comparison of the vector pairs in a matrix of asymmetric species similarity values (such as has been used by Swan 1970, see Goodall 1978b for the index which is usually ascribed to Dice (1945) but is in fact much older). Such an approach may be realistic in cases where some species occur within the range of other species, but with a much lower total frequency, e.g. in geographical distribution patterns (see Haeupler 1974 for an example of such distribution data for species in South Lower Saxony and also for a statistical treatment based on species associations).

192

The most commonly used species resemblance functions are the *chi* square measure or any of its derivates, such as the point correlation coefficient *phi*, the linear correlation coefficient *r* and the percentage co-occurrence (or 'spezifische Gemeinschaftskoeffizient', Ellenberg 1956, later called index of specific association by Mueller-Dombois & Ellenberg 1974) which is the exact equivalent of Jaccard's formula for relevé similarity. This measure is (like *phi*) only suitable for presence-absence data; therefore, the related similarity ratio could be suggested as an alternative for quantitative data.

### 3.3.3 Comparison of measures

It seems hardly possible to select any particularly preferable resemblance function from the many described in the literature. Goodall (1978b) discusses already 38 indices for species resemblance and 34 for relevé resemblance and his review is not exhaustive (it does, for instance, not include the similarity ratio). Goodall suggested some guidelines: (1) indices with statistical tests of the significance of association or correlation between species, or differences between samples, are of lesser concern. (2) Other things being equal, a simple index is to be preferred. (3) Appreciation of an index is easier if it has fixed upper and lower limits, such as + 1 and − 1 for a correlation coefficient, or + 1 and 0 for many similarity indices. (4) If sample size is varying and known to be of influence (which it usually is) an index is to be preferred which is less dependent on sample size. (5) In a discussion note to Goodall's list Whittaker (also Noy-Meir & Whittaker 1978) remarks that centred measures, i.e. with zero mean, may be most useful as a species correlation measure for narrow ranges of relevé variation, while they become less effective in cases of larger variation.

Goodall (1978b) suggested as suitable relevé similarity indices Sørensen's index for presence-absence data and *PS* and *ED* for quantitative data. For species resemblance he would recommend percentage co-occurrence as a non-centred measure for presence-absence data, and point correlation coefficient as a centred one; recommended for quantitative data are percentage similarity of distribution, and the correlation coefficient for presence-absence and quantitative data respectively.

Orlóci (1978a) presented a slightly different set of criteria: (1) meaningfullness in mathematical terms; (2) meaningfullness in ecological terms; (3) uniform scale of measure; (4) desired sampling properties; and (5) reasonable computational load.

For the average phytosociological study we would consider relevé resemblance as the primary basis for further multivariate analysis. Of the criteria mentioned above one may consider four as phytosociologically most relevant: (1) ecological meaningfulness, (2) easiness of calculation, (3) having fixed upper and lower limits and (4) mathematical meaningfulness. The first two criteria may be considered equally important for classification and ordination, the third criterion is more relevant for classification and the fourth one for ordination.

Fig. 13 presents an indicative scheme of appropriateness of the relevé resemblance functions mentioned here. With ecological meaningfulness we mean that the measure we use as a resemblance function takes both a likeness and an unlikeness

|                             | ED | ED* | DJ | DS | PD | DR | I |
|-----------------------------|----|-----|----|----|----|----|---|
| ecological meaningfulness   |    | *   | +  | +  | +  | +  |   |
| mathematical meaningfulness | +  | +   |    |    |    |    | + |
| easy calculation            | +  | +   | +  | +  | +  | +  | + |
| fixed upper and lower limits|    | *   | +  | +  | +  | +  |   |
| quantitative version possible| + | +  |    | +  | +  | +  |   |

Fig. 13. Comparison of some resemblance functions. For abbreviations of the coefficients see the text.

component into account (Orlóci 1978a). In the case of a Euclidean distance only the difference component is measured. This could lead to the situation that species-poor relevés with no species in common, but also few species occurring in both relevés, are situated closer to each other than species-rich relevés which differ in only a few species, but just a few more than in the former case.

Fig. 13 presents an indicative scheme of appropriateness of the relevé resemblance functions mentioned here. Indeed *ED* does not score on the criterion ecological meaningfulness. However, by applying *ED* to equalized species scores we introduce the likeness component in the calculations and make up for the disadvantage. Moreover, the measure becomes limited between 0 and 1. This is indicated in column *ED\** in Fig. 13.

In conclusion for phytosociological studies we would recommend the following set of resemblance functions:

(a)  For the classification of relevés, based up on a relevé resemblance matrix the (dis)similarity ratio is recommended, since it satisfies most of the criteria of Fig. 13 and it is preferable to other similarity measures because it is mathematically elegant and accepts both quantitative and presence-absence data.

(b)  For ordination of relevés the preferable resemblance function would be the Euclidean distance, but with equalized species scores.

(c)  For analyses based upon a species resemblance matrix we could recommend the same functions.

### 3.3.4  *Q*- and *R*-techniques

A usual distinction in multivariate analyses is between *Q*-techniques and *R*-techniques. Without entering upon differences in opinion and confusions (e.g. Dagnelie 1960, Greig-Smith 1964, Williams 1971, Kershaw 1973) we may adopt the solution of Noy-Meir & Whittaker (1978) and link *Q* with relevé resemblance and *R* with species resemblance matrices. This is accordance with the original scheme of Cattell (1952) which, however, includes a temporal dimension. According to this scheme we can distinguish six types of analyses (van der Maarel 1970b) which would all be worth being explored on the basis of the present ideas of transformation and resemblance functions:

*Q*-technique: comparison of relevés on the basis of their momentary species composition;

*R*-technique: comparison of species on the basis of their momentary occurrence in phytocoenoses;

*O*-technique: comparison of relevés of one phytocoenosis on the basis of their species composition at various moments;

*P*-technique: comparison of species in one phytocoenosis on the basis of relevé composition at various times;

*S*-technique: comparison of phytocoenoses on the basis of their relations as to the occurrence of one species in the course of time;

*T*-technique: comparison of occurrences of one species at various times in various phytocoenoses.

## 3.4  *Classification methods*

### 3.4.1  Principles of clustering

Clustering is the technical term for numerical classification, which in the context of phytosociology may be described as the division of a set of physociological entities (or a set of attributes) into a series of groups. Relevé clustering is indeed a technical step prior to the formal classification as described in the Introduction.

In all clustering procedures it is essential that the entities within a group show more likeness to each other than to entities of other groups. This relates to the situation in the field where phytocoenoses are delineated on the basis of a relative homogeneity towards the surroundings (see Westhoff & van der Maarel 1978). Discontinuities in the field should not be confused with those in abstract typologies. They are in fact independent and it is both possible that easily definable phytocoenoses are difficultly to classify because they do form a continuum in a multivariate analysis, or that relevés of difficultly definable phytocoenoses are easy to classify (see Lambert & Dale 1964 and Goodall 1978a for possible definitions of discontinuity and Westhoff & van der Maarel 1978 for a general discussion).

As in field delineation, in classification the emphasis may be on either aspect: internal homogeneity (similarity) of clusters and clear separation from other clusters (Williams 1971). Although a 'good' classification will usually satisfy both aspects for the clusters involved, different emphases are intrinsic in the clustering methods to be discussed. It should further be realized that in Braun-Blanquet classification, and more generally in the erection of types, the internal structure will be of primary importance.

Clustering is mainly concerned with relevés and so will be this review. Clustering of species is equally possible, either as a basis for subsequent relevé clustering or as a method of its own, leading to species groups which do attract the systematical attention of some phytosociologists, notably Doing (1962, 1969) and Scamoni & Passarge (1963), Scamoni, Passarge & Hofmann (1965). A two-way classification of both relevés and species is formalized in some methods, usually under the name nodal analysis (Lambert & Williams 1962, Williams & Dale 1965), after Poore (1955-1956). The part of the program dealing with relevé clustering is called normal analysis, the species-clustering part inverse analysis.

The idea of noda refers to the geometric model discussed in section 3.3.1. Williams & Dale (1965) defined a nodum as 'a set of points and the set of axes in which the points form a galaxy'. Indeed, we may visualize a clustering procedure as partitioning the phytosociological space in subspaces.

Results of clustering can be judged on a set of criteria, some of which are desirable from an ecological point of view (see Orlóci 1978a, partly according to Sneath & Sokal 1973):

(1) Stability. A classification should not be influenced too much by addition of new relevés. This is certainly a criterion to be fulfilled for the usual Braun-Blanquet approach to the description of comparatively unknown areas.

(2) Robustness. Small changes in the data should produce only small changes in the classification. This applies mainly to idiotaxonomy, but could also be relevant to phytosociology (identification problems).

(3) Predictivity. This refers to the success in predicting properties and place of relevés belonging to a data set which were not involved in the actual clustering procedure. Such a criterion, which is of course related to stability, is particularly relevant where data sets are too large to be handled at once and a classification is erected on the basis of part of the material.

(4) Objectivity. The method should be such that different workers arrive at the same classification. This is a very actual point of discussion in Braun-Blanquet syntaxonomy, which is said to be subjective (see Westhoff & van der Maarel 1978, Mueller-Dombois & Ellenberg 1974 and Werger 1974a, b).

Following Williams (1971, 1976) and Orlóci (1978a) we may distinguish a number of strategies, which can be discussed along a hierarchy of choices:

(1) Exclusive versus non-exclusive strategies. Exclusive strategies end up in groups with exclusive membership, non-exclusive strategies with overlapping groups. Although most of the clustering techniques and considerations about them deal with exclusive strategies, and also Braun-Blanquet syntaxonomy does so, the idea of overlapping groups does certainly not conflict with the typological concept of the plant community. Attempts such as by Yarranton et al. (1972) and Holzner & Stockinger (1973) towards non-exclusive groupings should therefore be extended particularly in view of spatial and temporal transitions between phytocoena, which are generally recognized in the Braun-Blanquet approach.

(2) Intrinsic versus extrinsic strategies. In an intrinsic classification all attributes used are regarded as equivalent. The resulting groups are usually of interest in their own right, and this is certainly the case in Braun-Blanquet syntaxonomy. In an extrinsic classification some external attribute is declared in advance and either the resulting groups, though based upon intrinsic attributes, are required to reflect discontinuities in the external attribute, or the groups are directly formed on the basis of that external attribute and then discontinuities in the internal attributes are checked. No examples of true extrinsic phytosociological classifications are known, but the duality is known, viz. with floristic-structural attributes as intrinsic and environmental characters as external attributes. In a way direct gradient analysis (Whittaker 1967, 1978d) is an extrinsic ordination and since ordination results are

196

used in some classification methods (section 3.4.2) this point may become actual.

(3) Hierarchic versus non-hierarchic strategies. Within the exclusive intrinsic strategies hierarchic classification is the common choice. It works along an optimized route between the entire group and the individual relevés. Consequently each group formed with hierarchic methods acts at the same time as a member of a group at a higher level in the hierarchy. The resulting structure is called a dendrogram (see Fig. 5). Clearly the Braun-Blanquet system of plant communities is the most explicit form of a phytosociological hierarchy known today. In a non-hierarchical strategy the structure of individual groups is optimized. The resulting classification may be called reticulate, and this can be optimized as well. Although Williams (1971) stated that the state of development of reticulate strategies is behind, we may consider the plexus techniques (McIntosh 1978), and notably the lineament technique of Polish workers (e.g. Matuszkiewicz 1948, Faliński 1960) as examples of elaborate reticulate systems. In Braun-Blanquet systematics reticulate systems are used informally, often in relation of external (environmental) attributes, such as in synecological schemes (e.g. Mueller-Dombois & Ellenberg 1974). In fact Fig. 3 implies a combination of hierarchic and reticulate systems by combining syntaxonomical classification with ordination.

(4) Polythetic versus monothetic strategies. Within the hierarchic strategies two further choices can be made, which are related to each other. Poly- versus monothetic refers to the number of attributes used in the formation of groups. Monothetic methods use only one species at the time, polythetic methods use more, usually all species involved.

(5) Agglomerative versus divisive strategies. In agglomerative strategies the hierarchical route between total group and individual members is developed from the latter up to the ensemble by progressive fusion; a divisive (also called subdivisive) strategy progressively divides the ensemble into smaller and smaller groups. In an agglomerative method individual relevés are fused on the basis of their resemblances, which are always determined with respect to all species involved, hence agglomerative methods are polythetic. In divisive strategies the subdivisions may be on presence or absence of one single species at the time. Such monothetic divisive methods are computationally attractive and used in various well-known clustering methods. Polythetic divisive methods are much more complicated and have been devised in operational form only recently.

Some current methods will now be discussed, which are representative for the various possible approaches. Fig. 14 shows how they are related to each other.

## 3.4.2 Divisive methods

(1). *Association analysis* (Goodall 1953a, Williams & Lambert 1959, see also Pielou 1969, Orlóci 1978a, Goodall 1978a). The method starts from a matrix of species association values, usually a *chi* square measure, e.g. $S_{kl} = \chi^2_{kl} / n$. The first division of the relevés is based on the presence or absence of the species with the highest sum of association values $\sum_k S_{kl}$. Association matrices are calculated a new

| | monothetic | polythetic |
|---|---|---|
| divisive | 1. Association analysis<br>2. Group analysis<br>3. Divisive information analysis<br>   DIVINF | 4. Indicator species analysis |
| agglomerative | | 5. Nearest neighbour clustering<br>6. Average linkage clustering<br>   HIERAR<br>7. Relocated group clustering<br>   TABORD<br>8. Serial clustering<br>   CLUSLA<br>9. Minimal variance clustering |

Fig. 14. Survey of clustering methods, discussed in sections 3.4.2 and 3.4.3.

for both subgroups and for each new species are found with satisfy the option. The procedure is terminated using a 'stopping rule', e.g. an arbitrary level of $\Sigma\,S$, or some levels of probability, e.g. if no single significant *chi* square values are left. The latter option is attractive in a way, but on the other hand a set of phytosociological relevés can usually not be considered as a sample from a universe for which a null-hypothesis can be adopted. Association analysis is applied both to normal and inverse analysis and thus to nodal analysis (Williams & Lambert 1962). Dutch applications of association analysis include Nijland & Wind (1973), and Werger (1973, Coetzee & Werger 1975).

Arguments mentioned in favour of association analysis (Orlóci 1978a) include (my comments between brackets): (a) easy measure of species performance, i.e. presence/absence (see objections mentioned below); (b) effective programming and suitable for large data sets (this is true for many other methods nowadays); (c) easy assignment of new relevés to the classification, i.e. stable and predictive (yes).

Objections against association analysis may be summarized as follows (see also Werger 1973, Coetzee & Werger 1975, Hill et al. 1975, Williams 1976): (a) argument-in-favour (a) is rather a drawback, because all information provided by the quantitative differences between species is neglected; (b) *chi* square measures are not suitable for relevé comparison, which would urge adaptations for inverse analysis; (c) there is a serious risk for misclassifications, particularly when relevés miss a common species or contain rare ones; (d) the method is not robust.

(2). *Group analysis* (Crawford & Wishart 1967, see Orlóci 1978a). This method is similar to association analysis, but works with a simpler measure for species association. Dutch applications include Baars & Baars-Kloos (1969) and de Lange (1972). The method suffers from the same drawbacks as association analysis but it deals with larger numbers of species.

(3). *Divisive information analysis* (Lance & Williams 1968, see also Williams 1976). The relevé set is subdivided into two subsets according to presence or

198

absence of that species that leads to subgroupings which show the largest decrease in the value of the information measure (formula 12) as compared with the parent ensemble. This method is now preferred to association analysis, mainly because it does not suffer so much from the misclassification risk described for association analysis (Williams 1976). The program, named DIVINF is well-known, also outside Canberra. From the objections mentioned for association analysis only objection (a) remains. In an improved version of this divisive program a re-allocation procedure is added, in which single relevés are replaced whenever this would lead to further decrease in information value.

(4). *Indicator species analysis* (Hill, Bunce & Shaw 1975). This polythetic divisive method seems to be the most elaborated attempt, working along ideas of Noy-Meir (1973b), to what is considered the most preferable approach (see Orlóci 1978a and Williams 1976 for previous attempts). The program starts with an ordination of relevés. A first division of relevés occurs according to their position left or right of the mean on axis 1. Then an indicator value is calculated for each species:

$$I_k = |n_{Lk} / n_L - n_{Rk} / n_R| \qquad (15)$$

where $n_{Lk}$ is the number of 'left' relevés containing species $k$, $n_L$ is the total number of 'left' relevés and so on. The five species with the highest $I$-values are used to calculated an indicator score for each relevé. Finally, relevés are divided over two groups with high $L$- respectively $R$-values. Some options for classification of relevés near the centre of axis 1 are included.

Evidently, the main objection against the monothetic procedures, viz. the non-robustness, is diminished with this method. Moreover, the limitation to presence-absence information has been overcome. Still, as in the first Dutch application by van Groenendael (1978) serious misclassifications may be found, mainly in the case of relevés lacking any of the indicator species, or possessing indicator species that are frequent but show large differences in cover-abundance values. Improvements can be expected when more indicator species are used — or rather all species —, and more axes.

Two other recent approaches to polythetic divisive clustering are that of Feoli (1977b) and that of Bouxin (1978). Feoli starts with an ordination of species together with a clustering, and divides the relevés on the basis of maximal intersection of groups based upon the presence of the dividing species. It is thus monothetic, but it could be extended to a polythetic method. Bouxin (after Berthet et al. 1976) applied an iterative method on the basis of a Euclidean distance matrix. 'Virtual points' are then introduced, the first one in the centroid of the ensemble, the second one at maximal distance. The position of the two virtual points are optimized by alternatively assigning relevés to one of the centres and replacing the virtual points to the centroid of the new clusters.

### 3.4.3 Agglomerative methods

(5). *Nearest neighbour clustering* (Sneath 1957, Jancey 1974, Orlóci 1976, see also Jardine & Sibson 1971, Sneath & Sokal 1973, Orlóci 1978a). This method is the best known of the single linkage clustering methods. Its principle is that progressive fusion of relevés, and clusters, occurs on the basis of the highest similarity (or lowest dissimilarity) between two relevés, or a relevé and a cluster, or two clusters. In the case a cluster is involved, the similarity with that relevé from the cluster with which the highest value occurs is taken, i.e. the nearest neighbour. Its simplest form, with a Jaccard index of similarity (formula 8) was introduced in idiotaxonomy.

Jancey (1974) and Orlóci (1976) devised a more elaborate program which is based upon a distance matrix (for which any symmetric measure of dissimilarity is admissible). The program fuses relevés as long as their distance is exceeded by a 'neighbour radius', which is constantly increased from 0 onwards with small increments. The procedure can be stopped a an arbitrarily fixed number of clusters. This approach is related to graph theoretical methods, such as that of van Groenewoud & Ihm (1974) − see also Hogeweg 1976 − and, of course, with plexus and lineament methods (McIntosh 1978, Romane et al. 1977 and Meyer 1977 for phytosociological examples). Dutch examples of nearest neighbour clustering include Thalen (1971).

Arguments in favour of the method were put forward by Jardine & Sibson (1971) and mainly concern theoretical clarity and efficient computation. Orlóci (1978a) adds that the method discerns discrete ('natural') groups if they exist, while it also finds groups at a specified minimum group size this is desirable. A major disadvantage of nearest neighbour clustering is its tendency to chaining, i.e. the joining of one group by relevés after another. In the terminology of Williams (1976) the method is space contracting, i.e. clusters when growing seem to move towards the centre of the model.

(6). *Average linkage clustering* (Sørensen 1948, Williams, Lambert & Lance 1966, see Orlóci 1978a, and also van der Maarel 1966a). The method starts with a relevé similarity matrix. The first fusion occurs between the relevé with the highest similarity. A new similarity matrix is calculated with new values for similarities between the cluster and each of the remaining relevés. This may occur by averaging values for a relevé with each of the relevés in the cluster, or, preferably by replacing the fused relevé by a synthetic relevé with average scores (it then becomes what is called centroid sorting). The method has no stopping rule.

It can also operate using the information measure (formula 12), but then only presence-absence data are possible. Orlóci (see 1978a) described quantitative information measures, but they have hardly been used yet. Van der Maarel (1966a) found that with average linkage on the basis of an information measure clusters formed at an $I$-level of c. 50% of the total $I$-value in the ensemble were optimal in terms of recognition in the field. Maybe there is a more general validity in such a relation, meaning that the investigator intuitively has a level of abstraction in mind that is related to the entire heterogeneity one is facing. This would lead to a natural stopping rule. A general and often used program for this method is the program HIERAR in the CLUSTAN package.

No systematical objections have been raised against the method. There is a risk of misclassifications, mainly in the case of many relevés having almost equal mutual similarities. For such situations a re-allocation procedure is realistic (e.g. the program RELOC in CLUSTAN). However, disadvantage of relocation is that the hierarchy is influenced: As compared with nearest neighbour clustering, average linkage is completely or nearby completely space conserving, or space indifferent, meaning that group sizes do not influence fusions. An example of Dutch applications is the numerical syntaxonomy of *Spartinetea* communities (Kortekaas, van der Maarel & Beeftink 1976, see Fig. 5).

(7). *Relocated group clustering.* A special form of average linkage clustering was adopted in the table rearrangement program TABORD (van der Maarel et al. 1978, see section 2.3.4). The program starts with clusters in stead of single relevés, which have been formed either by the investigator on the basis of previous knowledge, or automatically by the program. The clusters are homogenized through relocation. Outlier relevés may be placed in a residual group by applying a threshold similarity value, below which a relevé is not allowed to remain in a group (see Goodall 1969, 1978a for other approaches to outliers). Fusion occurs as in average linkage. The results are dependent to some extent on the initial clustering, in a complicated way depending on the relation between threshold value and fusion limit. However, the method is supposed to take advantage of the knowledge of the user, and then it may enable an effective start on a relatively high level of abstraction. In that case the only disadvantage mentioned for agglomerative strategies, viz. the inefficiency in calculating all fusion from the bottom when only the higher levels are of interest, would be overcome. Because the method might have some relevance on its own, it is added to this survey, with a new name: relocated group clustering.

(8). *Serial clustering.* Another derivation of average linkage clustering was devised for very large sets of relevés. The program was described by Janssen (1975) and it is now being extended (Louppen & van der Maarel, in prep.). It can handle thousands of relevés, and it takes previous knowledge into account in a rather special way: it builds up clusters by comparing series of relevés, which have been stored according to the Trieste system (Pignatti 1976), i.e. as they occur in phytosociological tables already composed to document plant community types. The first relevé is considered a cluster; relevé 2 is fused with 1 if their similarity exceeds a threshold value; if not, it becomes a cluster of its own. The groups formed in this way are relocated. Again it is supposed that this approach may have a wider practical significance (see Bouxin 1978) and therefore it is mentioned here under the name of serial clustering.

(9). *Minimum variance clustering* (Ward 1963, see Hogeweg 1976, Orlóci 1978a). This method minimizes variances within groups and maximizes variances between groups. It is based upon squared Euclidean distances weighted by group size. The fusion steps are similar to those in average linkage clustering. The method is space dilatating, meaning that as groups grow their dissimilarity with other relevés increase. Thus the method tends to form a clearly stratified dendrogram without chaining. It may depend on the material or the user's ideas whether this is an advantage or not. Generally spoken a space indifferent method is to be preferred. As compared with average linkage clustering it has the limitation of being restricted to Euclidean distances (see discussion in section 3.3.3), whereas average linkage clustering can be used with any similarity measure. Dutch examples of minimum variance clustering include Hogeweg (1976) and van Schaik & Hogeweg (1977).

### 3.4.4 Nodal analysis

The relevé clustering and the inverse species clustering are usually performed after each other in nodal analysis (Lambert & Williams 1962, see also Ivimey-Cook & Proctor 1966). Dale (1977, Dale & Anderson 1973, Dale & Webb 1975) devised an iterative two-way clustering method, named inosculate analysis, in which the following model is used:

$$x_{ik} = a_i b_k \left(1 + a_i b_k\right) \tag{16}$$

where $x_{ik}$ is the probability of finding species $k$ in relevé $i$, $a_i$ is a parameter measuring relevé richness and $b_k$ is a parameter measuring species abundance. The fit to the model is measured by the discrepancy between the observed $x_{ik}$ values and the estimates based on the $a$ and $b$ values. Two subsets are monothetically derived, using the species one after another and measuring the fit. Species are then ranked in order of effectiveness. The same is done for the relevés. Relevé grouping by division is continued until either the model fits adequately or the division modulates to the inverse approach.

Many of the table rearrangement programs are also two-way approaches (e.g. Moore 1966, 1973, Češka & Roemer 1971, Lieth & Moore 1971, Stockinger & Holzner 1972, Spatz & Siegmund 1973, van der Maarel et al. 1978). However, in none of them the two-way procedure is fully formalized. Two types may be distinguished (Westhoff & van der Maarel 1978): (a) concentrating on finding species groups, (b) concentrating on relevé groups as a first step. Type (a) programs are of course adapted to tables with many species, type (b) programs for many relevés. It is very difficult to compare these programs, because they are devised for different types of relevé sets and suppose different relations between the computerized and the manual subroutines.

### 3.4.5 Numerical classification and phytosociology

Many studies on numerical methods pay attention to the possible applications to and implications for phytosociology, notably the Braun-Blanquet approach. In the following summary of considerations and experiences with the various clustering methods described above this relation with phytosociology will be emphasized. This means that some aspects of which the relevance to phytosociology is not (yet) seen are not summarized here. (For such aspects, e.g. probability considerations, dendrogram structure, and discriminant analysis, see Goodall 1978a, Orlóci 1978a, Williams 1976, and also Hogeweg 1976). Our summary is condensed into Fig. 15.

202

| | association analysis | group analysis | divise information analysis | indicator species analysis | nearest neighbour clustering | average linkage clustering | relocated group clustering | serial clustering | minimum variance clustering |
|---|---|---|---|---|---|---|---|---|---|
| logical basis | + | | + | | + | + | | | + |
| stability | ± | + | + | + | | | + | + | |
| robustness | | | ± | ± | ± | + | + | + | + |
| predictivity | + | + | + | + | | | + | + | |
| objectivity | + | + | + | ± | + | + | ± | | + |
| resemblance function specified | + | + | + | + | | | | | + |
| resemblance function not specified | | | | | + | + | + | + | |
| presence-absence data only | + | + | + | + | | | | | |
| quantitative data possible | | | ± | | + | + | + | + | + |
| little risk of mis-classifications | | | ± | ± | | + | + | ± | + |
| re-allocation possibility | | | | ± | | + | + | + | ± |

Fig. 15. Comparison of some current clustering methods, as described in sections 3.4.2 and 3.4.3. + : yes/present; ± more or less so, depending on data structure.

Some remarks may be added to it:

(1) Polythetic methods are more useful and appropriate for phytosociological classification. Despite the emphasis on character and differential species in community diagnosis the Braun-Blanquet approach deals with the entire species composition and hence monothetic methods fall through in the establishment of clusters. This view is supported by reports on difficult phytosociological interpretation of results obtained with association analysis (Coetzee & Werger 1973, 1975, also Ivimey-Cook 1972, Moore & O'Sullivan 1970).

(2) Agglomerative methods are more useful when dealing with heterogeneous sets of relevés; divisive methods may be more optimal when dealing with homogeneous sets. Moore (1972) considered the table work of the Braun-Blanquet approach a typically polythetic divisive method (and his own PHYTO table rearrangement program reflects this view). This may be true at least for table work on lower syntaxonomical levels, e.g. an association with its subunits. On the other hand, when dealing with more heterogeneous material which is partly not yet well-known, the table worker would rather follow an agglomerative procedure, starting with some nuclei of better known relevés. Evidently TABORD (and the program of Spatz &

203

Siegmund) links up with this tendency. Experiences with agglomerative methods, notably nrs 6 and 9, in relation to phytosociology are all rather promising (e.g. Stanek 1973, Kortekaas, van der Maarel & Beeftink 1976, Adam 1977, Feoli-Chiapella & Feoli 1977, van Schaik & Hogeweg 1977).

Given these experiences we may look from a different viewpoint at the more technical discussion on agglomerative versus divisive methods (e.g. Williams 1971, Williams & Dale 1965, Goodall 1978a, Orlóci 1978a). The time-argument, i.e. waste of time in an agglomerative method if one is only interested in the higher hierarchical levels, can be contradicted with relocated group clustering and its results. The possible misclassifications in agglomerative methods (which are not more in number than in divisive methods) can largely be overcome through relocation (Wishart 1969, 1973).

(3) The two-way approach is essential in the Braun-Blanquet approach: both the distinction of homogeneous groups of relevés in a hierarchy and the distinction of species groups, so in a hierarchy, are aimed at it (see Moore et al. 1970, Češka & Roemer 1971 for a numerical underlining of this statement, and Moravec 1975, 1978 for an elaboration towards the treatment of synoptic tables, i.e. of types of plant communities already synthesized on a lower level). The two-parameter model advocated by Dale (1977) in his approach to associations should be considered with practical phytosociological examples before we can judge its full significance. In this connection the reciprocal averaging method as described by Hill (1973), although devised as on ordination method, can be considered as an additional classification strategy.

In view of the generally satisfying results with TABORD, with the exception of some uncertainties introduced in the initial clustering, a combination of TABORD with a divisive method may be envisaged. Combination with DIVINF as preclassification has been tried with success (Persson 1977, Jensén 1978) and combination with indicator species analysis (as heen been tried by van Groenendael 1978) seems even more promising. More generally table rearrangement programs will have a bright future, certainly if they continue to adapt to new clustering developments and if they are connected with display techniques in order to facilitate the final steps (Dale & Quadraccia 1973).

(4). A separate remark should be made on the judgement of the heterogeneity of accepted clusters. A further advantage of using agglomerative clustering with a similarity measure having fixed limits is the possibility of indicating levels in a syntaxonomical hierarchy with numerical values. This has been done by Kortekaas, van der Maarel & Beeftink (1976) and Neuhäusl (1977), partly along lines indicated by Češka (1966) and Moravec (1971). Westhoff & van der Maarel (1978) reviewed these and earlier attempts and suggested further use of such measures.

Another aspect concerns the relation between the level of homogeneity (or homotoneity) and size of the cluster involved. As Dahl (1957, 1960, see also Barkman 1958a) pointed out, there is a systematic increase in the total number of species in the cluster with increasing cluster size. Dahl found Williams' (1964) semi-logarithmic relation — we even found a double-logarithmic relation as the best approximation for many types, (Rijken-Kępczyńska & van der Maarel, in prep.) – despite repeated observations by Tüxen (e.g. 1977, also 1970) that something of a species saturation occurs. If we take the Raunkiaer (1934) frequency classes into account, it is clear that the increase in total species number is largely caused by species with a low frequency (class I). Thus indices of homogeneity can be based on the 'prevalent' species (Curtis 1959) and, as Dahl (1960) showed, his index of uniformity $\overline{m}$ / $\alpha$ is correlated with such homogeneity measures (see Westhoff & van der Maarel 1978).

## 3.5 *Ordination methods*

### 3.5.1 Principles of ordination

In terms of the geometric model outlined in section 3.3.1 the main objective of ordination is to simplify the multidimensional structure of the phytosociological space. Again we may deal with a species space as well as with a relevé space, but usually we deal with the latter one. Ordination principles and methods have been reviewed by Dale (1975), Williams (1976), Orlóci (1978a, b), Noy-Meir & Whittaker (1977, 1978) and Whittaker & Gauch (1978). The following survey will largely be based upon these reviews.

As Noy-Meir & Whittaker (1977) pointed out two related dichotomies are involved. The first one stems from Dale (1975), who considers two major types of pattern that are of interest. He distinguished between (1) the identification of sequences of compositional variation in plant communities (coenoclines, see section 2.3.1) which can be directly related to environmental gradients; and (2) the identification of related groups of relevés and species, without necessarily drawing discrete boundaries between them. Noy-Meir & Whittaker distinguished two types of ordination: (1) ordination sensu strictu (Whittaker e.g. 1975, 1978d, Austin 1976a, b): the arrangement of relevés in relation to environmental gradients or to abstract axes respresenting such gradients; (2) ordination sensu latu (Orlóci 1978a, b, Dale 1975): the arrangement of relevés along axes regardless of the interpretation of the axes. This distinction may seem artificial, but it may clarify some confusion in the discussion about preferable ordination methods. However, in the present survey ordination will be taken broadly, as in the opening sentence of this section, which is in accordance with the original ideas about ordination.

In all cases of relevé (species) ordination a resemblance matrix forms the basis for further calculations. Choices of transformation and resemblance functions both influence ordination results considerably, and as with classification, partly in relation to the ordination method used. Hence, much attention should be paid to these choices (see sections 3.2 and 3.3).

It is rather difficult to derive a clear system of ordination methods from the reviews mentioned (Whittaker & Gauch 1978). Moreover many names of methods are very similar and do not express specifications of the methods involved (what to think of principal component(s) analysis, principal coordinate analysis, principal axis analysis, principal axis ordination, position vectors ordination?). Whittaker & Gauch (1978) mentioned nine criteria for a classification of ordination methods, which are useful as an introduction to the objectives of ordination: (1) whether direct or indirect, i.e. on environmental or floristic-structural data; (2) the model by which relevés are arranged; (3) the mathematical procedure or algorithm; (4) whether relevés, species, or both at the same time are ordinated; (5) the kind of transformation and resemblance function; (6) the number of axes; (7) whether these are orthogonal or permitted to be oblique; (8) whether indirect techniques deal with species or other categories or structural characters; and (9) the degree to which the technique is informal. With Orlóci (1978a) we may add the goal to be achieved. From his list we adopt three different goals: summarization, ordering and trend seeking.

For our purposes the most relevant criteria are the general goals, the mathematical procedure and the underlying model. As to this model, our basic geometrical model, the phytosociological space, assumes a linear relationship between the axes and the original species attributes. Some ordination methods, notably principal component analysis (PCA) work on a matrix of correlation-, covariance-, or distance values in which a linear structure of the data is reflected. However, we know that the species performance parameters normally used, such as cover-abundance or biomass, are non-linearly distributed along environmental gradients and hence phytosociological data sets will usually not be linearly structured. It is commonly appreciated that such distribution curves are more or less of a Gaussian form (van Groenewoud 1965, 1978, Whittaker 1967, 1975). In some cases this may be doubted, however, and at least for the phytosociological gradients described in Europe (Louppen et al. 1978) the non-linearity and moreover the large zero-occurrence ranges are quite obvious.

In addition to PCA, which could be called a linear ordination type, one other type of ordination is explicitly based on Gaussian distribution curves: Gaussian ordination (Gauch, Chase & Whittaker 1974, Ihm & van Groenewoud 1975, see Noy-Meir & Whittaker 1977, 1978). It has been found succesful for simulated coencoclines and some real data, but no tests on phytosociological material is available. The usefulness of Gaussian ordination will be limited in cases where non-Gaussian species distribution types are in the majority.

A third type of ordination does not presume any particular species distribution type at all. Two of such non-parametric ordinations will be discussed here, viz. reciprocal averaging (RA) and non-parametric multidimensional scaling (NMS). Noy-Meir (1974, Noy-Meir & Whittaker 1977) proposed catenation as a general name for non-linear ordinations and applied one of these, continuity analysis, effectively to vegetation-environment data.

As to the mathematical procedure: through PCA as well as principal co-ordinate analysis and factor analysis the multidimensional relevé (species) space is replaced by a space with few dimensions each of which — in descending order — reflecting as much of the total variance as possible. Their goal is summarization. As was remarked in section 2.2 factor analysis seeks underlying fundamental factors which are difficultly to establish or interprete, especially in the relevé space. Principal co-ordinate analysis is related to PCA, but adapted to non-Euclidean resemblance functions (see Fig. 8 for an example). In this survey only PCA will be treated.

Polar ordination (PO), being a new name (see Cottam, Goff & Whittaker 1978) for the ordination developed by Bray & Curtis (1957) is a geometric procedure of arranging relevés (species) on axes defined by extreme, polar, points. Although the axes arrived at may reflect major parts of the total variance they need not do so. PO is rather an ordering procedure. It is basically a linear method, although this is not always explicitly stated. (Some non-linear forms have recently been suggested, see Noy-Meir & Whittaker 1978).

The most explicit ordering technique is reciprocal averaging (RA), but the mathematical procedure developed for it (Hill 1973, 1974) is identical with an eigen vector procedure also known as factorial analysis of correspondences (Benzécri 1966, 1973, Guinochet 1973, Bouxin 1976).

A completely different mathematics is used in NMS (Kruskal 1964a, b). It replaces the configuration of points in the phytosociological space by a new configuration in fewer dimensions and tries to maintain the original ordering of points as well as possible.

### 3.5.2 Comparison of ordination methods

As in classification, criteria have been listed for the evaluation of the large variation of ordination methods. Whittaker & Gauch (1978) present six criteria: (1) freedom from distortion, i.e. relative success in maintaining the relations between the ordinated relevés as following from the resemblance matrix; (2) range, i.e. the extent of community variation that is covered by the ordination without undue distortion. (This variation is usually expressed as beta diversity); (3) clarity, of the basis of calculations as well as of the ecological meaning of the results; (4) objectivity, i.e. freedom from influences by the choices in the application of the technique; (5) efficiency, in terms of information produced by the ordination as compared with the total information; (6) heuristic function, i.e. success in revealing new insights in the relationships between plant communities and their environments.

The most important criterion is the distortion, at least for Whittaker & Gauch. Distortion has different causes (Orlóci 1972b, 1974b, 1978a). Type A distortion refers to the curvilinear representation in ordination diagrams by linear methods, usually called the horseshoe effect (Swan 1970, Austin & Noy-Meir 1971, see Whittaker & Gauch 1978 for a review). Type B distortion refers to the projection of points from the phytosociological space into the new space, in methods such as polar ordination. Feoli's (1977a) measure for the resolving power of an ordination is in fact related to B-type distortion. Type C distortion refers to the distance function used (this technical point is treated by Orlóci). Type D distortion is linked with resemblance functions which do not fit with the mathematics; this may occur with non-Euclidean distance measures in geometrical ordination. Type E distortion occurs when unsuitable co-ordinates are used to predict levels of environmental influence along a gradient. It is related to type A distortion. There is difference of opinion as to the relative effects of various types of distortions, particularly between Whittaker and Orlóci. In this context only something can be said on the horseshoe distortion.

The four chosen methods will now be described very briefly.

(1). *Polar ordination (PO)* (Bray & Curtis 1957, see also van der Maarel 1966a, Orlóci 1974b, 1978a, Cottam et al. 1978). This method seeks a first axis on the basis of the most dissimilar pair of relevés. These relevés are given the minimal and maximal score on axis 1 and the other relevés are ordered in proportion to their dissimilarities to the polar relevés. This can be done by construction or through computation with the Pythagorean theorem:

$$c_i = (d_{ab}^2 + d_{ia}^2 - d_{ib}^2) / 2\, d_{ab} \tag{17}$$

where $c_i$ is position of relevé $i$, $d_{ab}$ the distance between polar relevés $a$ and $b$, $d_{ia}$ and $d_{ib}$ distances from relevé $i$ to the respective polar relevés. In the same way a second and possibly a third axis can be constructed. (see Fig. 2 for an example). The choice of polar relevés for these further axes may vary (and in fact that for the first axis as well (Orlóci 1966, 1974b, van der Maarel 1969, Swan et al. 1969, Gauch & Whittaker 1972, see Noy-Meir & Whittaker 1977 for a review). PO is attractive by its simplicity and clarity of method (Beals 1973). According to experiences of Whittaker an co-workers it is less sensitive to curvilinear distortion; in fact the axis construction has elements of a non-linear ordination (Gauch et al. 1977). Serious disadvantages are the possible type B distortion and the limited number of axes extractable.

(2). *Principal component analysis (PCA)* (van Groenewoud 1965, Orlóci 1966, Austin & Orlóci 1966, Williams 1976, Orlóci 1978a, b). The resemblance matrix with which calculations start should be a covariance matrix, but a Euclidean distance matrix is equally possible. Eigen vectors and eigen values are calculated according to one of the current iterative procedures. To give one example: the program ORDINA (Roskam 1971) is based on Orlóci's (1966) method with square root of the Euclidean distance normalized by the number of species as resemblance function. The principal components are calculated according to Torgerson with an eigen vector subroutine according to Householder. The percentages of explained variance are calculated, and a measure of success, relating the new distances to the original ones is given (see Fig. 7 for an example). PCA is sensitive to type A distortion which, however, in some cases can be reduced by species standardization (but see section 3.2.4!). The mathematical elegance and speed of the program and the minimum of other types of distortions make the program more attractive than PO for anybody having a large computer available.

(3). *Reciprocal averaging (RA)* (Hill 1973, 1974, Benzécri 1966, 1973, Bouxin 1976, Lacoste 1976, Lacoste & Roux 1971, Guinochet 1973). This program, which can best be used under this name, orders both the relevés and the species and reveals the correspondences between both sets. In one of the algorithms described

(Hill 1973) the structure of the method can be easily followed. In fact it can be calculated with desk computers, like PO. First the species are weighted by positions along an initial gradient with values between 0 and 100. The better these positions reflect actual gradient position the more rapid the further calculations will be. With these weights the relevé scores are calculated and from the relevé totals new species weights are derived. After rescaling so as to make the lowest value again 0 and the highest again 100, these species weights are used to calculate new relevé totals, etc. This iteration leads to a stable set of species and relevé weights, which does not depend on the initial weights. Finally, relevés and species are ordered according to the (rescaled) weights.

RA is an effective ordination with an elegant, clear mathematical procedure, but usually only one or two axes over a wide range of beta diversity can be obtained without distortion by curvilinear relationships. Moreover RA is sensitive to outlier relevés. Obviously the two-way procedure described includes double standardization, which may have its own disadvantages. Apart from that, the method can be of use in the ordering of tables (see del Moral & Watson 1978) especially of clusters. Gauch et al. (1977) explicitly preferred RA to PCA especially at high beta diversities. Bouxin (1976) and Werger, Wild & Drummond (1978), however, slightly preferred PCA to RA! Goff & Cottam (1967) suggested a procedure similar to RA but on a similarity matrix instead of a species score matrix. According to Whittaker & Gauch (1978) this may only improve some distortion in extremely varied coenoclines.

(4). *Non-parametric multidimensional scaling (NMS)* (Kruskal 1964a, b, Fasham 1977, Matthews 1978, see Noy-Meir & Whittaker 1978, Orlóci 1978a). The method starts with specifying the number of axes to be extracted. Next, a dissimilarity matrix is calculated. (A pecularity of the method is that it can cope with missing values; this will not normally occur in phytosociological resemblance matrices). A configuration of points in the new space with reduced dimensionality is chosen arbitrarily and distances are calculated. Now the original and the newly determined distances are compared through regression analysis, i.e. a least-squares regression (linear or polynomial) of ordination distances on observed dissimilarity values. After determining the regression coefficients the regression approximations of the ordination distances are computed. A stress function is determined measuring the distortion in approximated distances with chosen ordination distances. Then the chosen configuration is changed and a new stress value is calculated. The process continues iteratively until the stress decreases below a chosen threshold. The only demand concerning the original dissimilarities is that their order is maintained.

Evidently NMS can deal with non-linear trends and it is extremely flexible as to the resemblance function. It is also possible to impose a structure based on environmental relations which enables predictive versions of NMS. Weaknesses of the method are that the iteration may end up in a 'local minimum' which does not reflect the real phytosociological gradients, if present (Fasham 1977). As with many iterations this can be improved by adopting a realistic configuration, e.g. one based on another simpler ordination. The method is computationally heavy.

209

| | PO polar ordination | PCA principal components analysis | RA reciprocal averaging | NMS non-parametric multidimensional scaling |
|---|---|---|---|---|
| logical basis | weak, but related to vegetation | sound, but not related to vegetation | sound, but not related to vegetation | reasonable, can be related to vegetation |
| resemblance function | not specified, metric or semi-metric preferable | limited choice, metric preferable | specified | not specified, unlimited |
| sensitivity to choice of resemblance function | moderate | large | (no choice) | not large |
| sensitivity to distortion | little to fairly large | large | fairly large | little to not large |
| computation method | easy | moderately difficult | easy | difficult |
| single solution guaranteed | usually | yes | yes | no |
| number of axes extractable with ease | few | many | few | not many |
| interpretability of axes (pairs) in average situations | all axes (pairs) interpretable | few | very few | few |

Fig. 16. Comparison of four current ordination methods described in section 3.5.2.

Fig. 16 summarizes some comparisons between two or more of the four methods described here. In this summary part of the findings of a series of comparative studies are incorporated, notably Austin (1976a, b), Fasham (1977), Gauch & Whittaker (1972), Gauch, Whittaker & Wentworth (1977), Kessell & Whittaker (1976), which are again partly reviewed in Orlóci (1978a) and Whittaker (1978a). The first general conclusion from these comparisons is the complexity of the relations between the results. They are indeed very much dependent on the combination of a certain transformation with a certain resemblance function and the method used. One conclusion comes out rather universally: standardization, both by species and by relevé (introduced or implicite in the ordination method), improves the ordination result, i.e. decreases effects of distortion.

According to Whittaker and co-workers PO is always more successful than PCA in dealing with distortion caused by non-linear structures, especially in cases of high beta diversity. Since many results obtained with PCA have been judged satisfactory, the question is whether such long coenoclines as simulated for tests (e.g. Gauch & Whittaker 1976) are realistic, and if present whether they should be approached as a whole. The latter remark refers to re-interpretations of published PCA results showing clear horseshoe distortions (van der Maarel, in prep.). It seems possible, e.g. in the cases of Jensén (1978) and Werger, Wild & Drummond (1978), to detect ecologically distinct parts of overall gradients from the pattern shown in the diagram of dimensions 1 and 2, which could be subjected subsequently to new ordinations.

More generally it may be stressed that ordination axes should be interpreted in pairs rather than individually. Not only curvilinear but also diagonal trends can be traced and interpreted in relation to resultants of environmental factors (Loucks 1962, van der Maarel 1971, see also Figs. 2 and 7), or subjected to a specially devised polynomial ordination (Phillips 1978).

A final remark is concerned with the underlying model. So far we have used a phytosociological space, essentially a species-relevé space. However, with the general aim of environmental gradient detection in ordinations an 'ecological space' would be more appropriate (Whittaker 1967, Beals 1973, Austin 1976a). The problem here is, of course, how to link the species-relevé matrices on which we apply the ordination with the ecological space. The only straightforward solution here would be to apply direct gradient analysis (Whittaker 1978d).

### 3.5.3 Numerical ordination and phytosociology

Although the relevance of ordination methods in the proper framework of phytosociology may not be as clear as with classification methods, numerical ordination certainly has its perspectives for the development of theories in the synecology of plant communities and in syndynamics (van der Maarel 1969, Westhoff & van der Maarel 1978). We may mention here the use of isocenes and other isolines reflecting structural or ecological characteristics of plant communities (e.g. Londo 1971). Recently the use of ecological indicator values (Ellenberg 1974) has appeared to be informative as well. Such indirect interpretations of ordination diagrams are a welcome addition to the scarce systematically obtained real environmental data (see Figs. 4 and 7 and Werger, Smeets, Helsper & Westhoff 1978 for examples). Simultaneous ordination of real data on vegetation and the environment, e.g. in canonical co-ordinate analysis (see Werger et al. 1979), is also promising.

The interpretation of ordination results is facilitated by display techniques for three-dimensional pictures (Frazer & Kovats 1966, Kershaw & Shepard 1972).

Another perspective is the use of higher taxonomical categories in stead of species for obtaining overall ecological interpretations of large-scale variation in plant communities (van der Maarel 1972). In a way this approach also solves the problem of distortion (see Whittaker & Gauch 1978). As to the proper choice of ordination methods for phytosociological purposes the first thing to remember is the effect of data transformations.

Effects of standardization, especially by species, should be examined and intermediate weighting of cover-abundance estimations (such as with the 1-9 ordinal transform) should be tried.

As to the resemblance functions: the similarity ratio, preferred for numerical classification, has the disadvantage of being semimetric which makes it less suitable for PCA. Here the suggested form of Euclidean distance, i.e. with equalized ordinal transform, should be tested.

Apart from these aspects PCA could be recommended as a lucid, efficient summarization method which has proven its use in phytosociology (e.g. van der Maarel 1969, 1972, Pignatti & Pignatti 1975, Werger, Wild & Drummond 1978, Werger 1978, see also Nichols 1977). RA could well be considered a useful additional method for the more elaborate approach of ecologically more homogeneous parts of supposed phytosociological gradients revealed by PCA. The attraction of the double ordering of both relevés and species should be taken into account as well, also in view of the tabular representation of gradient structures.

### 3.5.4 Classification and ordination as complementary methods in phytosociology

The evident mutual benefits of classification and ordination for phytosociological studies have become clear from an increasing number of combined treatments of phytosociological data (e.g. Adam 1977, 1978, Bouxin 1975, 1976, Feoli-Chiapella & Feoli 1977, van Gils et al. 1977, van der Maarel et al. 1978, Moore et al. 1970, van Schaik & Hogeweg 1977, Werger 1973). In many cases ordination results have clarified obscure positions of relevés in classifications, and have deepened insights in the relations between community types and their environment. On the other hand ordination has been applied with success to clusters of relevés or even to already established phytocoena.

In this way the Braun-Blanquet approach has contributed to the 'reconciliation' between advocates of ordination and advocates of classification, which was reached during the International Botanical Congress at Edinburgh 1964. This reconciliation has been further promoted especially by Whittaker, who expressed his view most clearly in his contribution (1972) on convergences of ordination and classification. Convergence does not mean fusion; 'rather than fusion particular interlinkages, sharing of concepts and techniques even when applied within different perspectives, may be means of partnership for our two scientific cultures in exploring the phenomena of natural communities that are our common concern.'

## 4. Perspective

This review covers c. 250 publications dealing with multivariate methods in vegetation science, of which nearly 80% date from the last ten years. It includes a fairly complete survey of publications in which Dutch authors are involved (60, of which 50 are recent), but it is by far not complete for the world literature (see further the reviews mentioned in the introduction). We may conclude from these figures how strong recent developments of multivariate methods in vegetation science are, and particularly how young this branch of phytosociology is in the Netherlands. We may expect that this development will continue, since there is a continuous production of new original articles on the subject, notably in the journals *Journal of Ecology*, *Ecology*, and *Vegetatio*. In view of the interest in these trends and the participation in some of them by some Dutch centres and individuals a good future for numerical phytosociology in the Netherlands can be foreseen.

Five fields of research may be mentioned in which the Dutch contributions may play a part. Although this list tries to take both our traditions and limitations into account, it reflects my personal views and it is open to discussion. With this in mind, I see the following research activities as pertinent:

(1). extensive testing of a selection of classification and ordination methods, with emphasis on classification and on effects of transformation and resemblance functions, in a wide variety of community types;

(2). contribution to ordination theory by both descriptive and experimental research of species responses along environmental gradients, in combination with systematic soil description, and with emphasis on the influence of environmental dynamics (see also Grubb 1977);

(3). further exploration of the use of higher taxonomical categories and of structural characters in both classification and ordination;

(4). further development of numerical syntaxonomy in relation to the establishment of a data bank, or a network of local data banks;

(5). large scale comparative description of floristic composition, structure and environmental relations of carefully selected sites representing the main plant community types in the Netherlands on the basis of the results of the numerical syntaxonomy of the large phytosociological material already available.

## Acknowledgements

I thank colleagues of this department, notably Wil Kortekaas, Jan Janssen, Jo Louppen, Willem Schenk and Marinus Werger for many stimulating discussions. A special acknowledgement concerns Mike Austin who was a guest at our department while this contribution was drafted, and who stimulated the work very much! I also thank Dicky Clymo with whom I lectured at a course on numerical methods for the Nordic Council of Ecology, at the University of Lund, during which he gave me many stimuli, including ideas which led to the composition of Fig. 16. Finally I thank, again, Marinus Werger for his cheerful patience and for careful reading of the manuscript.

# References

Anon. 1974. The Canada geographic information system. Environment Canada, Ottawa.
Adam, P. 1977. On the phytosociological status of Juncus maritimus on British saltmarshes. Vegetatio 35: 81-94.
Adam, P. 1978. Geographical variation in British saltmarsh vegetation J. Ecol. 66: 339-366.
Adriani, M.J. & E. van der Maarel. 1978. Plant species and plant communities. An introduction. In: E. van der Maarel & M.J.A. Werger (eds.), Plant species and plant communities, p. 3-6. Junk, The Hague.
Austin, M.P. 1976a. On non-linear species response models in ordination. Vegetatio 33: 33-41.
Austin, M.P. 1976b. Performance of four ordination techniques assuming three different non-linear species response models. Vegetatio 33: 43-49.
Austin, M.P. 1977. Use of ordination and other multivariate descriptive methods to study succession. Vegetatio 35: 165-175.
Austin, M.P. & P. Greig-Smith. 1968. The application of quantitative methods to vegetation survey. II. Some methodological problems of data from rain forest. J. Ecol. 56: 827-844.
Austin, M.P. & I. Noy-Meir. 1971. The problem of non-linearity in ordination: experiments with two-gradient models. J. Ecol. 59: 763-774.
Austin, M.P. & L. Orlóci. 1966. Geometric models in ecology. II. An evaluation of some ordination techniques. J. Ecol. 54: 217-227.
Baars, J.A. & J.A. Baars-Kloos. 1969. Een classificatie van opnamen uit Thelypteris palustris-vegetaties volgens de methode van Crawford & Wishart (1967). Rapp. Hugo de Vries Lab. Amsterdam.
Barkmann, J.J. 1958a. Phytosociology and ecology of cryptogamic epiphytes. Van Gorkum, Assen.
Barkman, J.J. 1958b. La structure du Rosmarineto-Lithospermetum helianthemetosum en Bas-Languedoc. Blumea Suppl. 6: 113-136.
Barkman, J.J. 1978. Synusial approaches to classification, 2nd ed. In: R.H. Whittaker (ed.), Classification of plant communities. p. 111-200. Junk, The Hague.
Barkman, J.J., H. Doing & S. Segal. 1964. Kritische Bemerkungen und Vorschläge zur quantitativen Vegetationsanalyse. Acta Bot. Neerl. 13: 394-419.
Baum, B.R. 1977. Reduction of dimensionality for heuristic purposes. Taxon 26: 191-195.
Beals, E.W. 1973. Ordination: mathematical elegance and ecological naïveté. J. Ecol. 61: 23-35.
Becking, R.W. 1957. The Zürich-Montpellier school of phytosociology. Bot. Rev. 23: 411-488.
Beeftink, W.G. 1965. De zoutvegetaties van Zuidwest-Nederland beschouwd in Europees verband. Thesis. Wageningen. Meded. Landbouwhogeschool, Wageningen 65 (1): 1-167.
Beeftink, W.G. 1966. Vegetation and habitat of the salt marshes and beach plains in the southwestern part of the Netherlands. Wentia 15: 83-108.
Beeftink, W.G. 1972. Uebersicht über die Anzahl der Aufnahmen europäischer und nord-afrikanischer Salzpflanzengesellschaften für das Projekt der Arbeitsgruppe für Datenverarbeitung. In: E. van der Maarel & R. Tüxen (eds.), Grundfragen und Methoden der Pflanzensoziologie. Ber. Int. Symp. Rinteln 1970, p. 371-396. Junk, The Hague.
Beeftink, W.G., M.C. Daane & W. de Munck. 1971. Tien jaar botanisch-oecologische verkenningen langs het Veerse Meer. Natuur en Landschap 25: 50-64.
Beeftink, W.G. & J.-M. Géhu. 1973. Spartinetea maritimae. Prodrome des groupements végétaux d'Europe. Vol. 1. Cramer, Lehre.

Benninghoff, W.S. & W.C. Southworth. 1964. Ordering of tabular arrays of phytosociological data by digital computer. Abstr. 10. Int. Bot. Congr. Edinburgh: 331-332.

Benzécri, J.P. 1966. Leçons sur l'analyse factorielle et la reconnaissance des formes. Cours Inst. Stat., Univ. Paris.

Benzécri, J.P. et al. 1973. L'analyse des données. Vol. 2. L'analyse des correspondances. Dunod, Paris.

Berthet, P., E. Feytmans, D Stevens & A. Genette. 1976. A new divisive method of classification illustrated by its applications to ecological problems. Proc. 9 Int. Biometrics Conf. 2: 366-382.

Bie, S.W., J.R.E. Lieftinck, K.R. van Lynden. 1976. Computer-aided interactive soil suitability classification, a simple Bayesian approach. Neth. J. Agric. Sc. 24: 179-186.

Bouxin, G. 1975. Ordination and classification in the savanna vegetation of the Akagera Park (Rwanda, Central Africa). Vegetatio 29: 155-167.

Bouxin, G. 1976. Ordination and classification in the upland Rugege forest. Vegetatio 32: 97-115.

Bouxin, G. 1978. Etude comparative d'une nouvelle methode de classification des cendante polythetique avec reallocations. Biometrie-Praximetrie 18: 20-48.

Braun-Blanquet, J. 1964. Pflanzensoziologie, Grundzüge der Vegetationskunde. 3. Aufl. Springer, Wien-New York.

Braun-Blanquet, J. 1965. Plant sociology: the study of plant communities. Transl., rev. and ed. by C.D. Fuller & H.S. Conard. Hafner, London.

Bray, J.R. & J.T. Curtis. 1957. An ordination of the upland forest communities of southern Wisconsin. Ecol. Monogr. 27: 325-349.

Campbell, B.M. 1978. Similarity coefficients for classifying relevés. Vegetatio 37: 101-109.

Cattell, R.B. 1952. The three basic factor-analytical research designs — their interrelations and derivatives. Psychol. Bull. 49: 499-520.

Češka, A. 1966. Estimation of the mean floristic similarity between and within sets of vegetational relevés. Folia Geobot. Phytotax. 1: 92-100.

Češka, A. & H. Roemer. 1971. A computer program for identifying species-relevé groups in vegetation studies. Vegetatio 23: 255-276.

Coetzee, B.J. & M.J.A. Werger. 1973. On hierarchical syndrome analysis and the Zürich-Montpellier table method. Bothalia 11: 159-164.

Coetzee, B.J. & M.J.A. Werger. 1975. On association-analysis and the classification of plant communities. Vegetatio 30: 201-206.

Cormack, R.M. 1971. A review of classification. J. Roy. Stat. Soc. A-134, 3: 321-367.

Cottam, G., F.G. Goff & R.H. Whittaker. 1978. Wisconsin comparative ordination. 2nd ed. In: R.H. Whittaker (ed.), Ordination of plant communities, p. 185-213. Junk, The Hague.

Crawford, R.M.M. & D. Wishart. 1967. A rapid multivariate method for the detection and classification of groups of ecologically related species. J. Ecol. 55: 505-524.

Curtis, J.T. 1959. The vegetation of Wisconsin. An ordination of plant communities. The Univ. of Wisconsin Press, Madison Wisc.

Curtis, J.T. & R.P. McIntosh. 1951. An upland forest continuum in the prairie-forest border region of Wisconsin. Ecology 32: 476-496.

Dagnelie, P. 1960. Contribution a l'étude des communautés végétales par l'analyse factorielle. Bull. Serv. Carte Phytogéogr. C.N.R.S., B. 5: 7-71, 93-45.

Dagnelie, P. 1978. Factor analysis. 2nd ed. In: R.H. Whittaker (ed.), Ordination of plant communities, p. 215-238. Junk, The Hague.

Dahl, E. 1957. Rondane; Mountain vegetation in South Norway and its relation to the environment. Skr. Norske Vidensk-Akad; Mat.-Naturv. Kl. 1956 (3): 1-374.

Dahl, E. 1960. Some measures of uniformity in vegetation analysis. Ecology 41: 785-790.

Dale, M.B. 1975. On objectives of methods of ordination. Vegetatio 30: 15-32.

Dale, M.B. 1977. Planning an adaptive numerical classification. Vegetatio 35: 131-136.

Dale, M.B. & D.J. Anderson. 1973. Inosculate analysis of vegetation data. Austr. J. Bot. 21: 253-276.

Dale, M.B. & H.T. Clifford. 1976. On the effectiveness of higher taxonomic ranks for vegetation analysis. Austr. J. Ecol. 1: 37-62.

Dale, M.B. & L. Quadraccia. 1973. Computer assisted tabular sorting of phytosociological data. Vegetatio 28: 57-73.

Dale, M.B. & L.J. Webb. 1975. Numerical methods for the establishment of associations. Vegetatio 30: 77-87.

Dale, M.B. & W.T. Williams. 1978. A new method of species reduction in ecological data. Aust. J. Ecol. 3: 1-5.

Dice, L.R. 1945. Measures of the amount of ecologic association between species. Ecology 26: 297-302.

Dobben, W.H. van & R.H. Lowe-McConnell (ed.). 1975. Unifying concepts in ecology. Junk, Den Haag; Pudoc, Wageningen.

Doing, H. 1962. Systematische Ordnung und floristische Zusammensetzung niederländischer Wald- und Gebüschgesellschaften. Diss. Wageningen, Wentia 8: 1-85.

Doing, H. 1969. Sociological species groups. Acta Bot. Neerl. 18: 398-400.

Doing, H. 1974. Landschapsoecologie van de duinstreek tussen Wassenaar en IJmuiden. (English Summary). Meded. Landbouwhogeschool Wageningen 74 (12): 1-111.

Ellenberg, H. 1956. Grundlagen der Vegetationsgliederung. 1. Teil: Aufgaben und Methoden der Vegetationskunde. In: H. Walter, Einführung in die Phytologie 4 (1). Ulmer, Stuttgart.

Ellenberg, H. 1974. Zeigerwerte der Gefässpflanzen Mitteleuropas. Scripta Geobot. Göttingen 9: 1-97.

Faliński, J. 1960. Zestosowanie taksonomii wroclawskiej do fitosocjologii. (Anwendung der sogenannten 'Breslauer Taxonomie' in der Pflanzensoziologie). Acta Soc. Bot. Bol. 29: 333-361.

Fasham, M.J.R. 1977. A comparison of non-metric multidimensional scaling, principal components and reciprocal averaging for the ordination of coenoclines and coenoplanes. Ecology 58: 551-561.

Feoli, E. 1973. Un indici che stima il peso dei caratteri per classificazioni monotetiche. (English summary) Giorn. Bot. Ital. 107: 263-268.

Feoli, E. 1977a. On the resolving power of principal component analysis in plant community ordination. Vegetatio 33: 119-125.

Feoli, E. 1977b. A criterion for monothetic classification of phytosociological entities on the basis of a species ordination. Vegetatio 33: 147-152.

Feoli-Chiapella & E. Feoli. 1977. A numerical phytosociological study of the summits of the Majella massive (Italy). Vegetatio 34: 21-29.

Ferrari, Th.J., H. Pijl & D.T.N. Venekamp. 1957. Factor analysis in agricultural research. Neth. J. Agric. Sc. 5: 211-221.

Frazer, A.R. & M. Kováts. 1966. Stereoscopic models of multivariate statistical data. Biometrics 22: 358-367.

Fresco, L.F.M. 1969. Q-type factor analysis as a method in synecological research. Acta Bot. Neerl. 18: 477-482.

Fresco, L.F.M. 1971. Compound analysis: a preliminary report on a new numerical approach in phytosociology. Acta Bot. Neerl. 20: 589-59.

Fresco, L.F.M. 1972. A direct quantitative analysis of vegetational boundaries and gradients. In: E. van der Maarel & R. Tüxen (eds.), Grundfragen und Methoden in der Pflanzensoziologie. Ber. Int. Symp. Rinteln 1970. p. 99-111. Junk, Den Haag.

Gauch, Jr. H.G. 1977. Ordifex. Release B. Ecology and Systematics, Cornell University, Ithaca N.Y.

Gauch, Jr. H.G., G.B. Chase & R.H. Whittaker. 1974. Ordination of vegetation samples by Gaussian species distributions. Ecology 55: 1382-1390.

Gauch, Jr. H.G. & R.H. Whittaker. 1972. Comparison of ordination techniques. Ecology 53: 868-875.

Gauch, Jr. H.G. & R.H. Whittaker. 1976. Simulation of community patterns. Vegetatio 33: 13-16.

Gauch, Jr. H.G., R.H. Whittaker & T.R. Wentworth. 1977. A comparative study of reciprocal averaging and other ordination techniques. J. Ecol. 65: 157-174.

Gils, H. van, E. Keysers & W. Launspach. 1975. Saumgesellschaften im klimazonalen Bereich des Ostryo-Carpinion orientalis. Vegetatio 31: 47-64.

Gils, H. van & A.J. Kovács. 1977. Geranion sanguinei communities in Transsylvania. Vegetatio 33: 175-186.

Godron, M. et al. 1968. Code pour le relevé méthodique de la végétation et du milieu (principes et transcription sur cartes perforées). CNRS, Paris.

Godron, M. et al. 1969. Vade-Mecum pur le relevé méthodique de la végétation et du milieu. CNRS, Paris.

Godron, M. & J. Lepart. 1975. Sur la répprésentation de la dynamique de la végétation au moyen de matrices de succession. In: W. Schmidt (ed.), Sukzessionsforschung. pp. 269-287. Cramer, Vaduz.

Goff, F.G. & G. Cottam. 1967. Gradient analysis: the use of species and synthetic indices. Ecology 48: 793-806.

Goldstein, R.A. & D.F. Grigal. 1972. Computer programs for the ordination and classification of ecosystems. Ecol. Sci. Div. Publ. 417, Oak Ridge Nat. Lab., Oak Ridge.

Goodall, D.W. 1952. Quantitative aspects of plant distribution. Biol. Rev. 27: 194-245.

Goodall, D.W. 1953a. Objective methods for the classification of vegetation I. The use of positive interspecific correlation. Austr. J. Bot. 1: 39-63.

Goodall, D.W. 1953b. Objective methods for the classification of vegetation II. Fidelity and indicator value. Austr. J. Bot. 1: 434-456.

Goodall, D.W. 1954. Objective methods for the classification of vegetation. III. An essay in the use of factor analysis. Austr. J. Bot. 2: 304-324.

Goodall, D.W. 1961. Objective methods for the classification of vegetation. IV. Pattern and minimal area. Austr. J. Bot. 9: 162-196.

Goodall, D.W. 1969. A procedure for recognition of uncommon species combinations in sets of vegetation samples. Vegetatio 18: 19-35.

Goodall, D.W. 1970. Statistical plant ecology. Ann. Rev. Ecol. Syst. 1: 99-124.

Goodall, D.W. 1978a. Numerical classification. 2nd ed. In: R.H. Whittaker (ed.), Classification of plant communities, p. 247-286. Junk, The Hague.

Goodall, D.W. 1978b. Sample similarity and species correlation. 2nd. In: R.H. Whittaker (ed.), Ordination of plant communities, p. 99-149. Junk, The Hague.

Gounot, M. 1956. A propos de l'homogénéité et du choix des surfaces de relevé. Bull. Serv. Carte Phytogeogr. B 1: 7-17.

Gounot, M. 1961. Les méthodes d'inventaire de la végétation. Bull. Serv. Carte phytogeógr. Sér. B. 6: 7-73.

Gounot, M. 1969. Méthodes d'étude quantitative de la végétation. Masson, Paris.

Gounot, M. & M. Calléja. 1962. Coefficient de communauté, homogénéité et aire minimale. Bull. Serv. Carte Phytogéogr. B. 7: 181-210.

Greig-Smith, P. 1952. The use of random and contiguous quadrats in the study of the structure of plant communities. Ann. Bot. NS 16: 293-316.

Greig-Smith, P. 1964. Quantitative plant ecology. 2nd ed. Butterworths, London.

Groenendael, J.M. van 1978. The vegetation of Cors Goch, in relation to some environmental factors. Rapp. Afd. Geobotanie, Nijmegen.

Groenewoud, H. van. 1965. Ordination and classification of Swiss and Canadian forests by various biometric and other methods. Ber. Geobot. Inst. Rübel 35: 28-102.

Groenewoud, H. van. 1978. Theoretical considerations of the covariation of plant species along ecological gradients with regard to multivariate analysis. J. Ecol. 64: 837-847.

Groenewoud, H. van & P. Ihm. 1974. A cluster analysis based on graph theory. Vegetatio 29: 115-120.

Grubb, P.J. 1977. The maintenance of species-richness in plant communities: the importance of the regeneration niche. Biol. Rev. 52: 107-145.

Guinochet, M. 1973. Phytosociologie. Masson et Cie, Paris.

Haeupler, H. 1974. Statistische Auswertung von Punktrasterkarten der Gefasspflanzenflora Süd-Niedersachsen. Scripta Geobot. Göttingen 8: 1-41.

Hill, M.O. 1973. Reciprocal averaging. An eigen vector method of ordination J. Ecol. 61: 237-249.

Hill, M.O. 1974. Correspondence analysis: a neglected multivariate method. Appl. Statist. 23: 340-354.

Hill, M.O., R.G.H. Bunce & M.W. Shaw. 1975. Indicator species analysis, a divisive polythetic method of classification, and its application to a survey of native pinewoods in Scotland. J. Ecol. 63: 597-613.

Hogeweg, P. 1976. Topics in biological pattern analysis. Thesis Utrecht.

Hogeweg, P. & B. Hesper. 1972. BIOPAT. Program system for biological pattern analysis. Subfaculty Biology, University of Utrecht.

Holzner, W. & F. Stockinger. 1973. Der Einsatz von Elektronen-rechner bei der pflanzensoziologischen Tabellenarbeit. Oesterr. Bot. Z. 121: 303-309.

Holzner, W., M.J.A Werger & G.A. Ellenbroek. 1978. Automatic classification of phytosociological data on the basis of species groups. Vegetatio 38: 157-164.

Hopkins, B. 1955. The species-area relations of plant communities. J. Ecol. 43: 409-426.

Hopkins, B. 1957. The concept of minimal area. J. Ecol. 45: 441-449.

Hulst, R. van 1976. Theoretical aspects of vegetational change. Thesis London Ont. (will appear in three papers in Vegetatio 1978-1979).

Ihm, P. & H. van Groenewoud. 1975. A multivariate ordering of vegetation data based on Gaussian type gradient response curves. J. Ecol. 63: 767-778.

Ivimey-Cook, R.B. 1972. Association analysis-some comments on its use. In E. van der Maarel & R. Tüxen (eds.): Grundfragen und Methoden der Pflanzensoziologie. Ber. Int. Symp. Rinteln 1970, p. 89-97. Junk, Den Haag.

Ivimey-Cook, R.B. & M.C.F. Proctor. 1966. The application of association analysis to phytosociology. J. Ecol. 54: 179-192.

Jancey, R.C. 1974. Algorithm for detection of discontinuities in data sets. Vegetatio 29: 131-133.

Janssen, J.G.M. 1972. Detection of some micropatterns of winter annuals in pioneer communities of dry sandy soils. Acta Bot. Neerl. 21: 603-610.

Janssen, J.G.M. 1975. A simple clustering procedure for preliminary classification of very large sets of phytosociological relevés. Vegetatio 30: 67-71.

Jardine, N. & R. Sibson. 1971. Mathematical taxonomy. Wiley, New York.

Jensén, S. 1978. Influences of transformation of cover values on classification and ordination of lake vegetation. Vegetatio 37: 19-31.

Jong, W.W.W. de. 1964. Vloedmerkvegetaties van De Beer. Doct. Verslag Inst. Syst. Plantk. Utrecht.

Kershaw, K.A. 1957. The use of cover and frequency in the detection of pattern in plant communities. Ecology 38: 291-29.

Kershaw, K.A. 1973. Quantitative and dynamic ecology. 2nd ed. Elsevier, New York.

Kershaw, K.A. & R.K. Shepard. 1972. Computer display graphics for principal component analysis and vegetation ordination studies. Can. J. Bot. 50: 2239-2250.

Kessell, S.R. & R.H. Whittaker. 1976. Comparisons of three ordination techniques. Vegetatio 32: 21-29.

Knight, D.H. 1965. A gradient analysis of Wisconsin prairie vegetation on the basis of plant structure and function. Ecology 46: 744-747.

Knight, D.H. & O.L. Loucks. 1969. A quantitative analysis of Wisconsin forest vegetation on the basis of plant function and gross morphology. Ecology 50: 219-232.

Kopecký, K. & S. Hejný. 1974. A new approach to the classification of anthropogenic plant communities. Vegetatio 29: 17-20.

Kortekaas, W., E. van der Maarel & W.G. Beeftink. 1976. A numerical classification of European Spartina communities. Vegetatio 33: 51-60.

Kruskal, J.B. 1964a. Multidimensional scaling by optimizing goodness of fit to a non-metric hypothesis. Psychometrika 29: 1-27.

Kruskal, J.B. 1964b. Non-metric multidimensional scaling: a numerical method. Psychometrika 29: 115-129.

Kulczynski, S. 1928. Die Pflanzenassoziationen der Pieninen. Bull. Int. Acad. Pol. Sci., cl. Sci. Math. Nat. Ser. B. Suppl. 2: 57-203.

Laan, D. van der. 1971. Some aspects of vegetational and environmental research of the dune slacks of Voorne. Acta Bot. Neerl. 20: 717-718.

Laan, D. van der. 1974. Synecological research of the dune slacks on Voorne; analysis of vegetational and environmental data. In: Progress Report 1973 Institute of Ecological Research. Verh. Kon. Akad. Wetenschappen Afd. Natuurkunde. 2, 63: 93-96.

Laan, D. van der. 1979. Spatial and temporal variation in the vegetation of dune slacks in relation to the ground water regime. Vegetatio 39 (in press).

Lacoste, A. 1976. Relations floristiques entre les groupements prairiaux du Triseto-Polygonion et les Megaphorbiaies (Adenostylion) dans les Alpes occidentales. Vegetatio 31: 161-176.

218

Lacoste, A. & M. Roux. 1971. L'analyse multidimensionelle en phytosociologie et en écologie. Application à des données de l'étage subalpin des Alpes maritimes. 1. L'analyse des données floristiques. Oecol. Plant. 6: 353-369.

Lambert, J.M. & Dale, M.B. 1964. The use of statistics in phytosociology. Adv. Ecol. Res. 2: 59-99.

Lambert, J.M. & W.T. Williams. 1962. Multivariate methods in plant ecology. IV. Nodal Analysis. J. Ecol. 50: 775-802.

Lambert, J.M. & W.T. Williams. 1966. Multivariate methods in plant ecology. VI. Comparison of information analysis and association-analyses. J. Ecol. 54: 635-664.

Lance, G.N. & W.T. Williams. 1968. Note of new information-statistic classificatory program. Comput. J. 11: 195.

Lange, L. de. 1972. An ecological study of ditch vegetation in the Netherlands. Thesis Amsterdam.

Leeuwen, C.G. van. 1966. A relation theoretical approach to pattern and process in vegetation. Wentia 15: 25-46.

Leeuwen, C.G. van. 1974. Ekologie. Collegedictaat TH Delft. Afd. Bouwkunde

Leeuwen, C.G. van. 1977. Rangordebetrekkingen en landschapstechniek. Landbouwk. Tijdschr./pt. 89: 324-328.

Leser, H. 1976. Landschaftsökologie. Ulmer, Stuttgart.

Lieth, H. & G.W. Moore. 1971. Computerized clustering of species in phytosociological tables and its utilization for field work. In: G.P. Patil, E.C. Pielou & W.E. Waters (eds.), Spatial patterns and statistical distributions. Statistical Ecology Vol. 1, p. 403-422.

Londo, G. 1971. Patroon en proces in duinvalleivegetaties langs een gegraven meer in de Kennemerduinen. (with summary) Thesis Nijmegen.

Londo, G. 1974. Successive mapping of dune slack vegetation. Vegetatio 29: 51-61.

Londo, G. 1978. Möglichkeiten zur Anwendung von vegetationskundliche Untersuchungen auf Dauerflächen. Vegetatio 38: 185-190.

Looman, J. 1963. Preliminary classification of grasslands in Saskatchewan. Ecology 44: 15-29.

Looman, J. 1976a. Biological equilibrium in ecosystems. I. A theory of biological equilibrium. Folia Geobot. Phytotax. 11: 1-21.

Looman, J. 1976b. Biological equilibrium in ecosystems. 2. Parameters of ecosystems. Folia Geobot. Phytotax. 11: 113-135.

Looman, J. 1976c. Biological equilibrium in ecosystems. 3. Classification of ecosystems. Folia Geobot. Phytotax. 11: 337-365.

Loucks, O.L. 1962. Ordinating forest communities by means of environmental scalars and phytosociological indices. Ecol. Monogr. 32: 137-166.

Louppen, J.M.W., M.J.A. Werger & J.H.N. Eppink. 1978. Vegetation patterns and species performances along an environmental gradient. Abstracts 2nd. Int. Congr. Ecol. Jerusalem Vol. 1. p. 217.

Ludwig, J.A. & D.W. Goodall. 1978. A comparison of paired- with blocked quadrat variance methods for the analysis of spatial pattern. Vegetatio 38: 49-59.

Maarel, E. van der. 1966a. On vegetational structures, relations, and systems, with special reference to the dune grasslands of Voorne, The Netherlands. Thesis Utrecht.

Maarel, E. van der. 1966b. Dutch studies on coastal sand dune vegetation, especially in the Delta region. Wentia 15: 47-82.

Maarel, E. van der. 1969. On the use of ordination models in phytosociology. Vegetatio 19: 21-46.

Maarel, E. van der. 1970a. Vegetationsstruktur und Minimum-Areal in einem Dünen-trockenrasen. In: R. Tüxen (ed.), Gesellschaftsmorphologie. Ber. Int. Symp. Rinteln 1966, p. 218-239. Junk, Den Haag.

Maarel, E. van der. 1970b. Enkele begrippen en methoden in de kwantitatieve vegetatiesystematiek. Contactbl. Oecol. 6: 89-91.

Maarel, E. van der. 1971. Plant species diversity in relation to management. In: A.S. Watt & Duffey (eds.), The scientific management of animal and plant communities for conservation, p. 45-63. Blackwell, Oxford.

Maarel, E. van der. 1972. Ordination of plant communities on the basis of their plant genus, family and order relationships. In: E. van der Maarel & R. Tüxen (eds.), Grundfragen und Methoden der Pflanzensoziologie. Ber. Int. Symp. Rinteln 1970, p. 183-192. Junk, Den Haag.

219

Maarel, E. van der. 1974. Small-scale vegetational boundaries; on their analysis and typology. In: W.H. Sommer & R. Tüxen (eds.), Tatsachen und Probleme der Grenzen in der Vegetation. Ber. Symp. Int. Ver. Vegetationskunde 1968, p. 75-80. Cramer, Lehre.

Maarel, E. van der. 1975. The Braun-Blanquet approach in perspective. Vegetatio 30: 213-219.

Maarel, E. van der. 1976. On the establishment of plant community boundaries. Ber. Deutsch. Bot. Ges. 89: 415-443.

Maarel, E. van der. 1978. Experimental succession research in a coastal dune grassland. A preliminary report. Vegetatio 38: 21-28.

Maarel, E. van der. 1979. Transformation of cover-abundance values in phytosociology and its effects on community similarity. Vegetatio 39 (in press).

Maarel, E. van der & P.J. Dauvellier. 1978. Naar een globaal ecologisch model voor de ruimtelijke ontwikkeling van Nederland. Studierapp. Rijksplanologische Dienst 9. 2 Vols. (separate summary) Ministerie Volkshuisvesting Ruimtelijke Ordening, Den Haag.

Maarel, E. van der, J.G.M. Janssen & J.M.W. Louppen. 1978. TABORD, a program for structuring phytosociological tables. Vegetatio 38: 143-156.

Maarel, E. van der & J. Leertouwer. 1967. Variation in vegetation and species diversity along a local environmental gradient. Acta Bot. Neerl. 16: 211-221.

Maarel, E. van der & L. Orlóci & S. Pignatti. 1976. Data-processing in phytosociology, retrospect and anticipation. Vegetatio 32: 65-72.

Maarel, E. van der & M.J.A. Werger. 1979. On the treatment of succession data. Phytocoenosis (in press).

Maarel, E. van der & V. Westhoff. 1964. The vegetation of the dunes near Oostvoorne, Netherlands. Wentia 12: 1-61.

Matthews, J.A. 1978. An application of non-metric scaling to the construction of an improved species plexus. J. Ecol. 66: 157-173.

Matuszkiewicz, W. 1948. Roślinnośé lasów okolik Lwowa. (with English summary). Ann. Univ. Mariae Curie-Sklodowska Sect. C 3: 119-193.

May, R.M. 1975. Stability and complexity in model ecosystems. 2nd ed. Princeton University Press, Princeton.

McIntosh, R.P. 1978. Matrix and plexus techniques 2nd ed. In: R.H. Whittaker (ed.), Ordination of plant communities. p. 151-184. Junk, The Hague.

McNaughton, S.J. 1978. Stability and diversity of ecological communities. Nature 274: 251-252.

Meltzer, J. & V. Westhoff. 1942. Inleiding tot de Plantensociologie. Breughel, 's-Graveland.

Meyer, M. 1977. Vergleich verschiedener Chrysopogon gryllus-reicher Trockenwiesen des insubrischen Klimabereiches und angrenzender Gebiete. Vegetatio 35: 107-114.

Meyer Drees, E. 1949. Combined taxation and presence in analysing and comparing association tables. Vegetatio 2: 43-46.

Meyer Drees, E. 1951. Capita selecta from modern plant sociology and a design for rules of phytosociological nomenclature. Rapp. Bosbouwproefstation Bogor 52: 1-68.

Moore, J.J. 1966. Phyto. Re-arranging a phytosociological array according to the principles of Braun-Blanquet. Mimeogr. Paper, Dublin.

Moore, J.J. 1972. An outline of computer-based methods for the analysis of phytosociological data. In: E. van der Maarel & R. Tüxen (eds.), Grundfragen und Methoden in der Pflanzensoziologie. Ber. Int. Symp. Rinteln 1970, p. 29-38. Junk, Den Haag.

Moore, J.J. 1973. PHYTO. A suite of programs in FORTRAN IV for the manipulation of phytosociological tables according to the principles of Braun-Blanquet. Ms Dept. Botany, University College, Dublin.

Moore, J.J., P. Fitzsimons, E. Lambe & J. White. 1970. A comparison and evaluation of some phytosociological techniques. Vegetatio 20: 1-20.

Moore, J.J. & A. O'Sullivan. 1970. A comparison between the results of the Braun-Blanquet method and those of 'cluster-analysis'. In: R. Tüxen (ed.), 'Gesellschaftsmorphologie'. Ber. Int. Symp. Rinteln 1966, p. 26-29. Junk, Den Haag.

Moral, R. del & M.F. Denton. 1977. Analysis and classification of vegetation based on family comparison. Vegetatio 34: 155-165.

Moral, R. del & A.F. Watson. 1978. Gradient structure of forest vegetation in the Central Washington cascades. Vegetatio 38: 29-48.

Moravec, J. 1971. A simple method for estimating homotoneity of sets of phytosociological relevés. Folia Geobot. Phytotax. 6: 147-170.

Moravec, J. 1973. The determination of the minimal-area of phytocoenoses. Folia Geobot. Phytotax. 8: 23-47.

Moravec, J. 1975. Die Anwendung von Stetigkeitsartengruppen zur numerischen Ordnung von pflanzensoziologischen Tabellen. Vegetatio 30: 41-47.

Moravec, J. 1978. Application of constancy-species groups for numerical ordering of phytosociological tables — the synoptic table version. Vegetatio 37: 33-42.

Mueller-Dombois, D. & H. Ellenberg. 1974. Aims and methods of vegetation ecology. Wiley, New York.

Neuhäusl, R. 1977. Delimitation and ranking of floristic-sociological units on the basis of relevé similarity. Vegetatio 35: 115-122.

Nichols, S. 1977. On the interpretation of principal components analysis in ecological contexts. Vegetatio 34: 191-197.

Noordwijk-Puyk, K. van, W.G. Beeftink & P. Hogeweg. 1979. Vegetation development on saltmarsh flats after disappearance of the tidal factor. Vegetatio 39: (in press).

Noy-Meir, I. 1973a. Data transformation in ecological ordination. I. Some advantages of noncentering. J. Ecol. 61: 329-341.

Noy-Meir, I. 1973b. Divisive polythetic classification of vegetation data by optimized division on ordination components. J. Ecol. 61: 753-760.

Noy-Meir, I. 1974. Catenation: quantitative methods for the definition of coenoclines. Vegetatio 29: 89-99.

Noy-Meir, I. & D.J. Anderson. 1970. Multiple pattern analysis, or multiscale ordination, towards a vegetation hologram? Statistical Ecology 3: 207-225.

Noy-Meir, I., D.J. Walker & W.T. Williams. 1975. Data transformations in ecological ordination. II. On the meaning of data standardization. J. Ecol. 63: 779-800.

Noy-Meir, I. & R.H. Whittaker. 1977. Continuous multivariate methods in community analysis: some problems and developments. Vegetatio 33: 79-88.

Noy-Meir, I. & R.H. Whittaker. 1978. Recent developments in continuous multivariate techniques. In: R.H. Whittaker (ed.), Ordination of plant communities, p. 337-378. Junk, The Hague.

Nijland, G.O. & K. Wind. 1973. Een transsectstudie als oefenmodel voor de statische beschrijving van microdistributiepatronen in grasbestanden. Med. Vakgr. Landbouwplantenteelt Graslandcultuur Wageningen 33, 42 pp.

Oberdorfer, E. 1957. Süddeutsche Pflanzengesellschaften. Fischer, Jena.

Olschowy, G. 1975. Ecological landscape inventories and evaluation. Landscape Planning 2: 37-44.

Orlóci, L. 1966. Geometric models in ecology. I. The theory and application of some ordination methods. J. Ecol. 54: 193-215.

Orlóci, L. 1967. An agglomerative method for the classification of plant communities. J. Ecol. 55: 193-206.

Orlóci, L. 1972a. On objective functions of phytosociological resemblance. Amer. Midland Nat. 88: 28-55.

Orlóci, L. 1972b. On information analysis in phytosociology. In: E. van der Maarel & R. Tüxen (eds.), Grundfragen und Methoden in der Pflanzensoziologie. Ber. Int. Symp. Rinteln 1970, p. 75-88. Junk, Den Haag.

Orlóci, L. 1974a. Revisions for the Bray and Curtis ordination. Can. J. Bot. 52: 1773-1776.

Orlóci, L. 1974b. On information flow in ordination. Vegetatio 29: 11-16.

Orlóci, L. 1976. TRGRPS. An interactive algorithm for group recognition with an example from Spartinetea. Vegetatio 32: 117-120.

Orlóci, L. 1978a. Multivariate analysis in vegetation research. 2nd ed. Junk, The Hague-Boston.

Orlóci, L. 1978b. Ordination by resemblance functions. 2nd ed. In: R.H. Whittaker (ed.), Ordination of plant communities. p. 239-275. Junk, The Hague.

Persson, S. 1977. Data program for bearbetning av vegetationsdata. 1. Klassifikationsprogramdokumentation och handhavande. Meddn. Avd. Ekol. Bot. Lunds Univ. 33: 1-68.

Philips, D.L. 1978. Polynomial ordination: field and computer simulation testing of a new method. Vegetatio 37: 129-140.

Pielou, E.C., 1969. Association tests versus homogeneity tests: their use in subdividing quadrats into groups. Vegetatio 18: 4-18.

Pielou, E.C. 1974. Population and community ecology. Gordon and Breach, New York.

Pielou, E.C., 1975. Ecological diversity. Wiley Interscience, New York.

Pignatti, S. 1976. A system for coding plant species for data-processing in phytosociology. Vegetatio 33: 23-32.

Pignatti, S., G. Cristofolini & D. Lausi. 1968. Verwendungsmöglichkeiten einer elektronischen Datenverarbeitungsanlage für die Pflanzensoziologische Dokumentation. Discussion paper 12th Symposium Int. Soc. Vegetation Science Rinteln.

Pignatti, E. & S. Pignatti. 1975. Syntaxonomy of the Sesleria varia-grasslands of the calcareous Alps. Vegetatio 30: 5-14.

Poore, M.E.D. 1955-1956. The use of phytosociological methods in ecological investigations, Parts 1-4. J. Ecol. 43: 226-244, 245-269, 606-51. 44: 28-50.

Ramensky, L.G. 1930. Zur Methodik der vergleichenden Bearbeitung und Ordnung von Pflanzenlisten und anderen Objekten, die durch mehrere, verschiedenartig wirkende Faktoren bestimmt werden. (transl. from the Russian paper in Trudy sov. geobot.-lugov. 15-20 Jan. 1929, p. 11-36.) Beitr. Biol. Cbl. 18: 269-304. Breslau.

Raunkiaer, C. 1934. The life forms of plants and statistical plant geography. Oxford Univ. Press, Oxford.

Rejmánek, M. 1977. The concept of structure on phytosociology, with reference to classification of plant communities. Vegetatio 35: 55-61.

Romane, F. et al. 1972. Un example d'organisation du traitement des observations phyto-écologiques. Programme de phyto-écologie fondamentale et générale: Project écothèque. CNRS-CEPE Louis Emberger, Montpellier.

Romane, F., J.L. Guillerm & G. Waksman. 1977. Une utilisation possible de l'arbre de portée minimale en phyto-écologie. Vegetatio 33: 99-106.

Roskam, E. 1971. Programme ORDINA: Multidimensional ordination of observation vectors. Programme Bull. 16, Psychology Lab., Nijmegen.

Scamoni, A. & H. Passarge. 1963. Einführung in die praktische Vegetationskunde. Fischer, Jena.

Scamoni, A., H. Passarge & G. Hofmann. 1965. Grundlagen zu einer objektiven Systematik der Pflanzengesellschaften. Feddes Rep. Beih. 142: 117-132.

Schaik, C.P. van & P. Hogeweg. 1977. A numerical-syntaxonomical study of the Calthion palustris Tx. 37 in the Netherlands. Vegetatio 35: 65-80.

Schwickerath, M. 1931. Die Gruppenabundanz (Gruppenmächtigkeit); ein Beitrag zur Begriffsbildung der Pflanzensoziologie. Englers Bot. Jahrb. 64: 1-16.

Schwickerath, M. 1940. Die Artmächtigkeit. Fedde Rep. Beih. 121: 48-52.

Segal, S. 1969. Ecological notes on wall vegetation. Thesis Amsterdam. Junk, Den Haag.

Segal, S. 1970. Strukturen und Wasserpflanzen. In: R. Tüxen (ed.), Gesellschaftsmorphologie. Ber. Int. Symp. Rinteln 1966. p. 157-169. Junk, Den Haag.

Segal, S. & V. Westhoff. 1959. Die vegetationskundliche Stellung von Carex buxbaumii in Europa, besonders in den Niederlanden. Acta Bot. Neerl. 8: 304-329.

Sloet van Oldruitenborgh, C.J.M. 1976. Duinstruwelen in het Deltagebied. Thesis Wageningen. Meded. Landbouwhogeschool Wageningen 76 (8): 1-112.

Sloet van Oldruitenborgh, C.J.M. & M.J. Adriani. 1971. On the relation between vegetation and soil-development in dune shrub vegetations. Acta Bot. Neerl. 20: 198-204.

Smart, P.F.M., S.E. Meacock & J.M. Lambert. 1974. Investigations into the properties of quantitative vegetational data. I. Pilot study J. Ecol. 62: 735-759.

Smart, P.F.M., S.E. Meacock & J.M. Lambert. 1976. Investigation into the properties of quantitative vegetational data. II. Further data type comparison. J. Ecol. 64: 41-78.

Sneath, P.H.A. 1957. The application of computers to taxonomy. J. Gen. Microbiol. 17: 201-226.

Sneath, P.H.A. & R.R. Sokal. 1973. Numerical taxonomy. Freeman, San Fransisco.

Sobolev, L.N. & V.D. Utekhin. 1978. Russian (Ramensky) approaches to community systematization. 2nd ed. In: R.H. Whittaker (ed.), Ordination of plant communities, p. 71-97. Junk, The Hague.

Sørensen, Th.A. 1948. A method of establishing groups of equal amplitude in plant sociology based on similarity of species content. Biol. Skr. K. Danske Vidensk. Selsk. 5 (4): 1-34.

Spatz, G. & J. Siegmund. 1973. Eine Methode zur tabellarischen Ordination, Klassifikation und ökologischen Auswertung von pflanzensoziologischen Bestandsaufnahmen durch den Computer. Vegetatio 28: 1-17.

Stanek, W. 1973. A comparison of Braun-Blanquet method with sum-of-squares agglomeration for vegetation classification. Vegetatio 27: 323-338.

Stockinger, F.J. & W.F. Holzner. 1972. Rationele Methode zur Auswertung pflanzensoziologischer Aufnahmen mittels Elektronen rechner. In: E. van der Maarel & R. Tüxen (eds.), Grundfragen und Methoden in der Pflanzensoziologie. Ber. Int. Symp. Rinteln 1970, p. 239-248. Junk, Den Haag.

Swan, J.M.A. 1970. An examination of some ordination problems by use of simulated vegetational data. Ecology 51: 89-102.

Swan, J.M.A., R.L. Dix & C.F. Wehrhahn. 1969. An ordination technique based on the best possible stand-defined axes and its application to vegetational analysis. Ecology 50: 206-212.

Thalen, D.C.P. 1971. Variation in some saltmarsh and dune vegetations in the Netherlands with special reference to gradient situations. Acta Bot. Neerl. 20: 327-342.

Tjallingii, S.R. 1974. Unity and diversity in lanscape. Landscape Planning 1: 7-34.

Trass, H. & N. Malmer. 1978. North European approaches to classification. 2nd ed. In: R.H. Whittaker (ed.), Classification of plant communities, p. 201-245. Junk, The Hague.

Troll, C. 1968. Landschaftsökologie. In: R. Tüxen (ed.), Pflanzensoziologie und Landschaftsökologie. Ber. Int. Symp. Stolzenau 1963, p. 1-21. Junk, Den Haag.

Tuomikoski, R. 1942. Untersuchungen über die Untervegetation der Bruchmoore in Ostfinnland. I. Zur Methodik der pflanzensoziologischen Systematik. Ann. Bot. Soc. Zool. Bot. Fenn. Vanamo. 17 (1): 1-203.

Tüxen, R. (ed.). 1970. Gesellschaftsmorphologie. Ber. Symp. Int. Vegetationskunde Rinteln 1966. Junk, Den Haag.

Tüxen, R. 1977. Zum Problem der Homogenität von Assoziations-Tabellen. Doc. Phytosoc. NS 1: 305-320.

Tüxen, R. & H. Ellenberg. 1937. Der systematische und der ökologische Gruppenwert. Ein Beitrag zur Begriffsbildung und Methodik der Pflanzensoziologie. Mitt. Flor.-Soz. Arbeitsgem. 3: 171-184.

Utekhin, V. 1969. Vegetation changes in the Voronezh Reservation during 1936-1966 (In Russian). Mater. Moskov. Fil Geogr. Obshch. SSSR, Biogeography 3: 15-17.

Vestal, A.G. 1949. Minimum areas for different vegetations. Univ. Illinois Biol. Monogr. 20 (3): 1-129.

Vries, D.M. de. 1926. Het plantendek van de Krimpenerwaard I. Phytosociologische beschouwingen. Ned. Kruidk. Arch. 35: 215-275.

Vries, D.M. de. 1933. De plantensociografische rangorde-methode. Bot. Jaarb. Dodonaea Gent 24: 37-48.

Vries, D.M. de. 1938. The plant sociological combined specific frequency and other methods. Chron. Bot. 4: 115-117.

Vries, D.M. de. 1939. Zusammenarbeit der nördlichen und südlichen Schule ist zum Heil der gesammten Pflanzensoziologie unbedingt erforderlich. Rec. Trav. Bot. Neerl. 36: 485-493.

Vries, D.M. de. 1953. Objective combination of species. Acta Bot. Neerl. 1: 497-499.

Vries, D.M. de, J.P. Baretta & G. Hamming. 1954. Constellation of frequent herbage plants, based on their correlation in occurrence Vegetatio 5/6: 106-111.

Walker, B.H. 1974. The development and use of natural resource data banks in Europe and America: a report on a feasibility study for Rhodesia. Rep. Div. Biol. Sc. University of Rhodesia, Salisbury.

Walker, J., I. Noy-Meir, D.J. Anderson & R.M. Moore. 1972. Multiple pattern analysis of a woodland in south central Queensland. I. The original trees and shrubs. Aust. J. Bot. 20: 105-118.

Werger, M.J.A. 1972. Species-area relationship and plot size: with some examples from South-African vegetation. Bothalia 10: 583-594.

Werger, M.J.A. 1973. On the use of association-analysis and principal component analysis in interpreting a Braun-Blanquet phytosociological table of a Dutch grassland. Vegetatio 28: 129-144.

223

Werger, M.J.A. 1974a. On concepts and techniques applied in the Zürich-Montpellier method of vegetation survey. Bothalia 11: 309-323.
Werger, M.J.A. 1974b. The place of the Zürich-Montpellier method in vegatation science. Folia Geobot. Phytotax. 9: 99-109.
Werger, M.J.A. 1978. Vegetation structure in the southern Kalahari. J. Ecol. 66: 933-941.
Werger, M.J.A., J.W. Morris & J.M.W. Louppen. 1979. Vegetation-soil relationships in the southern Kalahari. Doc. Phytosoc (in press).
Werger, M.J.A., P.J.A.M. Smeets, H.P.G. Helsper & V. Westhoff. 1978. Ökologie der Subalpinen Vegetation des Lausbachtales, Tirol. Verh. Zool. Bot. Ges. Wien 117:
Werger, M.J.A., H. Wild & B.R. Drummond. 1978. Vegetation structure and substrate of the Northern part of the Great Dyke, Rhodesia: Gradient analysis and dominance-diversity relationships. Vegetatio 37: 151-161.
Werkgroep GRAN. 1973. Biologische kartering en evaluatie van de groene ruimte in het gebied van de stadsgewesten Arnhem en Nijmegen. Rapport Afd. Geobotanie, Nijmegen.
Westhoff, V. 1947. The vegetation of dunes and salt marshes on the Dutch islands of Terschelling, Vlieland and Texel. Thesis Utrecht.
Westhoff, V. 1951. An analysis of some concepts and terms in vegetation study or phytocenology. Synthese (Bussum) 8: 194-206.
Westhoff, V. 1954. Cline. Veenmans Agrarische Winkler Prins I, Wageningen.
Westhoff, V. 1967. Problems and use of structure in the vegetation. The diagnostic evaluation of structure in the Braun-Blanquet system. Acta Bot. Neerl. 15: 495-511.
Westhoff, V. 1969. Verandering en duur. Oratie Katholieke Universiteit Nijmegen. Junk, Den Haag.
Westhoff, V. & A.J. den Held. 1975. Plantengemeenschappen in Nederland. $2^e$ dr. Thieme, Zutphen.
Westhoff, V. & C.G. van Leeuwen. 1966. Oekologische und systematische Beziehungen zwischen natürlicher und anthropogener vegetation. In: R. Tüxen (ed.), Anthropogene vegetation Ber. Int. Symp. Stolzenau 1961, p. 156-172. Junk, Den Haag.
Westhoff, V. & E. van der Maarel. 1978. The Braun-Blanquet approach. 2nd ed. In: R.H. Whittaker (ed.), Classification of plant communities, p. 287-399. Junk, The Hague.
Whittaker, R.H. 1960. Vegetation of the Siskiyou Mountains, Oregon and California. Ecol. Mon. 30: 279-338.
Whittaker, R.H. 1967. Gradient analysis of vegetation. Biol. Rev. 42: 207-264.
Whittaker, R.H. 1970. The population structure of vegetation. In: R. Tüxen (ed.), Gesellschaftsmorphologie. Ber. Int. Symp. Rinteln 1966, p. 39-59. Junk, Den Haag.
Whittaker, R.H. 1972. Convergences of ordination and classification. In: E. van der Maarel & R. Tüxen (eds.), Grundfragen und Methoden der Pflanzensoziologie. Ber. Int. Symp. Rinteln 1970. pp. 39-57. Junk, Den Haag.
Whittaker, R.H. 1975. Communities and Ecosystems. 2nd Ed. Mac Millan, New York.
Whittaker, R.H. (ed.). 1978a. Ordination of plant communities. Junk, The Hague.
Whittaker, R.H. (ed.). 1978b. Classification of plant communities. Junk, The Hague.
Whittaker, R.H. 1978c. Approaches to classifying vegetation 2nd. In: R.H. Whittaker (ed.), Classification of plant communities. pp. 1-31. Junk, The Hague.
Whittaker, R.H. 1978d. Direct gradient analysis. 2nd. ed. In: R.H. Whittaker (ed.), Ordination of plant communities. pp. 7-50. Junk, The Hague.
Whittaker, R.H. & H.G. Gauch, Jr. 1978. Evaluation of ordination techniques. 2nd. ed. In: R.H. Whittaker (ed.), Ordination of plant communities. pp. 277-336. Junk, The Hague.
Williams, C.B. 1964. Patterns in the balance of nature and related problems in quantitative ecology. Academic Press, London-New York.
Williams, W.T. 1971. Principles of clustering. Ann. Rev. Ecol. Syst. 2: 303-326.
Williams, W.T. (ed.). 1976. Patterns analysis in agricultural science. CSIRO, Melbourne; Elsevier, Amsterdam.
Williams, W.T. & M.B. Dale. 1965. Fundamental problems in numerical taxonomy. Adv. Bot. Res. 2: 35-68.
Williams, W.T. & J.M. Lambert. 1959. Multivariate methode in plant ecology. I. Association-analysis in plant communities. J. Ecol. 47: 83-101.
Williams, W.T., J.M. Lambert & G.N. Lance. 1966. Multivariate methods in plant ecology. V. Similarity analyses and information analyses. J. Ecol. 54: 427-445.

Williams, W.T., G.N. Lance, L.J. Webb., J.G. Tracey & M.B. Dale. 1969. Studies in the numerical analysis of complex rainforest communities. III. The analysis of successional data. J.
Ecol. 57: 515-536.
Wishart, D. 1969. CLUSTAN Ia. User manual. St. Andrews Computing Laboratory, St. Andrews.
Wishart, D. 1975. CLUSTAN 1c User manual. Computer Centre University College, London.
Yarranton, G.A., W.J. Beasleigh, R.G. Morrison & M.I. Shafi. 1972. On the classification of
phytosociological data into nonexclusive groups with a conjecture about determining the
optimum number of groups in a classification. Vegetatio 24: 1-12.
Zonneveld, I.S. 1972. Land evaluation and land(scape) science. ITC Textbook of Photo-Inter-
pretation Vol. VII, Ch. VII-4. ITC, Enschede.

# 7. THE DEVELOPMENT OF PALYNOLOGY IN RELATION TO VEGETATION SCIENCE, ESPECIALLY IN THE NETHERLANDS

C.R. JANSSEN

# 7. THE DEVELOPMENT OF PALYNOLOGY IN RELATION TO VEGETATION SCIENCE, ESPECIALLY IN THE NETHERLANDS

C.R. JANSSEN

## 1. Introduction

From the way vegetation science developed in central Europe one unconsciously may get the impression that vegetation types are rather static phenomena. Of course this is not true: one of the important topics of vegetation study deals with vegetational change with time.

According to Westhoff (1970), these chronological aspects of vegetation study are threefold: (1) syndynamics or succession study is the science studying the development of (concrete) phytocoenoses from other phytocoenoses; (2) synepiontology is the science studying the development of vegetation as types (coenon) in the course of history; and (3) synchronology is the science studying the historical development of vegetation with palynological methods. In my opinion definition (3) is not on equal footing with the others, since palynological methods can be applied to the first two aspects as well. But it is not my aim to do hairsplitting over definitions. At any rate Westhoff distinguishes between the development of coena and coenoses and, although not explicitly stated, between successions over short periods of time (aspect (1)) and successions that deal with historic aspects and may last a long time (aspects 2, 3). Likewise, Major (1974), reviewing the concepts of kinds of changes in connection with duration, separates in a detailed way short-lived changes from the long to very long lasting developments in the vegetation.

Without going into the basic principles that underlie these kinds of changes, there is certainly an important methodological difference. Changes over short time intervals, more or less within the lifespan of man, can be studied by means of methods familiar to present day vegetation science. These changes can be followed in permanent quadrats from year to year on the basis of the complete species composition and structure of the vegetation. When dealing with long lasting changes however, the usual methods developed in vegetation science fall considerably short of our aims and we have to resort to methods that fall outside those of vegetation science properly. These methods have many concepts in common with the geological sciences, also studying biotic developments over long periods of time, especially those that deal with the development of life on earth: palaeontology and stratigraphy. In the context of this paper palaeontology will be narrowed down to palaeobotany, since palaeontology has, historically, mainly a zoological connotation. Furthermore, in this paper, only palynology will be discussed.

## 2. Scope of palynology

The term palynology was coined by Hyde and Williams in 1944 from the greek word 'Palunein' meaning dispersal; thus literally taken palynology is the science dealing with small particles that in nature are dispersed by either animals or by other agents such as water and wind, i.e. mainly pollen and spores, although sometimes other microorganisms are included.

Palynology today is a very diverse branch of science whose representatives seem to have only the object of study and a number of methods in common. It encompasses fields such as plant systematics, geological stratigraphy, honey research, chemistry, ontogeny, medicine, and morphology, to name a few.

A review of palynology in all these aspects is given in Janssen (1974). In this paper the relation with vegetation will be stressed and problems will be discussed that relate to what before World War-Two was called pollen analysis.

## 3. Origin of palynology

Although descriptions of pollen grains were published already in the late 18th century (cf. Jonker 1967), palynology arose in northern Europe at the end of the last century and in the beginning of this century in connection with research in peat bogs. Weber (1893) was the first to recognize pollen grains in peat deposits, whereas Lagerheim (1902) established pollen analysis as a quantitative method by counting pollen and by the calculation of their relative amounts. Thus, at a very early stage palynology was a quantitative method, taking advantage of the fact that pollen and spores, unlike many other plant remains are produced in huge quantities, dispersed, deposited and preserved in suitable habitats.

In 1916 the Swedish geologist L. von Post developed the so-called pollen diagram that showed on a relative time scale a reflection of forest history expressed in percentages of the various pollen types. Von Post was able to show, that going from south to central Sweden, the characteristics of the pollencurves changed not randomly, but according to trends that could be explained in a meaningful way (Fries 1967).

Pollen grains are laterally transported and this phenomenon has always been the weakness and at the same time the power of palynology. It is a weakness, because particles of different plant communities are mixed; it is a power because vegetational events in one place have a reflection somewhere else.

The recognition of similar events in pollen diagrams at various localities allows the establishment of synchronous levels. Von Post showed here the principle that by linking pollen diagrams along a line, a relative chronology could be established and that one could get a rough idea from where the various pollen types came from. Indeed, interpretation of pollen assemblages in terms of vegetation cannot been done unless there is some notion about the source area of the various pollen types. It may not be by accident that it had to be a geologist to show this principle. It is striking that in the following decades these principles were all to often ignored.

230

Since its invention the pollen diagram has proved to be the vehicle 'par excellence' to convey changes in the plant cover; it has become almost a sort of palynological esperanto. This is largely because of a rather remarkable high degree of standardization, which facilitates communication among palynologists. When studying a palynological paper, a palynologist will be inclined to study the diagram prior to the text. Thus even when a paper is in Russian, Japanese or, perhaps in the future in Arabic, it is still useful because of the pollen diagram.

Changes of vegetation over long periods of time depend on a plethora of environmental factors including man, climate, soil formation, etc. The success of pollen analysis is largely due to the fact that palynological results bear a strong relationship to other scientific disciplines, for which palynology is a method to solve essentially non-palynological problems. Sometimes it is used to such an extent that pollen values are treated not for what they are, but as indicators of palaeo-environmental parameters! The objects of study, however, are parts of plants and I fully agree with Faegri & Iversen (1964) that unless this botanical background is not well appreciated, conclusions may be very misleading. I shall return to this subject later.

## 4. Development of palynology

### 4.1 *The period before World War-Two*

In the early days, between 1916 and 1930, palynology was dominated by Scandinavians and Germans. As far as I know from the literature Weber recognized pollen in Dutch deposits (Kraantje Lek, Haarlem) already in 1910; the first counts from Dutch materials were carried out by him in his home in 1923 (van Baren 1927). No percentages were calculated and no pollen diagram was drawn up. At about the same time Erdtman, a pupil of Von Post began, after the completion of his Ph.D. thesis (Erdtman 1921), which dealt with South-West Sweden, to travel around the world, coring peatbogs including a core from the Peel, Limburg, in the Netherlands. And so the first published pollen diagram from a Dutch deposit was from the hand of a foreigner (Erdtman 1928; Fig. 1).

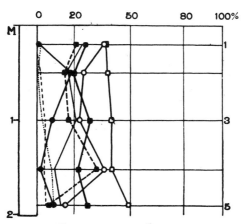

Fig. 1. The first Dutch pollen diagram, 'de Peel'. (Erdman 1928).

In the Netherlands two centres arose contemporaneously. In Amsterdam research in the western part of the Netherlands was carried out. In 1929 B. Polak published her thesis on the development of peatbogs in the western part of Holland, in which the presence and absence of certain pollen types was an important basis for interpretation. However no quantitative analyses were carried out.

A second thesis, that of Vermeer Louman (1934) formed temporarily an end of palynological research in Amsterdam.

A second palynological centre arose at Utrecht where F. Florschütz began research in the eastern part of the Netherlands in the late twenties and early thirties. This continued until a few years after World War-Two and resulted in a large number of publications from the northern, eastern and southern parts of the country.

When the Commission 'voor Biosociologie en het veenonderzoek van Nederland' was established in 1933 and began to organise the 'dagen voor Biosociologie en Palaeobotanie van het Holoceen' (after 1939 Holoceen was abolished, since older periods were dealt with as well; see Vroman, this volume) the Dutch palynologists took an active part in the Commission as well as in the deliberations at the meetings. Until 1945 Florschütz, Jonker, v. Oye, Polak, and Wassink were members of the Commission and took care of 21 talks (of which Florschütz delivered 8).

It is striking that at that time the marriage between vegetation science and palynology was considered a matter beyond discussion. The report of the first 'Nederlandse dag voor phytosociologie en palaeobotanie van het Holoceen' states that: 'the development of new concepts and methods for the study of the social relationships in the plant cover have initiated renewed field research. This applies also for the search of organic sediments'. Certainly there was in both fields a renewed interest in field work but although both dealing with vegetation, phytosociology and palynology were operating on different levels.

4.2. *Differences between vegetation science and palynology*

Although both in vegetation science and palynology assemblages of plant species are considered, palynology was compared with present day vegetation science in a disadvantageous position. The basis of phytosociology as vegetation science was called in the thirties, was the relevée in which taxa could be recorded down to species level, even to lower rank, because complete plants are available for identification. Moreover, phytosociology developed at a time in the Netherlands when the flora was, after more than 150 years of research in morphology and taxonomy, comparatively well known. Palynology came into existence at a time when pollenmorphology was virtually non-existent. The development of pollen analysis was very quick and soon began to outpace the more slowly developing pollen morphology. Palynology thus was, initially at least, based upon the few tree species whose pollen grains could be recognized. Vegetation history was in the thirties essentially forest history on a broad plant geographical scale.

Plant geography in these days excited perhaps more people interested in vegetation than it does today. J. Heymans for instance developed in those years the concept of accessibility which is of cours of prime importance to explain the distribution patterns of taxa.

The fact remains however that many plant associations were defined on the basis of the (often rare) herb species whereas palynologists could only consider the fate of tree species in the course of time, either in terms of the geography of single species and the development of their distributional areas, or, when some basic relationships between pollen assemblages and vegetation became known (Aario 1940, Firbas 1934), in terms of vegetation on a very broad scale viz. that of formations of very large areas.

The first pollen types of herbs that could be recognized were those of *Cyperaceae*, *Ericaceae*, *Poaceae* and the spores of *Sphagnum* and ferns. The remaining body of pollen types, often in considerable numbers, were listed as 'Varia' in the pollen diagrams (Fig. 2). But slowly the number of pollen types that could be ascribed to known plant taxa increased, often accompanied by much confusion. For some time, for instance, pollen of *Artemisia* was thought to be that of *Salix*.

An important step in the study of the morphology of pollen grains was marked by the invention of the acetolysis method (Erdtman 1934). Treatment with the acetolysis mixture gives pollen its familiar brown colour, facilitating the study of fine morphological detail.

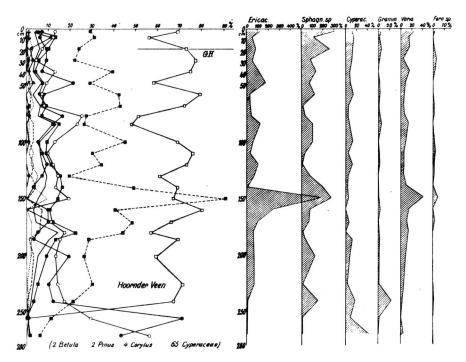

Fig. 2.   A pollen diagram from pre-World War-Two. Westerwolde. (Eshuis 1936).

233

## 4.3 Vegetation types within the formation

Although the relation between pollen assemblages and plant formations became gradually known, it was of course evident that within an area of a formation different vegetation types could be recognized, often dependent on the topography of the region.

In forested regions the difference was very striking between forest and the (often) treeless bogs, from where the samples for pollen analysis came from.

The early palynologist was aware of the fact that pollen types of herbs must be considered to originate from the local peat bog vegetation. Accordingly the pollen sum, the basis for calculation of percentages comprised tree pollen types only, even only those types that were assumed to originate from the upper forest layer. A calculation of percentages upon a basis that does not include all types is for statistical reasons a doubtful procedure. Also by handling the data this way the palynologist is pushing the data into a synecological mould that has to be proved. The alternative, namely no special pollen sum but calculation upon a basis including everything, as is done by some palynologists, for instance in Leuven, is, however, less attractive. Pollen diagrams are clearer when the local, often irregularly fluctuating pollen types are removed from the pollen sum.

Palynologists thus recognized that the plant cover of an area is often not homogeneous but can be delimited in areas covered by different vegetation types.

The establishment of a pollen sum, with the exclusion of others is in fact an indication that if the change in relative values in pollen diagrams are to be a reflexion of plant succession, only those elements must be considered that are actually competing. In applying a tree-pollen sum palynologists were saying in point of fact: 'we are interested in forest successions and inclusion in the basis of calculation of pollen and spores from bog species that do not compete with forest trees, because they are separated in space, 'distorts' the picture of forest succession that is reflected in the tree-pollen assemblages'.

For a long time the difference between bog and forest was the only distinction that had an impact on the pollen diagram. Although Tüxen already pointed out in 1931 that in the study of vegetation development successions on the valley floor, on the slopes, and on the plateau must be considered separately, this was not the time in palynology to put this in praxis, simply because of the poor resolution in pollen types.

In conclusion, the pollen sum thus is an expression of an assumed synecological identity. Its changing composition throughout the palynological literature is the tell-tale story of the changes in concepts of vegetation that any palynologist who does not follow dogmatic fashion, has in mind.

Of course comparison of pollen diagrams will be obstructed when pollen sums vary too much (Jonker 1952). But standardization of the diagrams would hamper ecological interpretation greatly. Ideally therefore, any pollen paper must include diagrams with standard pollen sums next to diagrams adapted to special problems.

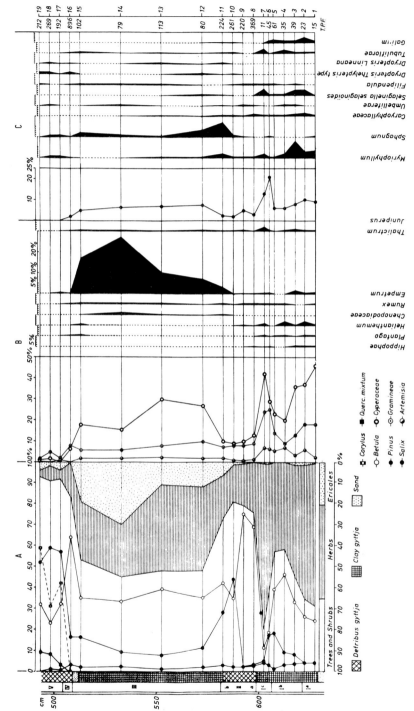

Fig. 3. The first 'Iversen' diagram from a Dutch deposit. (Van der Hammen's Hijkermeer, 1948, simplified).

235

An important development, initially largely unnoticed because of the war, took place in the late thirties and early fourties in Denmark where in connection with prehistorical research the identification of herb pollen types was developed far beyond the usual five plus 'varia'.

Already in 1935, Firbas had pointed out that the percentages of pollen from herbs could be an indication of the extent of forest of the landscape. The presence of many lakes in once glaciated Denmark prompted pollen analysis of lacustrine deposits. No longer did the majority of pollen from herbs originate from peatbogs. The consequence was the so-called 'Iversen Pollen Sum' (Iversen 1947), in which pollen of terrestrial trees and herbs together constituted the basis for calculation of the pollen percentages and in which the proportion of tree pollen and herb pollen was a rough measure of the degree of forest cover of an area.

When in 1948 Florschütz was appointed at the Leyden university, one of his students, van der Hammen, went to Denmark and in 1949 the first 'modern' pollen-diagram from a Dutch deposit was published (van der Hammen 1949; Fig. 3). It covered the lateglacial period in the northern Netherlands showing the two oscillations that were also present in Danish pollen diagrams. Next to the tree pollentypes, 19 herb and shrub types were shown in the diagram. The diagram was interpreted in terms of plant formations.

More than anybody else Iversen has established what may be called 'ecological palynology'. Foremost in his mind was the ecology of single species and the response of these species to environmental factors like light, snow cover, wind, etc. contrasting with, at that time, the more stratigraphical, plant distributional ('geological') character of palynology. The influence of Iversen is large. Although the Iversen diagram was strictly meant for lake sediments, it became gradually applied to other deposits. Today the majority of pollen diagrams is essentially styled in the fashion of Iversen. It is indeed difficult to imagine how the rapid increase of knowledge of vegetational and climatic events in the Quaternary in which Dutch scientists contributed considerably (Zagwijn 1960, 1961, 1973, van der Hammen et al. 1971) could have happened without the availability of this tool.

After 1950 palynology has expanded rapidly. It has been introduced in many parts of the world and it has been applied to almost any period of earth history. It is not the object of this paper to trace closely the developments of palynology in the various parts of the world. But if we look at the postwar period, then we note among the most important developments:

### 4.4.1 Radiocarbon dating

Before the invention of the radiocarbon dating method palynology was looked upon as an important means of dating. It has however not always been appreciated that because of migration of species, similar features in pollen diagrams could be of different age. The tremendous increase in the amounts of radiocarbon dates has shown that over large areas similar assemblages zones are often not synchronous. Indeed, synchroneity of events must be beyond any doubt if we will ever be able to show spatial differences in vegetation in a particular cross-section in time.

Because radiocarbon dating is only reliable up to ca 60,000 years B.P. palynology of the Late-Quaternary is in a much more favourable position than that of earlier times, where dating can be done in a relative way only.

### 4.4.2  Stratigraphical concepts

Of particular interest is the application of stratigraphical procedures (Cushing 1967) that geologists have developed in the fifties and sixties. Pollen diagrams can be zoned according to units based on rock, fossils, or time. The separation of these concepts in a formal way has been helpful in avoiding the confusions that existed in the comparison of the various zone systems throughout the world.

### 4.4.3  Incorporation of palynology in palaeo-ecology

In the reconstruction of past vegetation and the past environment palynology is but one approach. More and more palynology has become part of the highly interdisciplinary science of palaeo-ecology, the science that aims at the reconstruction of the total past environment. The emphasis on the total ecosystem has been accompanied by renewed study of other microfossils and of macrofossils. Thus in the last ten or twenty years pollen analysis has been complemented in many cases by analysis of other plant remains such as seeds, fruits, wood, leaves, etc. by the analysis of remains of animal life, and also by the study of geomorphological and pedological processes.

This multidisciplinary approach is especially noticeable in prehistoric research where in the so-called branch of 'ecological prehistory', palynology, however important it may be, is but part of an approach in which many other disciplines play their part. In the Netherlands palynology as a tool of archaeological research was introduced and developed by Waterbolk (1953) and van Zeist (1959, 1967). It is now an indispensable part of archaeology, being done at several archaeological institutes (see page 242).

Palaeo-ecology, literally 'ancient ecology', or ecology of the past, would imply perhaps similarity in approach to the problems. To the contrary, the approach in palaeo-ecology is quite different from that of present day ecology. In (syn)ecology* the relation between vegetation and environment is established by measuring all the attributes of vegetation and those of the environment, that do not include those of vegetation, followed by a determination of their relationships. In palaeo-ecology the approach is the other way around: the determination of the palaeo-environment is the major aim. It is reconstructed by studying the usually very imperfect remains

---

* The concept of synecology is used here in the usual anglo-american sense of the interrelations of species, not that of the ecology of discrete communities.

(transport and preservation are the main problems here) of a former vegetation and by the application of present day ecological relationships (Fig. 4). In this scheme knowledge of present day ecology is necessary to reconstruct the environment of the past. The present is the key to the past (Cain 1944).

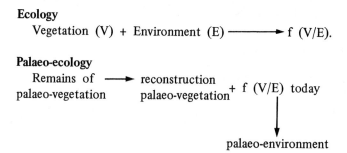

**Ecology**
Vegetation (V) + Environment (E) ⟶ f (V/E).

**Palaeo-ecology**
Remains of ⟶ reconstruction
palaeo-vegetation     palaeo-vegetation + f (V/E) today

palaeo-environment

Fig. 4. Differences in approach in ecology and palaeo-ecology.

I especially stress the necessity to recontruct former plant life before arriving at palaeo-environmental conclusions. Still, palynologists exist who treat pollen as independent parameters detached from the notion that they constitute parts of plants. It is especially in the application of numerical methods that people (often non-botanists) tend to forget that plant species are indicative of the environment and not pollen and spores as such. We agree fully with Faegri & Iversen (1964, p. 139): 'Pollen curves should not be regarded as phenomena per se, as a kind of index fossils, and then used independently of their botanical background. Such a procedure is bound, sooner or later, to lead to incongruous results'.

4.4.4 Conversion of pollen values into vegetation values

Throughout the history of palynology people have been aware of the fact that the pollen values of the dominant species in the vegetation do not bear a direct relationship with vegetational parameters because of strong differences in the production and dispersal of the various pollen types. There have been several attempts to determine conversion factors, but although the differences in pollen production of the various species could be quantified, the differences in dispersal escaped quantification because of the ever changing mosaic of vegetation types in the landscape. However, the use of conversion factors for local studies, where the distance of dispersal has been reduced to zero, looks more promising (Andersen 1970). Iversen (1969) has used Andersen's correction factors succesfully in his studies of the local vegetation at Draved forest in southern Jutland.

### 4.4.5  Spread of palynological studies to other parent materials

Palynology arose as a side branch of peat research. Today palynology has expanded to include almost any type of rock both marine and terrestrial.

Lake sediments were increasingly used as source of information after World War-Two. The special problem here is redeposition of material (Davis 1968). Soil pollen analysis was introduced in the Netherlands by Havinga (1962). The special problem here is percolation and preservation.

Marine deposits are most often used in pre-Quaternary palynology. The special problems here deal with transport by water.

### 4.4.6  Pollen influx diagrams

In the vast majority of pollen diagrams, the diagrams are relative, that is to say that the pollen values are expressed as percentages, not as numbers per volume sediment. In this way they certainly give an impression of competition in the vegetation: when the values of one type increase, the values of all the other types necessarily must decrease. However, the total output of pollen may vary considerably and the trends of the pollen curves based upon percentages may be quite misleading.

For a long time only pollen diagrams showing relative amounts of pollen could be prepared since no check was possible upon the time required for the formation of a certain amount of sediment. The easy availability of radiocarbon dates in the last ten years has changed this. The first pollen influx diagram (APF diagram) was produced by Davis (1967) in the USA, showing the total pollen influx per year per unit surface.

Recent accumulation data under various conditions are necessary, just like surface samples for the comparison of relative pollen data. APF diagrams therefore have incited an interest in pollen trapping and pollen dispersal studies on and in lakes (Tauber 1977, Peck 1973). The majority of pollen diagrams is, however, still of the relative kind, mostly the 'Iversen' version.

Essentially APF diagrams are only reliable for lake sediments where a more or less continuous sedimentation may be expected, without gaps in the record. Even then, the APF diagram can be quite irregular and seems to be most powerful for correction of overall trends in the pollen curves.

The full relative pollen diagram is still the most suitable tool to show minor vegetational fluctuations for which time control is difficult, unless the number of radiocarbon dates is increased beyond the financial means of most of the palynological institutes.

### 4.4.7  Expansion of pre-Quaternary palynology

Pre-Quaternary palynology, which developed largely independently from the Quaternary tradition, has experienced an enormous expansion after World War-Two, especially in connection with oil research. For a long time it was mostly qualitative, stratigraphic geology. Still stratigraphy is the most important aspect in pre-Quaternary palynology. But in recent years the quantitative aspects and phytogeographical and ecological interpretations have all become equally important approaches. The interpretations are often hindered by the long periods of time studied, by the fact that many taxa are extinct and that the distribution of habitats (even continents) cannot be compared with those of today. It is to be expected that pre-Quaternary palynology will benefit from experiences gained in the more recent past of the Quaternary.

### 4.4.8  Pollen deposition versus vegetation

Recent surface sample studies in many parts of the world have increased our knowledge on the relation between pollen assemblage and vegetation. This has been done on a regional scale to aid in the interpretation of pollen assemblages in terms of plant formations over large areas or on a more detailed local scale to interpret succession in single stands.

Regional pollen assemblages
The pioneer work of Aario (1940) has been complemented by many recent surface sample studies all over the world.

For most of the Quaternary palaeo-environmental interpretations are based upon regional interpretations of the pollen assemblages. For instance the palaeo-temperature curve of van der Hammen et al. (1967) is based mainly upon present day relationships between the average July temperature and plant formations like tundra, shrub tundra, and forest.

Although not explicitly expressed, the notion prevails in interpretations like these that formations are in equilibrium with the macroclimate, and thus are essentially climax vegetations. This is most clearly shown in a recent palynological study from the centre of climax theories, Quebec (Richard 1976). The principal plant formations along the north-south gradient in Quebec are called 'domaines climaciques' and the reconstruction of the palaeo-climate is treated accordingly.

However, we doubt today whether climax vegetations are realities. The time needed for their formation is too short and threshold and buffers in the environment, migration of species, competition, and many other factors spoil the climax fun.

Local pollen assemblages
The local pollen assemblages contain many pollen types from insectiferous species that are absent in the regional pollen deposition. Surface sample studies (Janssen 1973) have shown that these assemblages are indicative of local stands to such an extent that a rather detailed picture of the species composition can be obtained. To benefit from this information the pollen types must be identified to the lowest possible taxonomic level. Fortunately our knowledge of pollen types has been gradually widened after World War-Two, especially by Faegri & Iversen and by

Erdtman et al. (1961, 1963). At present a north-west European pollen flora (Janssen et al. 1974) is under way that plans to cover the total present flora of western and north-western Europe.

A drawback of the increase in types of identifiable pollen is that it makes the pollen diagrams unwieldy and unreadable. To avoid this (and often also the considerable publication expenses), a lot of the minor types, often not discussed in the paper, are sometimes not incorporated in the diagram, but listed on the last pages of the publication. I do not like this procedure because it makes the considerable information stored in these pollentypes less accessible. The readability of large diagrams can be improved when the pollen types are placed in some order. All too often there is no apparent order in the sequence of pollentypes, even not alphabetically. Clearly pollen types must be arranged according to a sequence that contributes to the solution of the problem at which the paper aims. When reconstruction of the vegetation is the aim, then pollen types may be arranged synecologically according to the present day affinities of the species involved. One should bear in mind however that in this way only the presence or absence of present day plant communities is shown, not those of communities with a species composition unknown today.

It is indeed well documented that species can be differently associated from one region to another because of different competition and differences in climate and flora. When this is true in space, it must be also true in time. Moreover, the more we go back in time the more there is the chance that we are dealing with different ecotypes.

Fortunately pollen assemblages can speak their own language: when it proves to be possible to separate allochthonous pollen from autochthonous pollen, then the latter can be arranged stratigraphically to visualize succession and species composition of past local stands (Janssen 1972).

## 5. Foundation of Palynological Circle

Many of the developments in palynology briefly described above are reflected in the reports of the 'Dagen voor Biosociologie en Palaeobotanie' of the Commission for 'Biosociologie en het veenonderzoek' of the Royal Botanical Society of the Netherlands (see Vroman, this volume). Palynologists regularly took an active part in the board of this commission, also following its name change in 1965.

In the midsixties palynology had moved much beyond the original peat research stage and although palynology, when dealing with spatial and temporal relationships between plants, is part of (micro)palaeobotany, just like the study of macrofossils, the word palaeobotany usually conveys a taxonomic aim. The simple name of the Commission since 1965 covers both researches of the present and the past and thus combines both recent vegetation research and many aspects of palynology. It was perhaps felt that although the methodologies of the study of past and recent vegetation are quite different, there was no neccessity to stress this, but in stead, to emphasis their common interests.

On the other hand, palynologists share a number of problems that are unfamiliar to the average vegetation scientist. In the midsixties palynologists felt that an own organization where 'palaeo' aspects could be discussed would be a useful instrument for the advancement of palynology. And so, October 4th, 1968 the Palynological Circle was established under the umbrella of the Royal Geological and Mining Society of the Netherlands. Today many of the discussions concerning palynology in all its interdisciplinary aspects are being held at the platform of the Palynological Circle.

The emergence of the Palynological Circle had a considerable impact on the number of talks by palaeo-oriented scientists at the meetings of the Commission. If we count the number of talks in the palynological or palaeo-ecological realm since 1933, we arrive at the following relative figures. 1933-1940: 25%, 1940-1950: 24%, 1950-1960: 17%, 1960-1970: 17%, 1970-mid 1978: 9%. Thus up to 1970 there is a slow relative decline. This does not reflect an absolute decrease (the absolute numbers are resp. 11, 22, 19, 23), but it merely indicates that other fields in vegetation science are gaining ground. But after 1970 there is a sharp decrease. After 1974 to mid-1978 there was only one 'palaeo-talk', no doubt the effect of the Palynological Circle.

## 6. Palynological centres in the Netherlands

After World War-Two the two centres Amsterdam and Utrecht were augmented by new ones in Groningen and Leyden, later also in Wageningen and Nijmegen.

Today palynology in the Netherlands has many aspects and is being done by representatives of many scientific fields, viz. by botanists, archaeologists, geologists, and geographers.

It would carry us too far to describe the origin and development of the palynological centres in the country. We suffice with a list of institutes, persons, and current research. Not mentioned is research in vegetation history in one way or another, since this is done in almost all the departments.

*Amersfoort*

(1) *State service for Archaeological Investigations in the Netherlands*: Postglacial palynology in connection with archaeology in the Netherlands (J. Buurman).

*Amsterdam*

(2) *Dept. of Palynology and Palaeo-ecology, University of Amsterdam*: Quaternary palynology of northern South America, the eastern Mediterreanean, and the Netherlands (T. van der Hammen, T.A. Wymstra, B. van Geel).
(3) *Dept. of Environmental Archaeology, University of Amsterdam*: Palynology and palaeobotany in connection with archaeology in the Netherlands, north-west Germany, Belgium, and Ireland (W. Groenman-van Waateringhe, J.P. Pals, M.J. Jansma, W. Voorrips).
(4) *Dept. of Physical Geography and Soil science, University of Amsterdam*: Palynology in connection with geomorphology in the Netherlands and Luxembourg (R.T. Slotboom).
(5) *Institute of Earth Science, Free University of Amsterdam*: Late-Quaternary palynology in connection with geomorphology and Quaternary geology in coastal northern France and the northern Netherlands (P. Cleveringa).

242

*Haarlem*

(6) *Dept. of Palaeobotany, Geological Survey of the Netherlands*: Caenozoic palynology and palaeobotany of the Netherlands and the adjacent continental shelf (W.H. Zagwijn, J. de Jong); Mesozoic palynology of the Netherlands, West Africa, and northern Brasil (G.F.W. Herngreen).

*Groningen*

(7) *Biological Archaeological Institute, University of Groningen:* Palynology and Palaeobotany in connection with archaeology in the northern Netherlands, south-east Europe, Carthago, the Near East, and Indonesia (W. van Zeist, W.A. Casparie, S. Bottema).

*Leyden*

(8) *State-Herbarium, University of Leyden*: Pollenmorphology of tropical families, mainly Malaysian, and palynology of the Cretaceous and Tertiary in the Malaysian Archipelago (J. Muller).
(9) *Institute of Prehistory, University of Leyden*: Postglacial palynology in connection with archaeology in the western and southern Netherlands and Bavaria (C.C. Bakels).
(10) *Dept. of Palynology and Palaeobotany, National Museum of Geology and Mineralogy*: Quaternary palynology and palaeobotany of the Netherlands, Belgium, and South Africa (H.J.W.G. Schalken).

*Nijmegen*

(11) *Dept. of Biogeology, section Biology, University of Nijmegen*: Late Quaternary palynology around Nijmegen, northern Limburg and adjacent Germany in connection with geomorphology and prehistory (D. Teunissen, H.G.C.M. Teunissen-van Oorschot).

*Utrecht*

(12) *Laboratory of Palaeobotany and Palynology, University of Utrecht*: Devonian-Jurassic Palynology of western and southern Europe (H.A. Visscher); Pollenmorphology of various plant families, north-west European Pollen Flora (W. Punt); Late-Quaternary palynology and palaeobotany of the southern Netherlands, eastern and central France, north-west Iberia, and Minnesota USA (C.R. Janssen); Palaeobotany of western Europe (J. van der Burgh, M. Boersma).

*Wageningen*

(13) *Dept. of Soils Science and Geology, Agricultural University of Wageningen*: Pollen corrosion and palynology of the river clay area in the central Netherlands and south-east Austria (A.J. Havinga).

*The Hague*

(14) *Section Palynostratigraphy, Shell Int. Petroleum Company*: Pre-Quaternary palynology (W.O. Tichler).

Summing up: Palynology today centres around archaeology (1, 3, 7, 9, 11), palaeo-ecology (2, 3, 6, 7, 9, 10, 11, 12), geomorphology and geology (4, 5, 6, 11, 13, 14), pollenmorphology (8, 12) and pre-Quaternary palynology (6, 8, 12).

In conclusion one may say that the two disciplines, palynology and recent-vegetation science have had their own separate developments. In the thirties there was the common background of plant geography. Then recent-vegetation research developed its detailed floristic approach, resulting in the detailed study of vegetation types. Palynology lagged behind because of pollenmorphological problems. The

243

fossil data could be interpreted only in terms of plant formations or in terms of the ecology and distribution of single species. Certainly palynology will continue to contribute to the science of plant geography in this narrow sense.

Still the overall regional vegetation history of many regions on earth is not known, even not for the most recent periods of earth history. But one day the regional developments will be known and will constitute a useful framework for local studies.

Palynology has now advanced to the stage that a comparison of lists of fossil species in single past local stands with present day sociological affinities is useful. It is here that vegetation science may benefit from palaeobotany. To turn Cain's phrase around: 'the past is the key to the present'. Structure and species composition of communities of the present day are not solely determined by the present-day biotic and abiotic factors but also by its historic past, an approach that unlike that of the 'B-sciences' may very well have included events that have occurred just once, and thus cannot be repeated in an experiment. It is this genetic bond with the past that makes that palaeobotany contributes to an understanding of the nature of plant communities, the way they came into existence and disappeared from the face of the earth. This proces is especially interesting in view of the immigration of new plant taxa that may have upset the ecological balance in a plant community.

Palaeobotany is potentially a science that may contribute to the detection of how, in the past, change was brought about, either gradually in the fashion of a continuum (Gleason 1926) or with jumps, separating discrete communities in time. Up to now discussions of this kind could be done only on the basis of a few dominant plant species in the vegetation (for instance, Mayer 1967, on the origin of the north-alpine community of the *Abieti-Fagetum*). I believe that we are now at the point to tackle these problems with the detailed information stored in local pollen assemblages.

It is indeed remarkable that despite more then 60 years of palynological research we still do not know much about the distribution and composition of stands in the landscape. This is especially true for the early Holocene and earlier periods, when competition between plants was different from that of today because of a different climate and a different flora. Local pollen studies under a variety of conditions in small sized areas may reveal the pattern of vegetation types in that area. They are particularly useful in the context of peat research, archaeology, and pedology. Integrated palaeo-ecological studies of small lakes and bogs and soils including the simultaneous analyses of all recognizable remains of life will throw light on past ecosystems.

The problem that faces us is that we do not have enough sites for pollen studies available in order to resolve all the problems. Worse, the number of suitable sites has decreased dramatically after World War-Two because of a variety of measures such as road building, re-allotment of agricultural plots, etc. Palaeo-ecology thus cannot be detached from present day vegetation science and it is my sincere wish that palynology will continue to play its role in the activities of the Commission for the Study of Vegetation.

Principally, however, it will always remain a matter of fact that fossil assemblages are less informative than those of the present day, simply because so much material has not been preserved at the site. This is a situation with which the palynologist, just like the palaeontologist, has to live with. But at the same time this makes the search into the past such an exhilarating experience.

## References

Aario, L. 1940. Waldgrenzen und subrezenten Pollenspektren in Petsamo, Lapland. Ann. Acad. Sci. Fenn. Helsinki, Ser. A, 54 (8): 1-120.

Andersen, Sv.Th. 1970. The relative pollen productivity and pollen representation of North European trees, and correction factors for tree pollen spectra. Danm. Geol. Unders., 2e Raekke Nr. 96.

Baren, J. van. 1927. Düne und Moor bei Vogelenzang. Beitrag zur Frage der Quartären Niveau-Veränderungen an der holländischen Nordseeküste. Mitt. Geol. Inst. der Landbouwhoge-school, Wageningen (Holland) 11: 1-38.

Cain, S.A. 1944. Foundations of plant geography. Harper & Row, New York.

Cushing, E.J. 1967. Late Wisconsin pollen stratigraphy and the glacial sequence in Minnesota. In: E.J. Cushing & H.E. Wright Jr. (eds), Quaternary Palaeo-ecology. pp. 59-88. Yale University Press, New Haven-London.

Davis, M.B. 1967. Pollen accumulation rates at Rogers Lake, Connecticut, during late and postglacial time. Rev. Palaeobotan., Palynol. 2: 219-230.

Davis, M.B. 1968. Pollen grains in lake sediments: redeposition caused by seasonal water circulation. Science 162: 796-799.

Erdtman, G.E. 1921. Pollenanalytische Untersuchungen von Torfmooren und marinen Sedimenten in Südwest Schweden. Arkiv f. Botanik 17 (10): 1-171.

Erdtman, G. 1928. Studien über die postarktische Geschichte der nordwesteuropäischen Wälder. 11. Untersuchungen in Nordwestdeutschland und Holland. Geol. Fören. Förhandl., 50 (3): 368-380.

Erdtman, G. 1934. Über die Verwendung von Essigsäureanhydrid bei Pollenuntersuchungen. Svensk Bot. Tidskr. 28: 354.

Erdtman, G., B. Berglund & J. Praglowski. 1961. An introduction to a scandinavian pollen flora. Almqvist & Wiksells, Uppsala.

Erdtman, G., J. Praglowski & S. Nilsson. 1963. An introduction to a scandinavian pollen flora II. Almqvist & Wiksells, Uppsala.

Eshuis, H.J. 1936. Untersuchungen an niederländischen Mooren. K. Westerwolde. Rec. Trav. Bot. Neerl. 33: 688-704.

Faegri, K. & Johs Iversen. 1964. Textbook of pollen analysis. Munksgaard, Copenhagen.

Firbas, F. 1934. Über die Bestimmung der Walddichte und der Vegetation waldloser Gebiete mit Hilfe der Pollenanalyse. Planta 22: 109-145.

Fries, M. 1967. Lennart von Post's pollen diagram series of 1916. Rev. Palaeobotan., Palynol. 4: 9-13.

Gleason, H. 1926. The individualistic concept of the plant association. Bull. Torrey Bot. Club 53: 7-26.

Hammen, T. van der. 1949. De Allerød-oscillatie in Nederland. Pollenanalytisch onderzoek van een laatglaciale meerafzetting in Drenthe I en II. Proc. Kon. Ned. Acad. van Wetenschappen 52: 69-75, 169-176.

Hammen, Th. van der, E. Gonzalez, G.C. Maarleveld, J.C. Vogel & W.H. Zagwijn. 1967. Stratigraphy, climatic succession and radiocarbon dating of the last glacial in the Netherlands. Geologie en Mijnbouw 46: 79-95.

Hammen, T. van der, T.A. Wymstra & W.H. Zagwijn. 1971. The floral record of the late Cenozoic of Europe. In: K.K. Turekian (ed.), The late Cenozoic Glacial Ages. pp. 391-424. Yale Univ. Press, New Haven-London.

Havinga, A.J. 1962. Een palynologisch onderzoek van in dekzand ontwikkelde bodemprofielen. Ph. D. Thesis, Wageningen. H. Veenman, Wageningen.

Iversen, Johs. 1947. Diskussionsindlaeg i: Nordiskt kvartärgeologiskt möte den 5-9 November 1945. Geol. För. Förhandl. 69: 205.

Iversen, Johs. 1969. Retrogressive development of a forest ecosystem demonstrated by pollen diagrams from fossil mor. Oikos (Suppl.) 12: 35-49.

Janssen, C.R. 1972. The palaeo-ecology of plant communities in the Dommel valley, North Brabant, Netherlands. J. Ecol. 60: 411-437.

Janssen, C.R. 1973. Local and regional pollen deposition. In: H.J.B. Birks & R.G. West (eds.), Quaternary Plant Ecology. 14th Symp. Brit. Ecol. Soc. pp. 31-42. Blackwell, Oxford.

Janssen, C.R. 1974. Verkenningen in de Palynologie. Oosthoek, Scheltema & Holkema, Utrecht.

Janssen, C.R., W. Punt & Tj. Reitsma. 1974. The Northwest European pollen flora, a new project. Geologie en Mijnbouw 53: 458-459.

Jonker, F.P. 1952. A plea for the standardization of pollen diagrams. Taxon 1: 89-91.

Jonker, F.P. 1967. Palynology and the Netherlands. Rev. Palaeobotan., Palynol. 1: 31-35.

Lagerheim, G. 1902. Metoder för pollenundersökning. Botaniska Notiser (1902): 75-78.

Major, J. 1974. Kinds and rates of changes in vegetation and chronofunctions. In: R. Knapp (ed.), Vegetation dynamics. Handb. Veg. Sc. 8: 7-18. Junk, The Hague.

Mayer, H. 1967. Entstehung des nordalpinen Abieti-Fagetum. In: R. Tüxen (ed.), Pflanzensoziologie und Palynologie. Ber. Int. Symp. Vegetationskunde, Stolzenau/Weser 1962. pp. 56-70. Junk, Den Haag.

Peck, R.M. 1973. Pollen budget studies in a small Yorkshire catchment In: H.J.B. Birks & R.G. West (eds.), Quaternary Plant Ecology. 14th Symp. Brit. Ecol. Soc. pp. 43-60. Blackwell, Oxford.

Polak, B. 1929. Een onderzoek naar de botanische samenstelling van het Hollandsche veen. Ph.D. Thesis, Amsterdam. Swets & Zeitlinger, Amsterdam.

Post, L. von. 1916. Forest tree pollen in south Swedish peat bog deposits. Translation of lecture to the 16th convention of Scandinavian naturalists in Kristiania (Oslo), 1916. Pollen et Spores 9: 375-399.

Richard, P. 1976. Contribution à l'histoire post Wisconsinienne de la végétation du centre du Quebec méridional par l'analyse pollinique. Ph. D. Thesis. Laval Univ., Quebec.

Tauber, H. 1977. Investigation of aerial pollentransport in a forested area. Dansk Bot. Arkiv 32: 1-121.

Tüxen, R. 1931. Die Grundlagen der Urlandschaftsforschung. Nachrichten aus Niedersachsens Urgeschichte 5: 59-105.

Vermeer-Louman, G.G. 1934. Pollenanalytisch onderzoek van den westnederlandschen bodem. Ph. D. Thesis, Amsterdam. De Westertoren, Amsterdam.

Waterbolk, H.T. 1953. De praehistorische mens en zijn milieu. Van Gorcum, Assen.

Westhoff, V. 1970. Vegetation study as a branch of biological science. Med. Bot. Tuinen Belmonte Arboretum, Landbouwhogeschool, Wageningen, 12: 11-30.

Weber, C.A. 1893. Über die diluviale Vegetation von Klinge in Brandenburg und Ihre Herkunft. Engler's Bot. Jb. 17. Beibl. 1.

Zagwijn, W.H. 1960. Aspects of the Pliocene and early Pleistocene vegetation in the Netherlands. Med. Geol. Stichting Ser. C-11-1, Nr. 5: 1-78.

Zagwijn, W.H. 1961. Vegetation, climate and radiocarbon datings in the Late Pleistocene of the Netherlands. 1. Eemian and early Weichselian. Mem. Geol. Found. in the Netherlands, 1961: 15-45.

Zagwijn, W.H. 1973. Pollenanalytic studies of Holsteinian and Saalian Beds in the northern Netherlands. Med. Rijks Geol. Dienst N.S. 24: 139-156.

Zeist, W. van. 1959. Studies on the post-Boreal vegetational History of south-eastern Drenthe (Netherlands). Acta Bot. Neerl. 8: 156-185.

Zeist, W. van. 1967. Archaeology and palynology in the Netherlands Rev. Palaeobotan., Palynol. 4: 45-65.

# 8. VEGETATION SCIENCE AND NATURE CONSERVATION

## P.A. BAKKER

1. The significance of nature reserves in vegetation science
2. The significance of the Royal Botanical Society of the Netherlands for the first steps in nature conservation
3. Rise and organization of nature conservation in the Netherlands
4. Terminology and theoretical backgrounds
5. Choice of nature reserves
6. Vegetation science as a basis for management plans
7. Some experiences from management practice
8. Creation of habitats
9. Wishes for the future
   References

# 8. VEGETATION SCIENCE AND NATURE CONSERVATION

## P.A. BAKKER

### 1. The significance of nature reserves in vegetation science

The Netherlands are very densely populated and as a result flora and fauna are seriously impoverished (Westhoff 1976). Quite a number of species and plant communities can only be studied in nature reserves nowadays. Just in time a network of these was chosen and established. This we owe to a small number of people who over and over emphasized the importance of nature conservation (e.g. van Dieren 1929, Thijsse 1946, Weevers 1938, 1951a, b, 1956, Mörzer Bruijns 1965a, 1967a, b, Westhoff 1952a, 1955, 1971a, b, d).

Nature reserves serve as open air laboratories for pure as well as for applied scientific research (Heimans 1956, Tüxen 1957, Weevers 1951a, Westhoff 1952a, 1955, 1971a). It is possible to carry out research undisturbedly only in well-guarded reserves. Extensive and costly equipment usually can be put up there without unacceptable risks. Furthermore, the influence of man can be controlled better. Especially in periodical analyses of vegetation samples on permanent plots it is absolutely necessary to have no undesired disturbance at all.

Nature reserves are essential also from an educational point of view. Field laboratories in or near nature reserves are indispensable in teaching biology, soil science, forestry, agri- and horticulture. Only here students can observe and examine themselves the complex interrelationships in our ecosystems which is necessary for a basic understanding. Biologists still 'germinate on waste grounds' as Thijsse wrote in 1938, and Westhoff (1971a) put it more or less like this: 'Only here they can acquire an attitude which makes them ask "where does it come from" instead of "how do I get rid of it" when confronted with a disturbance of the natural equilibrium, e.g. a plague'.

While nature reserves are important for scientific research, we are presently confronted with the fast growing problem of the rapidly increasing number of students. Therefore, in the near future we probaply will have to protect the reserves against too many research activities.

### 2. The significance of the Royal Botanical Society of the Netherlands for the first steps in nature conservation

Within the Royal Botanical Society of the Netherlands (K.N.B.V.) several commissions have been formed: one of them deals with floristic research (founded in 1914); a second dates from 1927 and deals with the protection of the wild flora (Weevers 1947); the Commission for the Study of Vegetation was founded in 1933 and its activities are documented by Vroman (this volume). All three of them cooperate closely, and are meaningful for nature conservation in the Netherlands.

At all 99 meetings of the Commission for the Study of Vegetation lectures have been presented which were directly or indirectly important for nature conservation. The first meeting of the Commission was held on November 26, 1933. There, J. Heimans lectured on the transport factor in sociology, J.W. van Dieren, one of the persons very much in favour of founding the Commission, talked about phyto-sociological studies on parabolic dunes, and Th. Weevers about forest communities in the Gelderse Vallei. During the fourth meeting in 1936, W.H. Diemont read a paper which is very much worth mentioning too. He stated that, in order to make a balanced choice of reserves, vegetation maps would be useful. The Dutch State Forest Service in the province of Drenthe had just started such mappings. Diemont (1937) mentioned calculations by W.C.A. Linn of Utrecht, based on the 'bottle-universe'-theory developed by Pearl in the U.S.A. According to these calculations the maximum population size the Netherlands can harbour is 14.4 million, a number that would be reached in 2100!

In 1929 van Dieren already had given a clear exposition of 'the significance of nature reserves to biologists' and had mentioned that nature reserves should be large enough to withstand the influences of land development in the vicinity. He also had pointed out that the main objective should be the preservation of biotic communities (biocoenoses), not of species. Selected areas should be studied and evaluated first in order to be able to make a firm recommendation for their preservation. Management of the areas should be based on biological studies too. Van Dieren certainly was ahead of his time.

The Botanical Society was active in nature conservation long before the three commissions mentioned earlier were founded (Anon. 1970, Smit & Margadant 1971). In 1901 Vuyck presented a lecture on the protection of the natural habitats of our native flora. He suggested that the government should purchase some sites in order to preserve them as botanical reserves (Anon. 1970).

Although the purchase in 1906 of the Naardermeer, the first nature reserve in the Netherlands, was considered a success, the prominent Society members Goethart and Lotsy stuck to their opinion that the Botanical Society should not have been involved (Wachter 1947). They preferred to protect the flora in a number of other places, notably some fens near Weert, an area where the culmination of the botanical richness of the Netherlands was found in the beginning of this century; however, the Botanical Society did not succeed in purchasing valuable sites near Weert.

In 1913 Th. Weevers and J. Heimans drafted an advice on areas to be preserved. In 1927 a list was published in the Netherlands Kruidkundig Archief, listing 255 rare vascular plants, whose distribution ought to be ascertained in detail. As a result florists were able to prepare a detailed report on the botanically important areas in some of the most threatened parts of the Netherlands. This report was presented to the Society for the Preservation of Nature Reserves ('Vereniging tot Behoud van Natuurmonumenten in Nederland', which was founded in 1905, and here will be called 'Natuurmonumenten').

Financial support was given to 'Natuurmonumenten' for the purchase of the wet heathland area Dwingelose Heide in 1930, and the raised bog Fochteloërveen in 1938, and to the Foundation Het Gelders Landschap for the purchase of the oakwood Wilde Kamp near Garderen (1939). In 1934 the Royal Botanical Society temporarily rented a boggy moorland of 0.5 ha, harbouring *Wahlenbergia hederacea*, in Harbrinkshoek near Tubbergen, but the species disappeared due to draining of the surroundings and lack of adequate management. In 1938 the slope with chalk grassland of the Bemelerberg was rented from the municipality of Bemelen but the agreement was cancelled soon by the municipality.

In 1939 action was taken by the Royal Botanical Society against large reclamation projects set up to provide work for the unemployed, e.g. in north-western Overijssel and the bogs near Vriezenveen. Also successful were actions to prevent the building of a dam between Epen and the Belgian border in the Geul valley, the lime diggings that would have destroyed the Heksenbos and the Willinks Weust near Winterswijk, and the building of a highway through the Oude Rijnstrangen. It proved to be impossible, however, to preserve the valley of the Beerse near Boxtel (1947) and the Sint Pietersberg near Maastricht (1949).

## 3. Rise and organization of nature conservation in the Netherlands

The nature conservation movement in the 19th century was started for the benefit of mankind, though this had a much more restricted meaning than at present. In that time one did not know anything about the importance of nature reserves for natural equilibria and as gene pools. The origin of actual nature conservation lies in France and dates from the reign of king Louis-Philippe (Westhoff 1955). The motives were purely esthetical. Painters from the Ecole de Barbizon, settling near the famous Forêt de Fontainebleau, prevented the cutting of large parts of their beloved woods. The French king intervened personally in 1837. This resulted in the creation of woodland reserves in the Forêt de Fontainebleau, which were enlarged later and nowadays amount up to 1692 hectares.

From that time onwards efforts to preserve nature were made at first in the USA and in Germany but now its importance has been recognized in almost all civilized countries in the world. Meanwhile, the aims have broadened. Westhoff (in Anon. 1978) defined nature conservation as follows: 'The efforts to preserve the largest possible diversity in geogenetic structures as well as plant and animal species living in ecosystems and resulting from biotic and abiotic developments, including human influence as far as this has enriched the total diversity of species and structures'. Van Leeuwen (1967) defined nature conservation as 'a social mechanism of feedback with a scientific basis, directed particularly at the relations of man with his natural environment'.

The term 'natuurmonument' (monument of nature) was introduced in Dutch usage by F.W. van Eeden in 1886 in connection with the destruction of the legendary Beekbergerwoud near Apeldoorn. Around 1900 the importance of nature conservation was more and more recognized in Europe, especially on the initiative of

Hugo Conwentz, Prussian State Commissioner for the Care of Monuments of Nature at Danzig. The term 'Naturdenkmal' was used by Conwentz, but in 1819 Alexander von Humboldt already spoke of old big trees as 'monuments of nature'. The term 'natuurmonument' is becoming old-fashioned now (Westhoff 1973a). It is suggestive of something romantic and invariable, whereas a nature reserve can be very dynamic and almost impossible to preserve in a densely populated country like the Netherlands.

In our country the rise of nature conservation is due first of all to E. Heimans and Jac. P. Thijsse. They brought about a up to then unknown interest in wild plant and animal life by publishing six little books about biocoenoses, founding a periodical 'De Levende Natuur' (1896), writing a flora and many other reports, and most of all by giving lectures.

Plans of the municipality of Amsterdam to fill up the Naardermeer with household refuses led to the foundation of 'Natuurmonumenten' in 1905. It started its activities by bying the Naardermeer in 1906. Nowadays 'Natuurmonumenten', which has over 250,000 members, is the largest private landowner in the Netherlands. It controls more than 35,000 hectares spread over about 150 nature reserves in all provinces of the country. They harbour representative examples of almost all ecosystems occurring in the Netherlands. This has been achieved primarily as a result of the unlimited dedication of Jac.P. Thijsse and P.G. van Tienhoven.

At the initiative of 'Natuurmonumenten' eleven provincial daughter-organizations were founded. Ten of them are foundations, of which the 'Utrechts Landschap' is the oldest; the eleventh is a society, 'It Fryske Gea'. These provincial organizations together control about 33,000 hectares spread over about 360 reserves.

On top of that there are a number of foundations set up by private persons which manage areas, like Twickel (4000 ha), Stichting Edwina van Heek (490 ha, a.o. Singraven) and Huis Bergh (about 2800 ha, including Koningsbelten/Sprengenberg). Also worth mentioning are the Hoge Veluwe (5450 ha), the Kennemerduinen (1240 ha), the Gooisch Natuurreservaat (about 1800 ha), the Royal Forestry Het Loo (10,500 ha), the Noordhollands Duinreservaat (4760 ha), the dune waterworks of The Hague (Meyendel, 1280 ha) and the waterwork dunes of Amsterdam (3400 ha), all of which were set up in cooperation with several governmental offices but which are not all real nature reserves (see page 254). Additionally, the Society for the Protection of Birds controls several bird sanctuaries which are partially of botanical interest. Characteristic of the situation of Dutch nature conservation is and always has been that there is no clear boundary between the concerns of the government and those of private organizations. The policy of the government depends largely on initiatives and advices of private organizations and persons.

Nature conservation by the government started with the activities of the Dutch State Forest Service. This organization was founded in 1899 to further forestry and afforestation of state-owned waste lands. In 1908 it first declared certain parts of its areas as nature reserves. The first ones were De Muy on Texel, the Kootwijkerzand on the Veluwe and the Lheebroekerzand near Dwingeloo, all saved from afforestated areas. The largest reserve in this category is the Boschplaat on Terschelling (4400 ha). In all about 22,500 ha of circa 120 state owned nature reserves are under the control of the Ministry of Agriculture and Fisheries under which the State Forest Service resorts.

On certain conditions, explained in the Natural Beauty Act (Natuurschoonwet), it is possible for estate owners to get tax facilities. This arrangement is used for over 100,000 ha.

Parts of the Crownlands, controlled by the Ministry of Finance, also have been declared nature reserves, e.g. the Kobbeduinen on Schiermonnikoog (2400 ha), the Zwarte Meer near Genemuiden (1500 ha), areas outside the dikes in the Haringvliet and the Grevelingen, and the domanial dunes on Schouwen (1000 ha). Parts of the Crownlands have been let to private organizations to manage them as nature reserves.

In 1929 the 'protection of nature' became one of the tasks of the State Forest Service. Up to the second world war the government only set apart nature reserves in areas which were already state-owned. In 1941 the government bought an area adjoining the Dwingelose Heide for the first time with the very purpose to make it into a nature reserve: the Kralose Heide, officially owned by the Ministry of Education, Arts and Sciences, as it was called at that time. After the war these activities were intensified, made possible by a permanent vote on the budget, first of this ministry, afterwards on that of the renamed Ministry of Cultural Affairs, Recreation and Social Welfare (CRM). This resulted in about 500 nature reserves with a total of about 35,000 ha. This does not include, however, the large seal reserve in the Wadden Sea between Texel and Vlieland (20,000 ha).

Important work was done by the Committee for advice on nature monuments of the State Forest Service, founded in 1928, and now called Scientific Committee of the Board for nature conservation. This committee consists of biologists and geologists and was presided by Th. Weevers for over twenty years. Therefore it often was called 'the Weevers' committee'. It soon became clear that all this work could not be done alone by scientists in their spare time. Thus, in 1947 a special division of the State Forest Service was formed, called 'Nature Conservation and Landscape'. Also in 1947 at 'Natuurmonumenten' a scientific department was started. Its workers carried out botanical research in their reserves for management purposes. Moreover, in 1957 the State Institute for Nature Conservation Research (RIVON) was founded; it was united with the Institute for Biological Field Research (ITBON) in Arnhem in 1969 into the Research Institute for Nature Management (RIN). Basic research concerning the choice and management of nature reserves is and was done primarily by RIVON and RIN. Supplementary research is also carried out by students of the ecological departments of the universities.

In accordance with the Nature Conservation Act (Natuurbeschermingswet) of 1969 areas which are of public interest from the point of view of natural beauty or scientific value can be declared protected nature reserves. The application of this law is impeded severely by financial and juridical problems. So far only 24,500 ha have been protected by means of this law.

## 4. Terminology and theoretical backgrounds

With the term 'natural area' we mean a part of the earth's surface where human influence, if not totally absent, in any case is relatively small, and where various kinds of spontaneous forms of life, which elsewhere are endangered in their existence by human technical activities or have already vanished, still can manifest themselves.

Technically speaking, natural areas can be used by man for various purposes, such as hunting or shooting, grazing, forestry, recreation and the supply or storing of water. Until the beginning of this century the largest part of the Netherlands consisted of natural areas and they were preponderantly used agriculturally. Among the few natural areas left today, many are used in this way. Most private estates, watershed areas, recreational areas, the state's forests resorting under the Ministry of Agriculture and Fisheries (except the nature reserves situated in those forests), and some of the properties of the Society for the Preservation of Nature Reserves in the Netherlands and the provincial organizations ('Provinciale Landschappen') belong to this category.

Apart from that function, natural areas provide also a technical means (and at that even the only practical one) by which we can realize or attempt to realize the conservation of all those kinds of organisms, which are endangered by the pressures of numerous, differently aimed technical measures. When we destinate and utilize a natural area for that specific goal, we call it a 'nature reserve'. Thus, the term nature reserve has both a technical and a legal aspect. A nature reserve is a natural area in use by the State or a private body, backed by the Law, with the specific aim to ensure as good as possible the survival of forms of life that have elsewhere vanished or are threatened in their existence. Such a reserve could be defined as a 'nature-saving device'.

Obviously a nature reserve needs its upkeep and, sometimes, needs improvement with respect to its purpose. It is also possible that a specific area first has to be developed before it can function properly as a nature reserve. Following P.G. van Tienhoven, the whole of measures aimed at the development, maintenance and improvement of nature reserves are called 'nature techniques'. Van Leeuwen (1973, 1978) distinguished between constructional aspects of nature techniques ('natuurbouw'), to which measures aimed at development and improvement belong, and management aspects of nature techniques ('natuurbeheer'), under which maintenance measures resort.

Management has two components: an external and an internal one. External or outward management of nature reserves comprises the measures aimed at a prevention of excessive human influence (Fig. 1). It is mainly directed against the effects of urban-, agro- and environmental technical developments in the vicinity of nature reserves. Internal or inward management, on the other hand, tries to ensure the required level of human influence necessary to maintain or to lead to the situation desired in the nature reserve. Interventions such as excavation, treading, grazing, burning, mowing, and also no interference at all ('doing nothing') are tools of

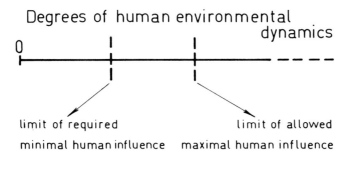

Fig. 1. Inward and outward management in relation to the degree of human environmental dynamics.

internal management. Those reserves, or the parts of such grounds, where the last-mentioned method of internal management is practised are called 'absolute' or 'strict nature reserves'.

The theoretical basis and the practical guidelines for ecological management of nature reserves were developed by Westhoff and van Leeuwen. Thanks to their efforts and their clear ideas the Netherlands occupy a honorable position among other nations in as far as the scientific approach to nature conservation and management is concerned.

In management of real nature reserves in the Netherlands (thus excluding natural areas which are used for water supply, military exercises, hunting, forestry and/or recreation) the following starting points are taken into account as much as possible:
1. Permanent conservation or development of a variety of ecosystems is the main objective. The use of nature reserves on behalf of the physical and psychic health of man is acceptable in as far as the effects stay within the boundaries of tolerance of the ecosystems concerned.
2. It is always necessary to apply external management, and this has priority to internal management.
3. In many cases the method of 'doing nothing' in internal management is wrong.
4. Contrary to the methods of agricultural techniques, management of nature reserves is restricted to taking measures in relation to environmental factors. The

organisms themselves are not manipulated, e.g. by planting, sowing, weeding. Planting is not allowed because in a nature reserve one just wants to have species which occur spontaneously. (For exceptional cases see Westhoff 1973b and Londo 1976).
5. Management is directed primarily on abiotic factors, secondly on the vegetation, thirdly on the fauna. By good management of the environmental factors important for the vegetation, niches of most animal species which are dependent on that vegetation are maintained. This generally goes for invertebrates and small vertebrates. Large mammals and a number of bird species (e.g. birds of prey, birds breading in colonies, meadowbirds, geese) which have a larger range of action, often need special measures. These can be contrary to the wishes from the botanical point of view. Also in respect to the small animals, however, the botanical criteria should be handled carefully (Mabelis 1977), for often little is known about their niches and way of life. This makes it difficult to know how to maintain their presence. For this reason we will confine ourselves primarily to botanical management.
6. Botanical management is conducted according to the six basic rules of internal (a, b, c) and external (d, e, f) control based on structure ecology, as stated by van Leeuwen (1966a). These are:
(a) The preservation of the botanical richness of a nature reserve is most assured if its treatment forms as good as possible a copy of, and is directly connected with, the methods applied formerly, and is later on subject to as little change as possible.
(b) In case the character and size of a nature reserve are suitable, its internal regulation can be amplified by furthering the development of ecoclines based on the degree of human influence within the area.
(c) Our controlling operations have to be done gradually and on a small scale.
(d) The external protection of nature reserves situated in landscapes with a coarse-grained pattern has to be based on their form and size. The more concentric the form and the larger the surface the more safe the areas will be against alterations threatening them from the outside.
(e) If the size of a nature reserve is sufficient, the controller has to take advantage of the ecoclines which are developing along the outskirts by interaction of internal and external influences. In this connection it is a condition that oligotrophic circumstances are predominating over eutrophic ones.
(f) If alterations are induced from the outside the internal botanical control has to be directed on delaying the processes evoked by them.
7. A management plan must be made for each nature reserve. It should be based upon thorough field studies and discussions between all people concerned.
8. The effects of the management should be studied regularly by means of periodical inventories and mappings, investigation of permanent plots and indicator species.
9. Meticulous accounts of everything taking place in and around the reserve is of great importance. These serve as a necessary feedback from the managers to the research-workers.

In the management plan of each reserve it should be stated whether or not this reserve can serve any other purposes besides nature conservation, and if so, which other functions and to what amount. In this, the vulnerability of the area or its future development have to be taken into account.

Up till the second world war among the people concerned with the protection of nature the view prevailed that management of nature reserves consisted of no intervention with nature itself. A nature reserve should be without human influence. At that time biologists did not realize that areas in which human influence is essential, like heathlands, grasslands and reedswamps, are of interest to nature conservation and can be maintained by adequate management (Westhoff 1970a, 1971c). Thanks to very practical managers like P.G. van Tienhoven and J. Drijver some of such areas nevertheless have been saved for the future. They let grass- and reedlands in nature reserves to farmers and argued that private organizations, which at that time were not yet subsidized by the government, needed the money obtained in that way. In this way the right management was guaranteed unknowingly! Furthermore these practical managers were first of all ornithologists, so they knew that meadowbirds require grasslands and reedwarblers reed. Therefore they listened politely to the non-intervention theory of the biologists of those times, but in reality had it all their own way. Nowadays it is more and more difficult to have farmer tenants in nature reserves, because the modern farmer wants to use modern agricultural methods which will destroy the value of the reserve. An increasing number of farmers is no more interested in using those lands now that modern methods are forbidden by the owner. The consequence is that the manager has to manage the land himself with the aid of employed labourers.

British and Dutch vegetation scientists independently formulated new views almost at the same time, notably in the report 'Nature Conservation and Nature Reserves' of the British Ecological Society (1943) and in the speech by V. Westhoff, called 'Biological problems of nature conservation', delivered at the congress of the N.J.N. (Dutch youth league for the study of nature) in Drachten in August 1945. Afterwards these insights have been further developed (Gabrielson 1947, Westhoff 1952b, 1955, van Leeuwen 1967). The non-intervention theory is now considered to be romantic, especially in densely populated countries where human impact has been working for centuries. Since the rise of environmental biology it is realized that man is one of the factors influencing existing ecosystems. This influence is not necessarily negative and impoverishing. Man also has enlarged the variety of ecosystems, in the first place by creating semi-natural landscapes. Many of our valuable areas which are rich in species belong to this category. In semi-natural landscapes structure and physiognomy of the vegetation have been thoroughly influenced by man: they belong to an other formation than the potential natural vegetation. Up to the 1930's the major part of our country consisted of semi-natural landscapes, and in the beginning of the century over 90% of our country was made up by semi-natural and sub-natural ecosystems of great ecological value. Nowadays only c. 12% of 'natural' areas remains, of which two thirds are woods, largely consisting of planted conifer forests of restricted biological value. Merely c. 4% of our area is managed as true nature reserves.

Westhoff (1952a, b) designed a classification of landscapes according to their

| | Flora and fauna | Development of vegetation and soil | Examples |
|---|---|---|---|
| Natural landscapes | spontaneous | not influenced by man | parts of the Wadden area (mudflats, coastal beaches and salt marshes) |
| Sub-natural landscapes | completely or largely spontaneous | to some extent influenced by man | parts of the dune landscape, most saltmarshes, inland drift sands, deciduous woods with some cutting, final stages of succession in hydroseries in fens |
| Semi-natural landscapes | largely spontaneous | drasticly influenced by man (other formation than the potential natural vegetation) | heathlands, oligotrophic grasslands, sedge swamps, reed swamps, inner dune grasslands, coppice, osier-beds, many woods in which the tree stratum is arranged by man |
| Agricultural landscapes | predominantly arranged by man | strongly influenced by man (soil often fertilized and drained; vegetation with ruderals, neophytes and garden escapes) | arable fields, sown grasslands, parks, conifer forests |

Fig. 2. *Degree of naturalness of landscapes*

degree of naturalness (Fig. 2). This classification was amplified by Sukopp (1972) who introduced a measure to indicate more or less the intensity of human influence, the so-called 'degree of hemeroby'. This is inferred from the floristic composition of the vegetation, and is measured as its proportion of neophytes. Van der Maarel (1975) adapted Sukopp's scheme to the terminology used in the Netherlands.

Thanks to studies by van Leeuwen (1967, 1973, 1978) we know that the connection between diversity and stability is of crucial importance in respect to the question whether human activity is favourable or unfavourable in nature conservation. Activities which enlarge the variety in space (e.g. peat cutting and winning of sand or clay on a small scale) are favourable, as are all influences which do not alter from year to year (e.g. mowing once a year, grazing). The former agricultural and mining systems induced enlargement of the variety in space within the landscape. This differentiation developed because

1. the methods did not or hardly change for centuries (stability of the methods);
2. the development of antropogeneous ecoclines as a result of the limited range of action which caused a decreasing intensity of human influence (isolation by distance). This induced the development of the landscapes named in Fig. 2.
3. our ancestors were operating gradually and on a small scale (dispersion of activities in space and time).

The old rural communities were based on a recycling economy. This is most obvious in the heathland farming system (Dutch: 'potstalsysteem') on Pleistocene sandy soils. Three gradients in space had developed there which run parallel to and intensified each other. These were (a) the gradient in scale (small arable fields near the village — larger hayfields in valleys of brooks — still larger heathlands, woods and peat bogs); (b) the gradient from private property to 'the commons'; (c) the gradient of intensity of cultivation. For centuries minerals have been transported

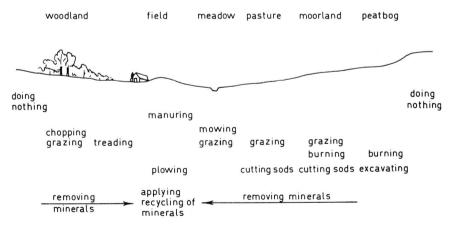

Fig. 3. Human influences on ecosystems at about the beginning of this century and their implication for the nutrient status.

from the vast hinterland to the relatively small centers of inhabitation. The former therefore became poorer, the latter richer in nutrients. Thus a landscape developed in which there were large, oligotrophic natural areas which were rich in species and not much influenced by man, and small settlements with relatively eutrophic and intensively used cultural fields (Fig. 3).

Fig. 4. Sketch of the most important subnatural and seminatural environments in the Netherlands.

Without human activity the low lands of north-western Europe would almost completely be covered by woods and be poorer in diversity than they have been up to the first part of this century. Factors of urban and agricultural techniques which level the variety in space or enlarge dynamics in the environment are unfavourable. Such techniques are of daily occurrence nowadays, which is catastrophic for our natural areas.

Inspired by the cybernetical approach of Ross Ashby and Margalef, van Leeuwen (1966b, 1973) developed his relation theory. This theory gives a sound basis for a new approach in the study of relations within ecosystems with special regard to nature conservation. The basis of the theory is the idea that each relation has aspects of 'open' (connection) as well as 'closed' (separation). A distinction is made between relations in space (pattern) and relations in time (process). Three basic relations are distinguished. Variation in space (differentiation) increases with levelling in time (regulation); variation in time (disturbance) leads towards a decrease in variation in space (equality). The relation between diversity and stability in ecosystems is complementary. Just a few people know that the development of this theory is based on 20 years of phytosociological investigation in permanent plots, primarily in the nature reserve 'De Beer' which has now been destroyed because of the expansion of Rotterdam.

As a next step van Leeuwen (1965) together with Westhoff marked out two very different types of border areas. A 'limes convergens' (ecotone) is allied to spatial concentration and characterized by sharp border lines, poverty in species and coarse granulation in pattern. A 'limes divergens' (ecocline) is allied to dispersion and characterized by faint lines of demarcation, richness in species, and fine granulation in its pattern. The first-named type occurs in areas with many fluctuations in environmental factors, in agricultural landscapes as well as in naturally unstable landscapes like the Wadden Sea, the outer coastal dunes, and parts of our river valleys. Ecocline border areas (gradient areas) can be found in landscapes without many fluctuations in environmental factors. Fig. 4 shows the distribution of the

---

A. Landscapes with gradual and stable environmental transitions.

Narrow zones with gradual transitions between mutually strongly contrasting habitats.

Lower course of large rivers: in the western part gradient of decreasing salt content, in the eastern part gradient of decreasing tidal movement.

Areas with many transitions between salt and fresh habitat.

Areas with concentration of rare plant species.

B. Landscapes without environmental transitions or with abrupt and (or) unstable ones.

Subnatural ecotones.

Large wildfowl areas (geese and water fowl).

Large areas with many grassland birds.

Areas with large heaths, moorlands, drift sands and monotonous forests.

261

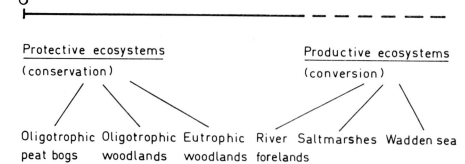

Fig. 5.  Degrees of non-human environmental dynamics.

most important natural areas in the Netherlands. In nature reserves examples of both types of border areas are preserved. The small, geomorphologically defined, gradual transition zones of the 'limes divergens'-type, as well as the dunes, are still partly characterized by a high percentage of rare plant species. Natural areas will be more susceptible to disturbance by man when they are less dynamical themselves (Fig. 5).

The arrangement in space of the ecotopes is very important too. Often it is caused by a difference in height or great distances between the extremes. If areas

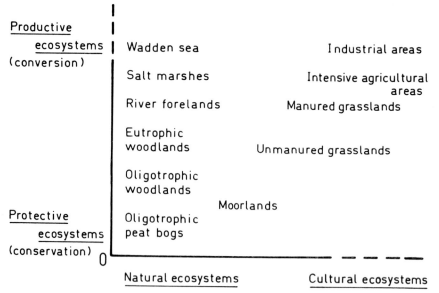

Fig. 6.  Relative position of ecosystems according to human and non-human environmental dynamics.

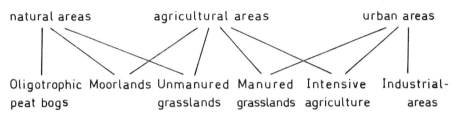

Fig. 7. Degrees of environmental dynamics added by man.

with a high degree of dynamics (i.e. eutrophic, basic, mineral and/or wet environments) dominate over those with a low degree of dynamics (i.e. oligotrophic, acid, organic and/or dry) a convergent border situation appears with few, mostly trivial species. If the reverse happens gradual transition zones can appear which are rich in rare species.

Not only landscapes (ecotopes) can be divided according to their requirements of the environment (Fig. 6), this can be done for separate plant species too. Every species needs or tolerates a certain amount of environmental dynamics. We can arrange all species into a series from those that need very little to those that need a great deal of environmental dynamics. On the minimal side of the scale for example *Eriophorum gracile* and *Coeloglossum viride* can be found, on the maximal side for example *Urtica dioica* and *Plantago major*. It should be quite obvious that nature conservation is interested mostly in the species on the minimal side for among them we find quite a few 'indicator' plant species. Each species requires a certain degree of environmental instability. If in a certain area the amount of environmental dynamics is too small or too large, i.e. outside the lower (minimal required) or higher (maximal allowed) border of tolerance, this plant species will not be able to survive there. Every species and every plant community regulates its environment. Human activities result in an amount of dynamics which is added to the ecosystems in the area, and is named A.T.D. by van Leeuwen (Dutch: 'Antropogeen Toegevoegde Dynamiek', in English 'Dynamics Added by Man', D.A.M.). This instability is added to the already present 'natural' amount in the abiotic environment (Fig. 7). Human activities essential for the ecosystems within a nature reserve should be continued and maintained in their own rhythm and way, just as they have been practiced for many generations. Consequently, in a number of cases certain traditional agricultural methods have to be continued. If this is not possible for technical reasons, we have to take measures simulating the former agricultural practice and coming as near as possible to it. The techniques of internal management can be

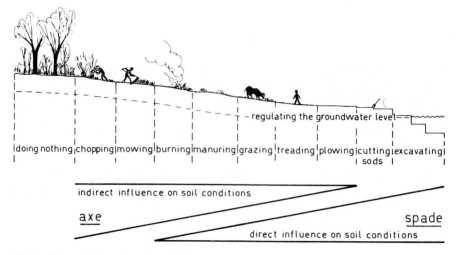

Fig. 8. Types of inward nature technical measures. Degrees of required minimal human influence.

listed in the sequence 'no action at all' ... 'excavating' (Fig. 8). In this sequence the rate of D.A.M. increases. The impact of D.A.M. is strongest in the three latter actions: they imply a direct influence on soil conditions ('spade effect'). It is weaker in the former actions, implying an indirect influence on soil conditions ('axe effect'). It is very important for the practice of management that van Leeuwen has pointed out the close relationships between the most important syntaxonomic units of higher rank and the types of required human activities (see van Leeuwen 1966a; in agreement with more recent opinions 'planting' is omitted from this scheme). Londo (1971) designed a similar scheme for the most important syntaxa of dune slack vegetations.

The most difficult task of nature technics is diminishing the amount of D.A.M., especially if we want to annihilate it in a strict nature reserve; only very gradually resp. with delay it can be realized successfully. Changes to a higher degree of dynamics ('disturbances') are easier and unfortunately occur already in many of our nature reserves.

## 5. Choice of nature reserves

The selection of nature reserves is made according to a well-balanced national planning of purchase, which is adjusted all the time to new insights and possibilities. This plan is made by the N.W.C. (Scientific Committee of the Nature Conservancy Council). Its members visit the sites and a more complete inventory is made by the State Forest Service. In the final decision six categories of points of view are taken into consideration (Westhoff 1970b, 1971b, c, 1973b): (1) the variety present in the Netherlands, including the relative scarceness of the type of landscape concerned; (2) the international importance; (3) the scientific points of view, which change from time to time; (4) the possibility to purchase; (5) the potential significance: (a) the possibility to improve a damaged area by effective management; (b) the possibility to create 'future nature reserves' by means of creation of habitats; and (6) the possibility to preserve the area in the future.

*Ad 1.* Syntaxonomical typologies have been made of quite a number of biocoenoses, e.g. former river beds, moorland pools, dry grasslands in river forelands, litter fens, spring vegetation, wall vegetation, raised bogs, heathlands and juniper scrubs. Unfortunately a usable typology of our forests, scrubs (except juniper scrubs) and coppice has not yet been made. There is no evaluation of the scientific value of the Dutch forests either. There does exist an evaluation of the river forelands (de Soet 1976), which takes into account geomorphology, landscape, and ornithology also. Of some parts of the Netherlands landscape ecological studies have been published, e.g. Zuidwest-Nederland (1972), Volthe-De Lutte (1971), the Kromme Rijn area (1974), the Veluwe (1977).

In several provinces a start has been made with the collecting of botanical data as a basis for physical and environmental planning (Hessel, Wildschut & Jansen 1975). A map on a scale of 1:200,000 of the potential natural vegetation of the entire country was produced, commissioned by the Physical Planning Service, and based on the soil map (Kalkhoven et al. 1976). Another ecological basis study for physical planning on national level is van der Maarel & Dauvellier (1978).

Rare habitats are: spring areas, non-canalized brooks, quaking sedge swamps, green beaches, limestone grasslands, dry grasslands in river forelands, and moist unmanured grasslands. Wet heathlands are scarcer than dry ones. Within the dry types, those on soil which is relatively rich in minerals are rarer, and richer in species, than those on very poor soils.

*Ad 2.* From an international point of view our most important natural areas are our dunes, tidal mudflats and salt marshes, the hydroseries in fens, inland drift sand areas, wet heathlands and oligotrophic moorland pools (van Donselaar 1970). Another important phenomenon is the fact that some plants in the Netherlands reach the edge of their distribution area, and here differ in properties from those in the centre. Specific examples are *Arnica montana, Vaccinium uliginosum*, some species of *Gagea, Euphorbia paralias, Euphorbia amygdaloides*. Examples on a coenotic level are the *Parietarion judaicae* and the *Koelerio-Gentianetum*. The *Taraxaco-Galietum maritimi, Anthyllido-Silenetum nutantis*, and *Pyrolo-Salicetum*, for example, are exclusively found in the Netherlands.

*Ad 3.* Ideas have changed in natural history. An example is the question whether some rare plants are indigenous or not. Due to a lack of knowledge it has long been believed, that some 'nice flowers', like *Fritillaria meleagris, Narcissus pseudonarcissus* ssp. *pseudonarcissus* and *Convallaria majalis*, were garden escapes, rather than wild species. A second example is the increasing attention that is given to gradual transition zones (gradient areas).

With the diminishing of natural areas, and the impoverishment of the flora and fauna, the standards for choosing reserves have changed. Nowadays even agricultural elements, like arable fields and cultivated grasslands have entered the scope of nature conservation. When around 1940 most of the litter fens (*Cirsio-Molinietum*) had already disappeared from our country, the marshy, slightly manured hayfields of the *Calthion palustris* type were still common. Most of our grasslands in those days would have been classified as *Lolio-Cynosuretum*. This association was found on different soils and showed a great deal of variation. As a result of agricultural intensification, the *Calthion* has been reduced to some remnants in nature reserves, while the *Lolio-Cynosuretum* is becoming rare. Due to intensive grazing, fertilizing, and lowering of the water level, it is gradually changing into the *Poo-Lolietum*. This association is poor in species and holds a relatively high number of nitrophilous therophytes, like *Stellaria media* and *Capsella bursa-pastoris*. In many areas the *Lolio-Cynosuretum* is nowadays mainly found along road verges and ditches, and it is high time to preserve some representative examples in nature reserves and national landscape parks. The same holds true for the *Agropyro-Rumicion crispi*. Many of its varieties have deteriorated or disappeared completely over large parts of the country, again mainly due to lowering of the water table (K.V. Sykora, personal communication).

*Ad 4.* The possibility of purchasing is a juridical and/or a financial problem, determined by economical, social, political and psychological aspects. In several parts of the country property is divided over many people who all have inherited just a small part. Sometimes it takes a lot of time to get each owner to sell his part in order to create a reserve of sufficient size. An extraordinary example is the case in which 'Natuurmonumenten' in 1966 bought 3811/2,016,000 part of a lot of 3/4 hectares in the Wieden which still had 50 other owners.

*Ad 5.* In the potential significance different factors play a part. With good management and if levelling influences from the outside can be excluded, differentiation in a nature reserve can increase. This not only concerns the establishment of new species but also the development of elements that were not, or hardly present. Examples of this are the development of certain vegetation types by sod-cutting and the furthering of the initial development of raised bog by yearly mowing in fens which have reached a certain stage in the hydroseries.

Equally important is the possibility of restoration of a damaged area by effective management. Before the Naardermeer was bought by 'Natuurmonumenten' in 1906, efforts had been made twice to drain the area (1629 and 1883-1886). The Korenburgerveen, bought in 1918, had been the property of a land developing company which by drainage had tried to make it into a meadow. In the Naardermeer as well as in the Korenburgerveen the former ditches are still visible. In former ages buckwheat was grown on certain plots in raised bogs, which were burned first. In considering buying a bog area it should be known beforehand to what extent such burning, or maybe former peat cutting, have damaged the area. In the case of fertilized grasslands the possibility and desirability to enlarge the differentiation by diminishing the amount of nutrients in the soil should be considered. Heathlands deteriorated by the ingrowth of trees and grasses can be improved by the removal of organic material and re-introduction of sheep. Sometimes restoration is possible

266

by adequate internal management to compensate failing external management (e.g. in moorland pools which have become eutrophic).

The potential significance also is important in cases of creation of habitats in so-called 'future nature reserves'. Examples can be found in the IJsselmeerpolders, the Grevelingen basin, the Veerse Meer and the Lauwersmeer. New reserves are created by differentiating the abiotic environment artificially.

*Ad 6.* The possibility to preserve an area in the future is determined by juridical factors and/or those of external management. It is a fiction to think that nature reserves are 'safe'. One just has to think about official reserves that nevertheless were destroyed by action of the state, like the Beer, the Rietput near Callantsoog, reserves in the so called Delta area, the Twiske in the Zaan region, the Hogt near Waalre, etc. The past shows us that generally state owned nature reserves were less safe than those of private organizations. The destiny of the first ones can be changed by just one signature of a Minister of State, while private organizations can resist more effectively because of proprietary rights and by means of actions. Private reserves can generally just be withdrawn from their destiny by expropriation. The situation in England is more favourable: ever since 1907 most properties of the National Trust are safeguarded unless parliament decides otherwise.

The future maintenance depends also on the size and boundaries of the reserve. The desired minimal size is not determined in the first place by the minimal area the biocoenosis needs to develop completely, but much more by the present and expected future situation in the area. Important are the hydrology and the eutrophication, by air (fertilizer blown in, air pollution) as well as by flowing in of polluted water. The relief and the extent of the future recreation are important too. If the ground water level in the area is going to be lowered the minimal size of the reserve should be many times larger than that of a reserve in an area with a consistent level. It is impossible to show figures applying to all situations, all depends on local circumstances. If the water level and the eutrophication are under control, a litter fen of 1 ha can be sufficient from a botanical point of view. If not, it is most likely that 50 ha will be too small to maintain the vegetation.

Ever since 1926 the choice of nature reserves is made by means of a nationwide inventory of what was called in those days 'natural beauty' (de Hoogh 1939). At first this was done at the initiative of the private committee 'Het Nederlandse Landschap' ('the Dutch landscape'), formed in 1929 by 'Natuurmonumenten' and the ANWB. The objectives proved to take a lot of time and work, and in 1928 the committee asked the government to have the inventory done by the State Forest Service. In 1929 The State Forest Service was commissioned to make up a systematic inventory, based on botanical, zoological, geological and esthetical (landscape) criteria. An integration of these data was out of the question in those days. The boundaries of the areas selected were plotted in different colours on ordnance survey maps, scale 1:25,000. By the time the project was finished in 1938, the data appeared to be out of date already. Therefore, a re-mapping project was started. Thanks to the cooperation of W.H. Diemont, G. Sissingh, and J. Vlieger phytosociological criteria could be added. After 1956 the inventory was continued by the RIVON, with important contributions of several institutes and persons (Westhoff 1978b).

One private initiative is worth mentioning: in 1939 the 'Contactcommissie inzake natuurbescherming' commissioned a survey named 'Het voornaamste natuurschoon in Nederland' (most important natural beauty in the Netherlands). According to the compilers (G.A. Brouwer, H. Cleyndert Azn., W.G. van der Kloot, Jac.P. Thijsse and Th. Weevers) this list only contained the most important objects, totalling 747 sites. Under the auspices of the Nature Conservancy Council the 'Bolwerkgroep' realized the mapping of the valuable natural and cultural sites in the rural parts of the Netherlands, on two maps (scale 1:250,000), each dealing with different aspects and with a seperate explanation (Ministry of Cultural Affairs, Recreation, and Social Welfare 1979). Apart from nature, the map dealing with natural values also shows sites of geological importance, special landscapes, areas important for meadow birds and wild geese, and agricultural areas with a high density of biologically important objects. The map dealing with cultural values shows important archeological monuments, old farms, patterns of parcelling, castles and estates, windmills and historical fortifications.

For an excellent British example of this type of survey it is referred to Ratcliffe (1977).

The Ministry of Cultural Affairs, Recreation and Social Welfare (C.R.M.) decides on the national plan of purchase made up by the N.W.C. (Scientific Committee of the Nature Conservancy Council). In each province the three organizations which purchase land to create nature reserves (the State Forest Service, 'Natuurmonumenten' and the provincial private organization, the 'Landschap') confer regularly, often under the chairmanship of the deputy of the provincial government who deals with matters of nature conservation. In most provinces the areas of purchase have been divided beforehand by the above mentioned. Since 1954 the purchases of the private organizations are subsidized by the C.R.M.-ministry. C.R.M. mostly pays 50%, as is done by the provinces. Sometimes nationwide actions are necessary to complement the amount of money needed to buy very large areas as was the case in the purchase of the Deelerwoud on the Veluwe (1967, 1000 ha). Since about 1964 the C.R.M.-ministry also contributes towards the costs of management of private nature reserves. In 1977 around Hfl.8,800,000.– was turned over to 'Natuurmonumenten', the provincial organizations and a few others.

In 1941 the first state nature reserve was purchased, the Kralose Heide. After 1946 other purchases were made regularly, thanks to a consistent vote on the budget. These state nature reserves come under the jurisdiction of the Ministry of C.R.M. and are managed by the State Forest Service.

Besides purchasing (the most effective) other possible methods to preserve natural areas are according to the civil law: tenure by long lease, tenancy, rent, and agreements on management. It is also possible according to public law to have an area denominated as a protected nature reserve (the above mentioned Nature Conservation Act).

268

## 6. Vegetation science as a basis for management plans

In the Netherlands it becomes common use to draw up management plans for nature reserves to guarantee the continuation in management in case of staff changing, to have a good account of all public rights and obligations, and to plan well the required manpower and material for the reserve management. These management plans have generally an operating period of ten years and the management for all different parts of the reserve concerned are described in detail from year to year. As the draw up of a management plan after the method Hoogenhout (1972) consumes quite a lot of time, it will be necessary to start for many reserves with a provisional plan, so-called management directives.

Since 1974 the Ministry of Cultural Affairs, Recreation and Social Welfare required of the state-aided private organizations on nature conservation to draw up management directives containing information on the following details (compare Stamp (1974) and Usher (1973) for the situation in Britain):

I. *Orientation and aims*
  1. General orientation
  1.1 Description of the reserve area
  1.2 Town and country planning
  1.3 Menaces and external developments
  2. Aims

II. *Inventory and evaluation*
  1. Area in property, buildings and cadastral details
  2. Abiotic environmental features
  2.1 Geology and geomorphology
  2.2 Soils
  2.3 Hydrological details
  3. Biotic environmental features
  3.1 Flora and vegetation
  3.2 Fauna
  4. Cultural history and archeology
  5. Public admission and ways of access
  6. Recreation
  7. Public rights and obligations
  8. Staff, guard and supervision
  9. Account and evaluation of the management up till present
  10. Incomes and costs
  11. Documents

III. *Planning*
  1. General planning vision on long term
  2. Planning on half long and long term, specified after details, mentioned under II

For drawing up a management plan general rules are followed (see pag. 255). As local conditions are always different, a plan needs to be worked out for every reserve. Every organization has its own particular way to realize these management plans or directives. Mutual arrangement between research workers and managers is necessary. The choice of the aims for a particular reserve is based on the integration of biological inventory, local (psychological-social) circumstances, and the policy of

269

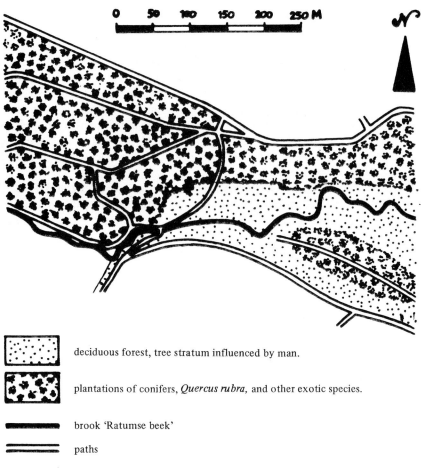

0    50    100    150    200    250 M

deciduous forest, tree stratum influenced by man.

plantations of conifers, *Quercus rubra,* and other exotic species.

brook 'Ratumse beek'

paths

Fig. 9. Dottinkrade. Situation in 1969.

the conservation organization. The selected management methods should agree with the aims and should be based on an evaluation of the previous management. If it is decided to change the management, the possibilities for a continued effective management along the new lines should be ensured. Financial and social aspects are important in this respect (Hessels 1978).

Field inventories and the use of existing documentation help to ascertain the intrinsic value of a nature reserve. These usually include the abiotic environment, flora and vegetation, fauna, cultural-historical features, and archeology. Depending on the vulnerability of the ecosystems present in the reserve and/or those to be developed, it can be established which functions the reserve may exert (Anon. 1978). In many cases it is important to decide to what degree a particular function will be acceptable. Because of public rights or agreements according to civil law the realization of the desired functions may be restricted.

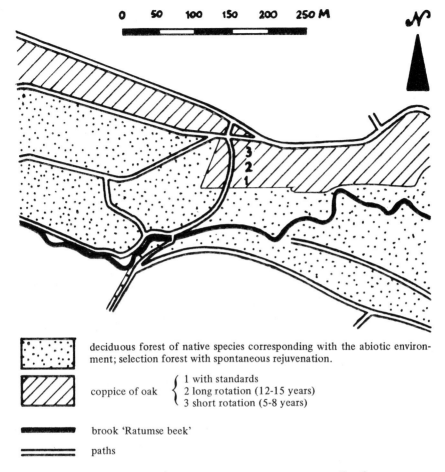

0    50    100    150    200    250 M

deciduous forest of native species corresponding with the abiotic environment; selection forest with spontaneous rejuvenation.

coppice of oak { 1 with standards
                 2 long rotation (12-15 years)
                 3 short rotation (5-8 years)

brook 'Ratumse beek'

paths

Fig. 10. Dottinkrade. Situation in future according to the management directives.

For a real nature reserve the aim will be the conservation and development of diversity of ecosystems. In the case of more than one use of a natural area it is no real nature reserve. Then a multiple use strategy must be developed, or a segregation of areas applied, which means that parts of the natural area will have different main functions. In the case of recreational use, one may try to concentrate people in certain parts of the area to save other parts.

For an integrated management plan of a nature reserve a vegetation map will be necessary (Tideman 1963, Diemont 1970, Westhoff 1954). For a simple management device a short description of flora and vegetation or maps of 'indicator species' may be sufficient. In the case of the reserve Dijkmanshuizen on the Island of Texel, maps of only three species, viz. *Dactylorhiza maculata, Orchis morio* and *Armeria maritima* (made in one day) revealed the existing microgradients and features of the local hydrology. In many cases maps based on Londo's (1974) simplified system of syntaxonomical-structural vegetation units were sufficient for a proper management plan. This will be illustrated with two reserves of the Society for the Preservation of Nature Reserves.

The reserve Dottinkade (Fig. 9 and 10), situated 3 km north-east of Winterswijk, is a woody area of 21 ha along the brook 'Ratumse beek'. On both sides of the meandering brook, we find fertile loamy, calcareous deposits. Further away from this brook there is a gradual change into poor, acid, podzolized, loamy sand. The areas near to the brook are periodically inundated. This gradient situation is expressed in a gradual change from an *Alno-Padion* wood towards a *Quercion robori-petraeae* wood. The area directly influenced by the eutrophic water contains a mosaic of *Pruno-Fraxinetum* and *Fraxino-Ulmetum* woods, with many species, including *Primula elatior, Impatiens noli-tangere, Equisetum hiemale* and *Campanula trachelium*, and with a well developed shrub layer. Further away from the brook a *Fago-Quercetum* wood occurs with *Pteridium aquilinum Teucrium scorodonia, Solidago virgaurea* and *Maianthemum bifolium*. On the high, podzolized soils we find a *Querco roboris-Betuletum* wood and some relics of the former heathland, both with *Calluna vulgaris, Erica tetralix, Vaccinium myrtillus, Vaccinium vitis-idaea* and *Molinia caerulea*. The remainder of the area contains planted woods. By the time of the purchase in 1969 about half the area consisted of pure or mixed conifer stands.

The management aim for the reserve was chosen as follows: conservation and development of those environmental conditions as to obtain a biologically optimally diverse woodland. To achieve this situation the present gradient situation will be reinforced and the exotic conifers and broad-leaved trees (*Quercus rubra*) will be removed after they have matured. For avifaunistic purposes a few pine groups will be maintained. In the clearings spontaneous growth of native species is awaited. In well-developed woods small clearings (5-20 are) will be made ('selection forest') as has been done traditionally by the local farmers. These clearings will be spread in time to obtain a varied age structure of the trees. In the woods on the former heathland organic material will be removed through heathland exploitation and coppice management with varying rotation times.

Thus, management of the major part of the reserve will be largely the same (Fig. 10). By increasing environmental differentiation and decreasing human disturbance a considerable ecosystem variety can be obtained, even in such a small reserve. In this case a superficial inventory of flora and vegetation, in combination with some old maps and records appeared to be sufficient for the drawing up of the management directives.

The second example concerns the 160 ha semi-natural wetland reserve 'Kierse Wijde' near Wanneperveen (cf. Duffey 1971 for a similar but less elaborated British example). This is an old peat excavation area in the about 4200 ha fenland reserve 'The Wieden'. A vegetation map, scale 1:5000, of the reserve part 'Kierse Wijde' was drawn after false colour aerial photographs and checked through field reconnaissance in 1972 by H. Piek (Fig. 12). 'Kierse Wijde' contains a small-scale mosaic of overgrowing peat cuttings, reedbeds, hayfields and pastures, 'floating fens', forb and shrub vegetation, and carr. The area is situated in the macro-gradient from the

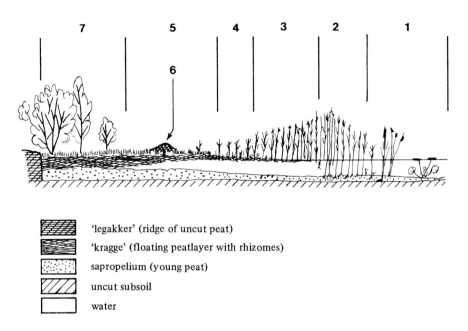

| | legakker (hatched) | 'legakker' (ridge of uncut peat) |
| | kragge | 'kragge' (floating peatlayer with rhizomes) |
| | sapropelium | sapropelium (young peat) |
| | subsoil | uncut subsoil |
| | water | water |

| | Vegetation types | Required internal management |
|---|---|---|
| 1 | { *Potameto-Nupharetum*<br>*Scirpetum lacustris*<br>*Typhetum angustifoliae* | doing nothing |
| 2 | *Scirpo-Phragmitetum caricetosum pseudocyperi*<br>or *solanetosum (*or *Cicution)* | mowing only in<br>severe winters on the ice |
| 3 | { *Scirpo-Phragmitetum ranunculetosum*<br>*Thelypterido-Phragmitetum* | annually mowing in winter |
| 4 | { *Scorpidio-Caricetum diandrae*<br>*Pallavicinio-Sphagnetum* | annually mowing in<br>autumn |
| 5 | { *Scorpidio-Caricetum diandrae*<br>*Sphagno-Caricetum lasiocarpae*<br>*Pallavicinio-Sphagnetum* | annually mowing in<br>late summer |
| 6 | *Sphagno palustri-papillosi* | annually mowing in late<br>summer or doing nothing |
| 7 | { *Salicion cinereae*<br>*Alnion glutinosae* | doing nothing |

Fig. 11. Succession in a peathole in the 'Kierse Wijde' near Wanneperveen and required inward nature technical measures (After H. Piek 1972).

Pleistocene coversand plateau of Drenthe towards the more eutrophic peat area of North-west Overijssel. At many places in 'Kierse Wijde' where coversand ridges occur near the surface hardly any peat was formed.

From the Middle Ages until the 19th century the *Sphagnum* peat was excavated for fuel. The peat was dug from long narrow strips which resulted in canal-like lakes

273

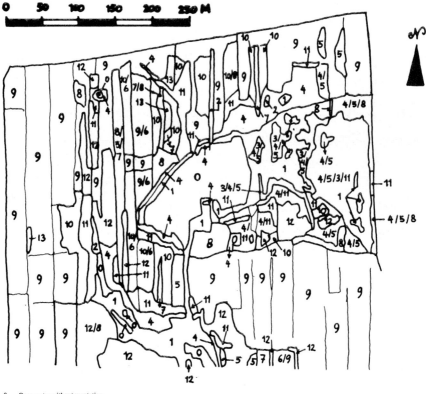

| | | |
|---|---|---|
| 0 | Open water without vegetation | |
| 1 | *Nymphaeion* | a *Potameto-Nupharetum*<br>b *Nymphoidetum peltatae* |
| 2 | *Hydrocharition* | a *Hydrocharito-Stratiotetum*<br>b sociation of *Utricularia vulgaris* |
| 3 | *Cicuto-Caricetum pseudocyperi* | |
| 4 | *Phragmition* | a *Scirpetum lacustris*<br>b *Typhetum angustifoliae*<br>c *Scirpo-Phragmitetum*<br>d *Thelypterido-Phragmitetum* |
| 5 | *Magnocaricion* | a *Caricetum paniculatae*<br>b veg. with dominance of *Carex acutiformis, Carex riparia* or *Phalaris arundinacea* |
| 6 | *Molinietalia*-communities. a.o. *Senecioni-Brometum racemosi* (*Calthion palustris*) | |
| 7 | *Cirsio-Molinietum* | |
| 8 | *Valeriano-Filipenduletum* | |
| 9 | | a *Lolio-Cynosuretum*<br>b *Poo-Lolietum* |
| 10 | *Caricion curto-nigrae* | a *Sphagno-Caricetum lasiocarpae*<br>b *Pallavicinio-Sphagnetum*<br>c *Caricetum curto-nigrae* |
| 11 | *Scorpidio-Caricetum diandrae* | |
| 12 | *Carici elongatae-Alnetum, Alno-Padion, Myricetum gale, Salicetum pentandro-cinereae, Frangulo-Salicetum auritae* or *Alno-Salicetum cinereae* | |
| 13 | *Prunetalia spinosae* | |

Fig. 12. Vegetation map of the northern part of the 'Kierse Wijde' near Wanneperveen (Nature reserve 'De Wieden' in North-West Overijssel) (After H. Piek 1972).

274

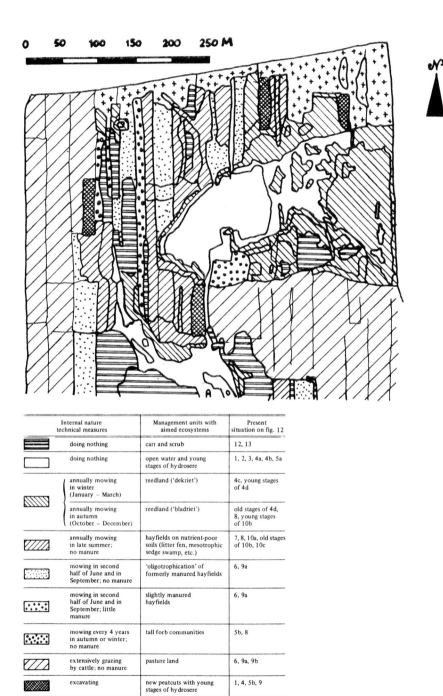

| Internal nature technical measures | Management units with aimed ecosystems | Present situation on fig. 12 |
|---|---|---|
| doing nothing | carr and scrub | 12, 13 |
| doing nothing | open water and young stages of hydrosere | 1, 2, 3, 4a, 4b, 5a |
| annually mowing in winter (January – March) | reedland ('dekriet') | 4c, young stages of 4d |
| annually mowing in autumn (October – December) | reedland ('bladriet') | old stages of 4d, 8, young stages of 10b |
| annually mowing in late summer; no manure | hayfields on nutrient-poor soils (litter fen, mesotrophic sedge swamp, etc.) | 7, 8, 10a, old stages of 10b, 10c |
| mowing in second half of June and in September; no manure | 'oligotrophication' of formerly manured hayfields | 6, 9a |
| mowing in second half of June and in September; little manure | slightly manured hayfields | 6, 9a |
| mowing every 4 years in autumn or winter; no manure | tall forb communities | 5b, 8 |
| extensively grazing by cattle; no manure | pasture land | 6, 9a, 9b |
| excavating | new peatcuts with young stages of hydrosere | 1, 4, 5b, 9 |

Fig. 13. Management units in the 'Kierse Wijde' with types of internal technical measures (After H. Piek 1972).

275

('petgaten'), separated by small ridges of uncut peat, used for drying the excavated peatmud ('legakkers'). Later on the ridges disappeared locally through erosion by wave action and larger lakes were formed, while at other places they persisted and new peat was formed (Gorter 1964, Segal 1966).

A duck decoy in the area contains *Carici elongatae – Alnetum* woods, while locally on sandridges fragments of *Alno-Padion* wood and *Prunetalia spinosae* scrub are found. Along ditches in manured pastures in the gradient from coversand ridges towards the lower peat filled basins, a *Prunetalia* scrub developed spontaneously. In many of the peat cuts the succession starts with *Nymphaeion* and *Phragmition* communities (Fig. 11). It often happens that submerse peat with rhizomes of *Phragmites* and *Typha* 'jumps upwards' and becomes floating (end of phase 2). Peat forming continues by reed and mosses and in wintertime the reed is cut as soon as the peatlayer supports the weight of one man (phase 3). With increasing isolation from the eutrophic water the vegetation on this growing peatlayer gradually develops towards a *Sphagnum* reedbed when cut in wintertime or autumn, or towards a so-called 'floating fen' ('trilveen') when cut in summertime (phase 4 and 5). The Dutch 'trilveen' refers to a vegetation type resembling that of the 'lagg' around raised bogs on a floating peatsoil. It is a quaking sedge swamp. If the topsoil becomes acid and poor in nutrients, an oligotrophic vegetation succession towards a bog may start (phase 6). When the yearly mowing is stopped before an ombrotrophic vegetation has become established, a rapid regression towards carr may occur (phase 7).

From a conservation point of view, the most interesting phases of the sere are the quaking sedge swamps (*Scorpidio-Caricetum diandrae* and *Sphagno-Caricetum lasiocarpae*), the initial bog vegetation (*Sphagnetum palustri-papillosi*), the eutrophic marsh vegetation (*Calthion palustris*) and the litter fen (*Cirsio-Molinietum*), the latter type occurring locally on the uncut peat ridges. In this situation, where a micro-gradient has developed between old peat and floating young peat, *Carex buxbaumii* occurs. The ecology of this very rare species was studied by Segal & Westhoff (1959) in this particular area.

In order to draw up the management plan for this area the 13 vegetation types distinguished on the vegetation map (Fig. 12) were grouped into 10 management units, characterized by the same management type (Fig. 13). (On this scale not more than 13 vegetation types could be drawn on the map, but in the field 30 syntaxa were distinguished which were noted down in the detailed descriptions of every management unit. Mowing of reedland in autumn or in winter could not be separated in Fig. 13). A separate management method useful for this area consists of excavating new peat cuts to maintain the complete sere. In fact this is a form of habitat creation. Estimating that the succession from open water towards initial bog vegetation will take about a century, we can calculate the area which must be yearly excavated to maintain all phases of the sere. Places where new peatcuts are planned, are indicated on the management map. The management plan also gives the indicated times of excavation.

The surrounding grasslands on the coversand ridges in the reserve area (*Lolio-*

*Cynosuretum* and *Poo-Lolietum*) will be safegarded from heavy exploitation as soon as they will be free from tenancy. In future they will be extensively pastured with cattle, starting with 1 cow per ha, later on decreasing to 1 cow per 3 ha. It is expected that the *Prunetalia* scrub will spread. The unmanured hay fields (litter fen, quaking sedge swamp, etc.) are mown in late summer, whereas the reedbeds will be cut in wintertime ('dekriet' for thatched roofs) or in autumn ('bladriet': cover for flower beds, etc.). The *Calthion* marsh and the adjoining *Lolio-Cynosuretum* grasslands are mown twice a year (second half of June and in September) and slightly manured with ditch mud, raw manure or following the traditional manuring with collected plants of *Stratiotes aloides*. The forb vegetation (*Filipendulion* and *Magnocaricion*) is cut once every four years in autumn or winter. Open water, scrub and carr will not be treated by management work in this area. The hayfields which are well manured now and situated in the area of vulnerable meso- or oligotrophic communities, will be 'oligotrophicated' by removing organic material through mowing them twice a year.

## 7. Some experiences from management practice

A good reporting system is essential for long term management planning. All management work performed has to be noted down with care, just as all changes arising at and around the reserve which may effect the ecosystems to be maintained.

By means of periodical floristic and phytosociological investigations, preferably including permanent plot studies and mapping, at least of indicator species, changes in the reserve can be accurately identified. Westhoff (1969) distinguished six different ways of changes in vegetation. Usually the changes are caused by disturbances from outside the reserve (e.g. lowering of the watertable in the surroundings of the reserve, eutrophication and heavy recreation) or by sudden alterations in the internal management (stopping of hay-making, grazing or chopping as traditionally done by the local farmers, or the introduction of new methods of management such as burning, heavy trampling, and the use of fertilizers).

Mapping the vegetation every five or ten years appeared to be sufficient in cases of slow changes. We have such series of vegetation maps available now for several nature reserves, but their number could be much enlarged. Examples are (Nijland 1974): the island of Griend in the Wadden Sea (1932, 1936, 1941, 1963, 1969, 1972), the Oosterkwelder on the Wadden island of Schiermonnikoog, parts of the Kwade Hoek, a coastal reserve on Goeree (1965, 1977), Wooldse Veen near Winterswijk (1952, 1978), pine plantations on drift sand near Kootwijk, parts of the Biesbosch estuary, Kil of Hurwenen, Broekhuizerbroek and Baronie Cranendonck.

Unfortunately investigations by means of permanent plots reached a deadlock by lack of manpower. Apart from this, there are also cases of neglecting the management in or outside the permanent plot, which makes further investigation useless. A summary of the situation up to 1951 is given by Westhoff & Van Dijk (1952).

Lately, increasing attention is given to the use of plant species or plant communities as environmental indicators. Fundamental research about this subject is done particularly by the Research Institute for Nature Management at Leersum (Anon. 1974). Assuming that the presence of a given plant species at a particular place is an expression of the local environmental dynamics[1], observations on the local distribution of ecologically related species may produce information about the processes taking place in the particular locality. In situations where a series of ecologically related species occurs it will be possible to trace an environmental factor at several degrees of intensity. In this way the following series of *Filipendulion* species of damp habitats is used to indicate the degree of decomposition of organic material and nutrient status. From a low to high degree of environmental dynamics we list: *Peucedanum palustre* – *Filipendula ulmaria* – *Lysimachia vulgaris* – *Valeriana officinalis* – *Epilobium hirsutum* – *Eupatorium cannabinum* – *Urtica dioica*. If we find some of these species in a spatial zonation, a fertility gradient may be expected. If a succession of species is observed corresponding to this sere, an increase in its nutrient level may be deducted.

Similarly series of species of the genera *Carex, Plantago, Rumex, Trifolium, Ranunculus, Taraxacum*, and of the family *Orchidaceae* can be used as management indicators. A useful series in damp hayfields on peaty soils reflecting increasing anthropogenous environmental dynamics[2] are the following *Carex* species: *Carex dioica*– *C. pulicaris* – *C. hostiana* – *C. panicea* – *C. nigra* – *C. disticha* – *C. otrubae*. A critical level in management is indicated by *Carex hostiana*, where the differences in haymaking by skite or by light tractor may be of relatively great ecological importance. On the other hand, *Carex nigra* and the other species lower in the sere are thought to indicate haymaking by a four-wheel tractor.

Unfortunately it turned out that some species and some plant communities, doomed to disappear from landscapes with highly intensified agricultural use, can either not at all or with difficulty be maintained in nature reserves. Examples are *Nanocyperion flavescentis* communities depending on irregular trampling of tracks in heath, the complex vegetation of wood verges (*Trifolio-Geranietea sanguinei*), and the vegetation of ruderal places (a.o. *Onopordion acanthii* and *Arction*). These types of plant communities were obviously maintained by the traditional way of farming, depending on a particular degree of D.A.M.

The most important external factors effecting natural areas are draining, eutrophication and excessive recreation. The lowering of the water table everywhere in the

1. environmental dynamics (Dutch: milieudynamiek) = the sum of the dynamical factors such as heat, nutrients, solar energy, activity of men and animals and also their fluctuations in intensity (see pag. 263).
2. antropogenously added environmental dynamics (Dynamics Added by Man, D.A.M., see pag. 263) implies in this case: activity of men during haymaking, the removal of nutrients by harvesting of the sward, and trampling during cutting and collecting, or, as will happen nowadays, riding by tractors.

Netherlands destroyed many habitats. Research on the influence of lowering the ground water table has started here only about 1973 (Tüxen & Grootjans 1978a, b). Besides the obvious disappearance of phreatophytes (taxa occurring under permanent influence of the ground water table; Londo 1975) there may be indirect effects, such as increased mineralization, and desalination of formerly brackish areas. Peaty soils may dry out irreversibly and become more mineral. This all leads to eutrophication in habitats which were originally poor in nutrients. Species like *Glyceria maxima, Iris pseudacorus, Juncus effusus,* and *Urtica dioica* may then become established. Even periodical lowering of the water table for a short time in wet hayfields, e.g. *Cirsio-Molinietum* to facilitate haymaking, has already adverse effects (Bink 1978). Examples of desalination are found in the polders of the Island of Texel (water table lowering due to agricultural amelioration) and in the Southwest Netherlands (enclosure of estuaries within the scope of the Delta statute). Small reserves are definitely lost in cases of lowering the water table in the surroundings (cf. Smeets et al. 1979). Larger reserves may be saved by establishing a buffer zone around them. In this context it is worth mentioning that P.G. van Tienhoven, a president of the Society for the Preservation of Nature Reserves for many years, strived after the establishment of large reserves from the beginning. A good example is his 'own' achievement — the establishment of the large and outstanding heathland reserve Kampina.

Eutrophication is possible by air (dust drift of fertilizers, sulphuric acid pollution from industrial areas) and by water (run off from agricultural areas, etc.). A natural way of eutrophication will take place when colonies of birds start breeding in an oligotrophic moorland pool, while foraging outside the area, e.g. the black-headed gull. A well-known example of eutrophication in a reserve, i.e. the lake 'Choorven' near Oisterwijk, was caused by a clandestine discharge of sewage from a well-visited pub. This originally oligotrophic lake had become completely overgrown by about 1947 by eutraphent plant species. In 1950 the discharge was stopped and the water was pumped out of the pool. After that it was possible to dig out the mud by hand. The work was carried out carefully to avoid damaging the sandy soil. The pool was filled again in a natural way and the originally occurring *Eleocharitetum multicaulis* became re-established (Van Dijk & Westhoff 1960). However, later on the pool appeared to be more oligotrophic than it was before by the eutrophication by the sewage discharge, probably because of the general acidification of the $SO_2$-polluted rain (Van Dam & Kooyman-van Blokland 1978).

The effects of water pollution are most pronounced in wetlands; even very large reserves are affected nowadays. The considerable increase of outdoor recreation in these areas is certainly one of the major factors. In polders where the water table is regulated by waterboards problems with the water management occur as well. In summer, when the authorities promote a higher water table, polluted water is led into the polder; in winter the water table is lowered and large quantities of rather clear water are pumped out. In this way the lake reserves 'Loosdrechtse and Kortenhoefse Plassen' are polluted. The only solution in this case would be to establish very large reserve areas, but even then there may be problems. E.g., the lake reserve Naardermeer covers 750 ha and includes the outlet to the dirty river Vecht in the neighbourhood. From 1959 onwards no polluted water has been admitted from the river into the reserve. Nevertheless, pollution occurs in the reserve because of seepage of heavily polluted water from the drain 'Karnemelksloot' which borders the reserve. Another factor is that the former natural seepage of iron-rich water from the high sandy area 'het Gooi' has nearly stopped now. Thus there is less precipitation of phosphate nowadays and that means new eutrophication. Another unforeseen change is the lowering of the water table with 30-40 cm in dry summers, because of seepage towards adjoining polders where the water table was lowered in 1962.

Another important aspect is the geomorphological constitution. When an oligotrophic habitat borders an eutrophic one, e.g. a heavily manured field, the oligotrophic habitat will certainly be lost when it is situated at a lower level. The opposite situation may be rather safe, however. This happens for instance in South Limburg on the slopes of the chalk hills and river valleys. In most situations the top layers are rather poor, sandyloamy sediments. Still, there is a threat here, viz. in the use of fertilizers and manure in adjoining fields on the plateau which affects the slopes. Thus there is a need for a buffering zone of woodland at the edge of the plateau.

The internal management of nature reserves will now be exemplified with grazing and woodland management. The Research Institute for Nature Management has been studying the perspectives of grazing as a management technique. A first outcome are directives about the kind of animal and the breed to be used, the number per area and the total area needed (Oosterveld 1975, 1977). The 'Rare Breeds Survival Trust' (Stichting Zeldzame Huisdierrassen), established in 1976, aims at preserving 'primitive' breeds of domestic animal. These breeds generally need little care and can be kept in the field the year round. Grazing of heathland was traditionally done by flocks of sheep with shepherds, but this is very expensive nowadays. One solution has been tried with some success by fencing the area of sheep grazing, and thus not needing a shepherd any more.

Experiments for 'oligotrophication' (i.e. decreasing the amount of nutrients in the soil) by mowing and grazing are carried out by the Laboratory for Plant Ecology, University of Groningen (Bakker 1976a, b), the Division of Geobotany of the University of Nijmegen (van der Maarel in prep.) and the Centre for Agrobiological Research (CABO), Wageningen (Oomes 1976).

Until recently there was hardly any research on nature technical methods in forests, (except by Diemont and Sissingh, see Sissingh 1977). Diemont (1956) stated that in the forest reserves in South Limburg rejuvenation by means of planting should take place only on a very small scale. The soil should be spared as much as possible and only individual trees or small groups of trees should be felled at the same time. At some places there should be no human interference at all.

Westhoff studied the position of *Fagus sylvatica* in the forests of Middle and

West Europe, with special reference to the Forêt de Fontainebleau. Parts of this forest were made strict reserves in 1854 and the succession has since proceeded from *Querco-Carpinetum* to a pure *Fagetum*. The *Querco-Carpinetum* was maintained outside the absolute reserves by normal forestry (Doing Kraft & Westhoff 1959). The spontaneous succession leads towards a forest type that the painters of Ecole de Barbizon would not have appreciated!

Since 1976 forestry methods after nature-technical principles are in the focus of interest in management circles, particularly after the large-scale damages by gales in November 1972 and April 1973. In most cases the felled woodlands were conifer plantations (van Herwerden 1975).

In South Limburg Diemont developed plantation methods on a phytosociological basis, by replacing the species that were considered alien to the respective woodland communities. Now this principle has been abandoned for various reasons and only spontaneous establishment of indigenous species will be accepted as a source for rejuvenation of forests in nature reserves. Plantations will only be accepted in cases of buffer zones, screens and restoration of decaying wooded banks and coppices.

Too little research has been done on the effects of exotic species in reserves. Obviously, species like *Prunus serotina, Amelanchier lamarckii, Robinia pseudoacacia, Quercus rubra, Aronia spp.* have spread explosively in many forests at the expense of native species. The status of *Pinus sylvestris* is different. From palynological research it is clear that this species has been present from the last glacial period onwards. In historical times *Pinus sylvestris* persisted with low density as a pioneer on the poorest soils: on shifting sands and in bogs. However, the recent extension in its distribution in our country is largely due to plantation which, moreover, often occurred with foreign races or ecotypes.

The promising establishment of real coniferous woodland species as *Goodyera repens, Listera cordata*, and *Linnaea borealis* in some of our pine plantations in former heathland areas probably just depends on a particular phase in the succession due to accumulation of organic material in the topsoil of these plantations. In some of these pine plantations these 'conifer forest neophytes' occurred for a short period only. In many of these plantations more trivial understorey species become dominant.

To turn a production-plantation into a real forest with its typical self-regulation mechanisms is a process that can proceed only very slowly and would need at least a century. By means of selective cutting we try to improve the vertical and horizontal structure of these forests (larger variation in tree ages; more variation in stem densities, from nearly unaccessible parts to open spaces of varying sizes). In the course of time the intensity of the cuttings must be lowered (Londo 1977b).

## 8. Creation of habitats

By means of habitat construction following the principles of nature techniques we can try to create suitable environmental conditions for particular communities.

Well-known examples are to be found in the South-west Netherlands (Veerse Meer, Grevelingen basin), in the new polders in the IJsselmeer, and in the Lauwersmeer. Recently some small-scale habitat constructions have been carried out at some places in the dunes and on the Pleistocene sandy soils.

Beeftink (1973) drew up six principles for habitat creation analogous to the basic rules for botanical management by van Leeuwen (1966). The first three rules concern the internal, the next three the external management:

(1) introduce environmental differences by contrasting environmental factors: acid should dominate over alkaline; nutrient-poor over nutrient-rich; fresh over salt; dry over wet;

(2) situate the environmental extremes at such distances that stable gradients may develop;

(3) carry out operations in the area in a spatially varied and temporally spread manner;

(4) adjust differences in environmental features to the natural features of soil profile and topography;

(5) introduce contrasting environmental factors, but spread them in time in such a way that environmental changes will proceed gradually;

(6) adapt human activities as much as possible to locally traditional forms of agricultural and urban techniques.

The reclamation or embankment of salty or brackish areas mean a disturbance (disappearance of the tidal influence, desalination, etc.). On the other hand, it is possible to create a large variation in the newly formed environments, e.g. by placing temporary wind screans on former sandbanks. The subsequent establishment of vegetation should be spontaneous; sowing of cereals or grasses is wrong from the point of view of nature technique. On mud flats in the former tidal area which are now permanently dry, a good turf will establish through rough pasture (Beeftink 1975).

Londo (1971) gives a good account of the possibilities of habitat creation through excavation in dunes. According to Londo it is of no use to enlarge the number of artificial lakes in the dunes, but instead shallow dune slacks should be created. One should take some general hydrological and geomorphological features into account, such as the phreatic level and the relief (both shallow and steep slopes are important). One should avoid disturbance of soil structure and prevent mixing of organic material or raw humus with the sandy soil. As some shifting of sand may be useful, the newly formed bare slopes should not be planted with shrubs.

Examples are provided by the small scale excavations which were carried out according to the rules mentioned above near the 'Brede Water' in the dunes of Voorne in 1966. In these artificial dune slacks vegetation types belonging to the *Caricion davallianae* have developed, with rare species such as *Parnassia palustris*, *Blackstonia perfoliata* subsp. *serotina*, *Epipactis palustris* and *Liparis loeselii*.

An important form of habitat creation is renovation, as applied to eutrophicated oligotrophic moorland pools. If the source of eutrophication has already been abolished and the local hydrology is still suitable, excavation down to the mineral soil and removal of all detritus will lead to regeneration of *Littorellion uniflorae*. Examples are the 'Brunstinger plassen' near Hijken (Dr.) and the 'Kampina' near Boxtel.

A number of other examples can be mentioned briefly.

In the area of the riverbeds of the Rhine-system some botanical reserves could be established in old claypits; if they have a calcareous bottom and a favourable hydrology the *Caricion davallianae* may develop.

Paths and tracks may have a positive effect on variation in vegetation when they are running more or less perpendicular to soil boundaries or isohypses. Along unhardened paths a trampling gradient may develop which favours species of such micro-gradients.

The creation of new peatcuts in our fenlands is now urgent because we lack the early stages in the succession series.

Finally, it is worth drawing attention to the artificial gradients in experimental plots at the Research Institute for Nature Management at Leersum and in some private experimental gardens. Londo (1977a) gives detailed directives on how to create and maintain 'nature gardens' and 'nature parks'. By creating relief and using suitable soils it is possible to obtain vegetation types of relatively high diversity which need little maintenance.

## 9. Wishes for the future

In the future more and more difficulties in the management of nature reserves will arise because of increasing external influences such as drainage, heavy manuring, pollution, excessive recreation, etc. Furthermore, large scale amelioration programs are still going on, even in ecologically vulnerable landscapes. The agricultural policy is still aimed at scale enlargement, intensification and mechanization. In addition there is still some loss of natural areas by urbanization, industrial development, road construction, etc. I would like to end therefore with some wishes for our work in the future:

1. A good recording system of management measures should be developed. For this purpose it will be necessary to further develop the method of mapping environmental and management indicators in addition to regular vegetation mapping and detailed vegetation analyses in transects or permanent plots.

2. Managers and research people should meet regularly. The issue of a periodical 'Technical management bulletin' is very desirable.

3. The local managers ought to have an adequate schooling in management and conservation principles and techniques. Supplementary education will be necessary in many cases.

4. Alternative management methods should be developed and modified farming methods for practical use in semi-natural landscapes should be tested. Methods to effectively convert abandoned arable land and pine plantations into nature reserves should be developed, e.g. effective and cheap management of large areas in reserves

or national parks should be realized and the possibilities for restoring heavily eu-trophicated or polluted areas should be studied.

5. Research on fenland ecosystems should be carried out (particularly on the effects of mineralization of the peaty soils, and on ways of restoration of raised bogs).

6. Research on long-term changes caused by low level human impact, e.g. in wood-land and fenland (strict reserves), should be encouraged.

7. The actual and potential botanical value of our wood plantations should be investigated in order to be able to select the best parts for the establishment of forest reserves.

8. Research into the ecology of exotic species invading our natural communities should be intensified.

9. The actual and potential value of macro-gradients in the Netherlands for large-scale planning programs should be investigated.

10. A multi-use habitat typology, based on structural features of the vegetation should be developed, for a better integration of botanic and faunistic information for management and research purposes.

11. Completion of an ecological flora of the Netherlands is desired.

12. A 'Handbook for the management of nature reserves' for the average reserve manager should be produced.

It will be important for nature conservation when some of the wishes listed here can be realized soon. The last wish will even be fulfilled very soon: the first part of the Handbook for the management of nature reserves will appear within some months.

## Acknowledgements

The author would like to express his gratitude to Dr. Chr. G. van Leeuwen, Prof. Dr. V. Westhoff and Mr. H. Piek for critically reading the manuscript. He thanks the Research Institute for Nature Management and Mr. H. Piek for their kind permission to use some figures. He is much indebted to Drs. R.J. Beintema-Hietbrink, Drs. F.A. Bink and Dr. E. van der Maarel for translating and revising the English text.

## References

Anon. 1953. Keuze en Beheer van Natuurmonumenten. Koninklijke Nederlandse Natuurhisto-rische Vereniging, Amsterdam.

Anon. 1970. Inleiding tot de Tentoonstelling 'Van Floristiek tot Moleculaire Biologie: 125 Jaar Koninklijke Nederlandse Botanische Vereniging'.

Anon. 1974. Jaarverslag 1974. Rijksinstituut voor Natuurbeheer. pp. 15-18.

Anon. 1977. Structuurvisie Natuur- en landschapsbehoud. Ministerie van Cultuur, Recreatie en Maatschappelijk Werk, 's-Gravenhage.

Anon. 1978. Doelstellingen en hoofdlijnen beleid terreinbeheer Natuurmonumenten. 12 pp.

Anon. 1979. Natuur- en Cultuurwaarden in het landelijk gebied. Inventarisatie verricht onder auspiciën van de Natuurbeschermingsraad. Ministerie van Cultuur, Recreatie en Maatschap-pelijk Werk, 's-Gravenhage (in press).

Bakker, J.P. 1976a. Botanische onderzoek ten behoeve van natuurtechnisch beheer in het Stroomdallandschap Drentsche A. Natuur en Landschap 30: 1-12.

Bakker, J.P. 1976b. Tussentijdse resultaten van een aantal beheersexperimenten in de madelanden van het Stroomdallandschap Drentsche A. Contactblad voor Oecologen 12: 81-92.

Beeftink, W.G. 1973. Ecologie en vegetatie met betrekking tot het Deltaplan. In: De Gouden Delta. pp. 81-109. Pudoc, Wageningen.

Beeftink, W.G. 1975. The ecological significance of embankment and drainage with respect to the vegetation of the South-West Netherlands. J. Ecol. 63: 423-458.

Bink, F.A. 1978. Voorlopige richtlijnen voor het beheer van blauwgraslandreservaten. RIN-rapport, Leersum.

Dam, H. van & H. Kooyman-van Blokland. 1978. Man-made changes in some Dutch moorland pools, as reflected by historical and recent data about diatoms and macrophytes. Int. Revue Ges. Hydrobiol. 63: 589-609.

Diemont, W.H. 1937. Plantensociologie en natuurbescherming. Ned. Kruidk. Arch. 47: 93-96.

Diemont, W.H. 1956. Over het wetenschappelijk verantwoord beheer van de Zuidlimburgse bosreservaten. Jaarboek 1956 K.N.B.V. 27: 10-11.

Diemont W.H. 1970. Botanisch beheer van reservaten. In: J.C. van de Kamer et al., Het verstoorde evenwicht. pp. 184-192. Oosthoek, Utrecht.

Diemont, W.H., G. Sissingh & V. Westhoff. 1954. Die Bedeutung der Pflanzensoziologie für den Naturschutz. Vegetatio 5-6: 586-594.

Dieren, J.W. van. 1929. De beteekenis van het natuurmonument voor den bioloog. 1. Voor den botanicus. Vakblad voor Biologen 10: 165-174.

Doing Kraft, H. & V. Westhoff. 1959. De plaats van de beuk (Fagus sylvatica) in het Midden- en West-Europese bos. Jaarboek Ned. Dendrol. Ver. 21: 226-254.

Donselaar, J. van. 1970. De Nederlandse natuurbescherming gezien in internationaal verband — Botanie. In: J.C. van de Kamer et al., Het verstoorde evenwicht. pp. 231-244. Oosthoek, Utrecht.

Duffey, E. 1971. The management of Woodwalton Fen: a multidisciplinary approach. In: E. Duffey & A.S. Watt (eds.), The Scientific Management of Animal and Plant Communities for Conservation. pp. 581-597. Blackwell, Oxford.

Duffey, E. 1974. Nature Reserves and Wildlife. Heinemann, London.

Dijk, J. van & V. Westhoff. 1960. Situatie en milieu van Choorven, Witven en van Esschenven in het licht de wijzigingen, die zich in het decennium 1946-1956 daarin hebben voltrokken. De veranderingen in de vegetatie van het Choorven van 1948-1955. In: Hydrobiologie van de Oisterwijkse Vennen, publ. no. 5 der Hydrobiol. Vereniging. pp. 9-24. Amsterdam.

Gabrielson, I.N. 1957. Management of nature reserves on the basis of modern scientific knowledge. Proc. and Papers 6th Technical Meeting I.U.C.N.: pp. 27-35. London.

Gorter, H.P., J. van Dijk, D.A. Docter & V. Westhoff. 1964. Conservation and management of the Netherlands lowland marshes.- Proceedings MAR Conference, I.U.C.N. Publ. New Series, 3, I-C: 248-259.

Heimans, J. 1956. Het belang van natuurreservaten voor de wetenschap. Gedenkboek Natuurmonumenten. pp. 187-192. Amsterdam.

Herwerden, P.J. van. 1975. De bebossing van stormvlakten. Benadering vanuit het natuurbehoud. Nederlands Bosbouw Tijdschrift 47: 58-71.

Hessel, P., J.T. Wildschut & T.R. Jansen. 1975. Milieukartering provincie Utrecht. Inventarisatiegegevens flora en vegetatie. Gorteria 7: 148-160.

Hessels, E.P.L. 1978. Het beheer van natuur en landschap. Landbouwkundig Tijdschrift 90: 49-53.

Hoeve, J. ter. 1973. Gedachten over ontwikkeling en behoud van natuurgebieden bij landinrichting. Cultuurtechnisch Tijdschrift 12: 193-206.

Hoogenhout, H. 1972. Beheersplannen voor natuurterreinen. Nederlands Bosbouw Tijdschrift 44: 8-18.

Hoogh, J. de. 1939. De inventarisatie van het natuurschoon in Nederland. In: Gedenkboek 40-jarig bestaan Staatsbosbeheer. pp. 143-149. 's-Gravenhage.

Kalkhoven, J.T.R., A.P.H. Stumpel & S.E. Stumpel-Rienks. 1976. Landelijke Milieukartering. Staatsuitgeverij, 's-Gravenhage.

Leeuwen, Chr.G. van. 1965. Het verband tussen natuurlijke en anthropogene landschapsvormen, bezien vanuit de betrekkingen in grensmilieu's. Gorteria 2: 93-105.

Leeuwen, Chr.G. van. 1966a. Het botanische beheer van natuurreservaten op structuur-oecologische grondslag. Gorteria 3: 16-28.

Leeuwen, Chr.G. van. 1966b. A relation theoretical approach to pattern and process in vegetation. Wentia 15: 25-46.

Leeuwen, Chr.G. van. 1967. Tussen observatie en conservatie. In: Tien jaren RIVON. pp. 38-58. Zeist.

Leeuwen, Chr.G. van. 1973. Oecologie en natuurtechniek. Natuur en Landschap 27: 57-67.

Leeuwen, Chr.G. van. 1978. Dertig jaar onderzoek voor het natuurbeheer. Rede bij het afscheid van prof. dr. M.F. Mörzer Bruyns (in press).

Londo, G. 1971. Patroon en proces in duinvalleivegetaties langs een gegraven meer in de Kennemerduinen. Thesis, Nijmegen.

Londo, G. 1974. Karteringseenheden op vegetatiekundige basis (herziene voorlopige lijst). Leersum.

Londo, G. 1975. Nederlandse lijst van hydro-, freato- en afreatofyten. RIN, Leersum.

Londo, G. 1976. Uitgangspunten en ideeën betreffende het natuurbeheer. Contactblad voor Oecologen 12: 77-81.

Londo, G. 1977a. Natuurtuinen en -parken. Thieme, Zutphen.

Londo, G. 1977b. Bossen en natuurbeheer. Nederlands Bosbouw Tijdschrift 49: 219-228.

Maarel, E. van der. 1975. Man-made natural ecosystems in environmental management and planning. In: W.H. van Dobben & R.H. Lowe-McConnell (eds.), Unifying concepts in ecology. pp. 263-274. Junk, 's-Gravenhage; Pudoc, Wageningen.

Maarel, E. van der. (in prep.). Ontwikkeling van graslandvegetaties in een voormalige boomgaard onder invloed van verschillende beheersmaatregelen.

Maarel, E. van der & P.L. Dauvellier. 1978. Naar een Globaal Ecologisch Model voor de ruimtelijke ontwikkeling van Nederland. Studierapporten Rijks Planologische Dienst 9, 2 dln. 's-Gravenhage.

Mabelis, A.A. 1977. Inventarisatie en onderzoek van ongewervelde dieren ten behoeve van het natuurbeheer. De Levende Natuur 80: 204-210.

Mörzer Bruyns, M.F. 1965a. Natuurbehoud als gemeenschapsbelang. Rede, Wageningen.

Mörzer Bruyns, M.F. 1965b. Onderzoek voor het natuurbehoud. De Levende Natuur 68: 193-201.

Mörzer Bruyns, M.F. 1967a. Wat moeten wij verstaan onder natuurbehoud? Natuur en Landschap 21: 33-49.

Mörzer Bruyns, M.F. 1967b. Value and significance of nature conservation. Nature and Man 1967: 37-47.

Nijland, G. 1974. Nieuw overzicht van de Nederlandse vegetatiekaarten. Mededelingen Landbouwhogeschool Wageningen 74-20.

Oomes, M.J. 1976. Vegelijkend beheersonderzoek aan marginale graslanden. Contactblad voor Oecologen 12: 92-99.

Oosterveld, P. 1975. Beheer en ontwikkeling van natuurreservaten door begrazing. Natuur en Landschap 29: 161-171.

Oosterveld, P. 1977. Beheer en ontwikkeling van natuurreservaten door begrazing (III). Bosbouwvoorlichting 16 (7): 94-98. Staatsbosbeheer, Utrecht.

Piek, H. 1972. De beheersproblematiek van de Wieden, in het bijzonder van het Kiersche Wijde. Voordracht 81e Dag voor het Vegetatie-onderzoek (not published).

Ratcliffe, D.A. (ed.). 1977. A nature conservation review. 2 Vols. Cambridge Univ. Press, Cambridge.

Segal, S. 1966. Ecological studies of peat-bog vegetation in the northwestern part of the province of Overijsel. Wentia 15: 109-141.

Segal, S. & V. Westhoff. 1959. Die vegetationskundliche Stellung von Carex buxbaumii Wahlenb. in Europa, besonders in den Niederlanden. Acta Bot. Neerl. 8: 304-329.

Sissingh, G. 1977. Bosbouw en natuurbeheer. Nederlands Bosbouw Tijdschrift 49: 229-238.

Smeets, P.J.A.M., M.J.A. Werger & H.A.J. Tevonderen. 1979. Vegetation changes in a moist grassland following draining. J. appl. Ecol. (in press).

Smit, P. & W.D. Margadant. 1971. Lijst van archivalia aangaande personen, boeken en documenten aanwezig op de tentoonstelling te Nijmegen ter gelegenheid van het 125 jarig bestaan der K.N.B.V. Acta Bot. Neerl. 20: 690-716.

Soet, F. de. (ed.). 1976. De waarden van de uiterwaarden. Een milieukartering en -waardering van de uiterwaarden van IJssel, Rijn, Waal en Maas. Pudoc, Wageningen.

Stamp, D. 1974. Nature Conservation in Britain. 2nd. ed. Collins, London.

Sukopp, H. Wandel von Flora und Vegetation in Mitteleuropa unter dem Einfluss der Menschen. Ber. Landwirtschaft 50: 112-139.

Thijsse, Jac.P. 1946. Natuurbescherming en Landschapsverzorging in Nederland. Versluys, Amsterdam.

Tideman, P. 1963. Vegetationskartierung als Grundlage für Verwaltungspläne in Naturschutzgebieten in den Niederlanden. In: R. Tüxen (ed.), Bericht über das Internationale Symposion für Vegetationskartierung 1959. Cramer, Weinheim.

Tüxen, R. 1957. Die Bedeutung des Naturschutzes für die Naturforschung. Mitt. flor.-soz. Arbeitsgem. N.F. 6/7: 329-334.

Tüxen, R. & A.P. Grootjans. 1978a. Bibliographie der Arbeiten über Grundwasserganglinien unter Pflanzengesellschaften II. Excerpta Botanica, Sectio B 17: 50-68.

Tüxen, R. & A.P. Grootjans. 1978b. Bibliographie der Arbeiten über Vegetation und Wasserhaushalt des Bodens. Excerpta Botanica, Sectio B 17: 69-80.

Usher, M.B. 1973. Biological Management and Conservation. Chapman & Hall, London.

Wachter, W.H. 1947. De Nederlandsche Botanische Vereeniging 1845-1945. Ned. Kruidk. Arch. 55: 12-116.

Weevers, Th. 1938. De betekenis van natuurreservaten voor de botanische wetenschap. Ned. Kruidk. Arch. 48: 83-88.

Weevers, Th. 1947. Commissie tot Bescherming der wilde flora. Ned. Kruidk. Arch. 55: 126-130.

Weevers, Th. 1951a. De betekenis der natuurbescherming voor de biologische wetenschappen. In: Natuurbescherming, Kunsten en Wetenschappen. Contactcommissie voor Natuur- en Landschapsbescherming. pp. 3-6. Amsterdam.

Weevers, Th. 1951b. Natuurbescherming in verleden, heden en toekomst. Natuur en Landschap 5: 49-56.

Westhoff, V. 1945. Biologische problemen der natuurbescherming. Inleiding gehouden voor het congres van de N.J.N. op 15 augustus 1945 te Drachten. 13 pp.

Westhoff, V. 1952a. De betekenis van natuurgebieden voor wetenschap en practijk. Contactcommissie voor Natuur- en Landschapsbescherming, Amsterdam.

Westhoff, V. 1952b. The management of nature reserves in densely populated countries considered from a botanical viewpoint. Proc. and Papers technical meeting I.U.C.N. pp. 77-82.

Westhoff, V. 1953. Het botanisch beheer van natuurreservaten. In: Jaarboek van de Vereniging tot Behoud van Natuurmonumenten 1950-1953. pp. 104-113. Amsterdam.

Westhoff, V. 1954. Die Vegetationskartierung in den Niederlanden. Angewandte Pflanzensoziologie. Festschrift Aichinger, pp. 1223-1231. Springer, Wien.

Westhoff, V. 1955. Hedendaagse aspecten der natuurbescherming. Wetenschap en Samenleving 9: 25-34.

Westhoff, V. 1969. Verandering en duur. Beschouwingen over dynamiek van vegetatie. Oratie Katholieke Universiteit Nijmegen. Junk, Den Haag.

Westhoff, V. 1970a. New criteria for nature reserves. New Scientist 16 April 1970: 108-113.

Westhoff, V. 1970b. Botanisch onderzoek als grondslag voor de keuze van natuurreservaten. In: J.C. van de Kamer et al., Het verstoorde evenwicht. pp. 111-124. Oosthoek, Utrecht.

Westhoff, V. 1971a. De wetenschappelijke betekenis van het natuurbehoud. In: J.Th.J.M. Willems (ed.), De noodzaak van natuur- en milieubeheer. pp. 22-41. Bruna, Utrecht.

Westhoff, V. 1971b. Botanische criteria. In: A.P.A. Vink et al., Criteria voor milieubeheer. pp. 28-42. Oosthoek, Utrecht.

Westhoff, V. 1971c. Choice and management of nature reserves in the Netherlands. Bull. Nat. Plantentuin Belg. 41: 231-245.

Westhoff, V. 1971d. Het natuurbehoud in Nederland. Jaarboek 1970 K.N.B.V.: 100-116.

Westhoff, V. 1973a. Natuurbehoud en natuurbeheer. Natuurkundige Voordrachten N.R. 52: 71-84.

Westhoff, V. 1973b. Vegetatie-ontwikkeling; Reservaten. In: A.D. Voûte & J.F. de Vries Broekman (eds.), Natuurbeheer in Nederland. pp. 46-54; 180-189. Samson, Alphen aan de Rijn.

Westhoff, V. 1976. Die Verarmung der niederländischen Gefässpflanzenflora in den letzten 50 Jahren und ihre teilweise Erhaltung in Naturreservaten. Schriftenreihe für Vegetationskunde 10: 63-73.

287

Westhoff, V. 1977. Botanical aspects of nature conservation in densely populated countries. In: A. Miyawaki & R. Tüxen (eds.), Vegetation science and environmental protection. Proc. Int. Symposium Tokyo: 369-374. Maruzen, Tokyo.

Westhoff, V. 1978a. Natuurbeheer en het agrarische landschap. Rede bij het afscheid van prof. dr. M.F. Mörzer Bruyns (in press).

Westhoff, V. 1978b. Een halve eeuw wisselwerking tussen wetenschap en natuurbehoud. In: Wetenschap in dienst van het natuurbehoud (voordrachten ter gelegenheid van het vijftigjarig bestaan van de Natuurwetenschappelijke Commissie van de Natuurbeschermingsraad. pp. 13-25.

Westhoff, V. & J. van Dijk. 1952. Experimenteel successie-onderzoek in natuurreservaten, in het bijzonder in het Korenburgerveen bij Winterswijk. De Levende Natuur 55: 5-16.

Westhoff, V. & J. van Dijk. 1953. Overzicht van het wetenschappelijk beheer van een negental bezittingen der Vereniging. In: Jaarboek van de Vereniging tot Behoud van Natuurmonumenten 1950-1953. pp. 114-123. Amsterdam.

Westhoff, V. & A.J. den Held. 1969. Plantengemeenschappen in Nederland. Thieme, Zutphen.

9. LIST OF LECTURES HELD AT THE HUNDRED MEETINGS OF
THE COMMISSION FOR THE STUDY OF VEGETATION OF THE
ROYAL BOTANICAL SOCIETY OF THE NETHERLANDS

JAN VROMAN

# 9. LIST OF LECTURES HELD AT THE HUNDRED MEETINGS OF THE COMMISSION FOR THE STUDY OF VEGETATION OF THE ROYAL BOTANICAL SOCIETY OF THE NETHERLANDS

## JAN VROMAN

In the following list each meeting is labeled by a caption listing the number, date and locality of the meeting and a source reference to a brief report on the meeting. Whenever the meeting dealt with a special theme, this also is indicated. Only the captions of the most recent meetings have no source references as these have not yet been published. The titles of the lectures given at these meetings have been taken from widely circulated convocations.

Following the titles there is a reference to a published summary of the lectures. In cases where no summary or an extremely brief one was available, references to other publications on the same topic (taken from NKA) sometimes have been included. In a number of years the summaries apparently have not been published.

In nearly all cases the first cited author delivered the lecture.

With very few exceptions all lectures were presented in Dutch. The lecture titles in this list, however, are given in the same language as their summaries. When summary titles were available both in Dutch and in a foreign language, the titles are given in the foreign language. In some cases the title of a lecture presented and that of its summary were not identical, even though they were in the same language. In those cases the title giving most information is listed.

Abbreviated references to meeting reports and summaries are as follows:

| | | |
|---|---|---|
| ABN | ..... | Acta Botanica Neerlandica |
| CO | ..... | Contactblad voor Oecologen |
| JB (with year) | ..... | Jaarboek van de Koninklijke Nederlandse Botanische Vereniging (year-book of the Royal Botanical Society of the Netherlands) |
| NKA | ..... | Nederlandsch Kruidkundig Archief |
| VB | ..... | Vakblad voor biologen |

*Nederlandsche Dagen voor Phytosociologie en Palaeobotanie van het Holoceen*

1.  26 November 1933. Utrecht (NKA 44: 93-105)
J. Heimans. De transportfactor in de sociologie (NKA 44: 96-98)
Th. Weevers. Enkele boschassociaties in de Geldersche vallei (NKA 44: 98-99)
J.W. van Dieren. Plantensociologische studies aan paraboolduinen (NKA 44: 99-101)
K. Zijlstra. Plantengroei en landaanwinning (NKA 44: 101-102)
J. Hofker. Diatomeeën als indicatoren voor facies-verschillen (NKA 44: 102-104)
W. Beijerinck. Mikropalaeontologisch onderzoek der post-Riss-afdeeling (NKA 44: 104-105)

2.  16 December 1934. Utrecht (NKA 45: 208-217)
B. Polak. De vegetatie van het zandsteengebied bij Mandor (West-Borneo) (NKA 45: 208-209)

M.J. Adriani. De oekologie van enkele mediterrane halophyten-associaties (NKA 45: 209-211)

P. van Oye. Plankton-spectra (NKA 45: 211)

W. Feekes. Snelle veranderingen in vitaliteit van pionierplantengezelschappen op maagdelijken bodem en de oorzaken daarvan (NKA 45: 211-213).

J. Vlieger. Eenige waarnemingen omtrent de degradatie van het Querceto-Betuletum (NKA 45: 213-214)

J.G. ten Houten A.H. zn. Vorming en vegetatie van het Korenburgerveen (NKA 45: 214-215)

D.M. de Vries. Verloop der aanwezigheidsverdeelingsdiagrammen bij twee floristisch verschillende vormen van een gezelschap, dat gekenmerkt is door de overheersching van een bepaalde plantensoort (NKA 45: 215-216)

J.W. van Dieren. De vegetatie van het eiland Griend en haar verandering onder invloed van de afsluiting van de Zuiderzee (NKA 45: 217)

3. 1 December 1935. Utrecht (NKA 46: 397-407)

D.M. de Vries. Werkwijzen, gebruikelijk bij het plantkundig graslandonderzoek aan het Rijkslandbouwproefstation voor Akker- en Weidebouw te Groningen (NKA 46: 398-401)

J. Vlieger. Over enkele bosch-gezelschappen van de hooge Veluwe-gronden (NKA 46: 401-402)

M.J. Adriani. Zuurgraadbepalingen aan den bodem van enkele Nederlandsche bosschen, tevens een bijdrage tot het climaxvraagstuk (NKA 46: 402-403)

F. Florschütz & E.C. Wassink. Over de geschiedenis van bosch en heide in Drenthe (NKA 46: 404-405)

J. Lanjouw. De vegetatie van de Surinaamsche savannen en zwampen (NKA 46: 405-407)

4. 15 November 1936. Utrecht (NKA 47: 89-96)

F. Florschütz. Het nut van makroskopisch onderzoek van venen met beperkten horizontalen omvang (NKA 47: 90)

P. van Oye. De plantengroei der kalkformaties van de Belgische Jurastreek, Crons genoemd (NKA 47: 90-91)

B. Polak. Veenanalyse uit het IJselmeer (NKA 47: 91-92)

G.W. Harmsen. De onder invloed van de afsluiting der Zuiderzee ingetreden veranderingen in het plantendek van het eilandje Griend (NKA 47: 92-93)

E. Kolumbe. Biologische Fragen bei der Landgewinnung (NKA 47: 93)

W.H. Diemont. Plantensociologie en natuurbescherming (NKA 47: 93-96)

5. 14 November 1937. Utrecht (NKA 48: 47-60)

D.M. de Vries. De vereenigde aanwezigheids- en rangorde-methode (NKA 48: 48-51)

E.C. Wassink. De tegenwoordige vegetatie in de Engbertsdijkvenen te Vriezenveen (NKA 48: 51-52)

J. Vlieger. Algemeene opmerkingen over de hoogere plantensociologische eenheden in Nederland (NKA 48: 53-54)

M.A.J. Goedewaagen. Akkeronkruiden in verband met den zuurgraad van den grond (NKA 48: 54-58)

B. Polak. Het veenlandschap aan deze en aan gene zijde van de Noordzee (NKA 48: 58-59)

F. Florschütz. Wederkeerig dienstbetoon van palaeophytosociologie en archaeologie (NKA 48: 59-60)

6. 13 November 1938. Utrecht (NKA 49: 67-78)

J. Wasscher. Die Verbreitung der Getreideunkräuter in Groningen und Nord-Drente (NKA 49: 69-70)

V. Westhoff. Die Vegetation der Muschelkalkinsel von Winterswijk (NKA 49: 70)

W. Feekes. Botanische Untersuchung in Bezug auf den Nord-Ost-Koog (NKA 49: 70-73)

P. van Oye. Le p.H. de l'eau, facteur biosociologique (NKA 49: 73-74)

Th. Weevers. The dynamic view of a flora (NKA 49: 74-77)

F.P. Jonker & F. Florschütz. Die stratigrafische und pollenanalytische Untersuchung der Lehmlagen bei Wijk bij Duurstede (NKA 49: 77-78)

F. Florschütz. Ueber den Pflanzenwuchs in den Niederlanden während des Pleistocens (NKA 49: 78)

*Nederlandsche Dagen voor Biosociologie en Palaeobotanie*

7.  26 March 1939. Utrecht (NKA 50: 71-77)
E.M. van Zinderen Bakker. Recherches botaniques au lac de Naarden (NKA 50: 71-73)
G. Kruseman & J. Vlieger. Iets over Nederlandsche akkergezelschappen (NKA 49: 327-386)
F. Florschütz. Het veen op grootere diepte (NKA 50: 73-74)
J. Heimans. Accessibiliteit en plantenverspreiding (NKA 50: 74-75)
W.H. Diemont. Enkele merkwaardige plantengezelschappen langs de Drenthsche beekdalen (NKA 50: 75-77)
J. Vlieger. Vegetatie en podsolprofiel (NKA 50: 77)

8.  10 March 1940. Leiden. Theme: Onderzoekingen betreffende het natuurreservaat Griend (NKA 51: 51-60)
G.W. Harmsen. Inleidend woord (NKA 51: 51-52)
A. Scheygrond. Geschiedenis van Griend (NKA 51: 52-53)
W. Feekes. Plantengroei in verband met den bouw van het eiland (NKA 51: 53-56)
A. van der Werff. De bodemalgen van Griend (NKA 51: 56-57)
V. Westhoff. Systematik der Pflanzengesellschaften (NKA 51: 57-59)
G.A. Brouwer. De beteekenis van Griend voor de vogels (NKA 51: 59-60)

9.  10 November 1940. Utrecht (NKA 51: 61-71)
F.P. Jonker & F. Florschütz. Palaeobotanisch onderzoek van quartaire afzettingen nabij Utrecht (NKA 51: 61-62)
F. Florschütz. De bosschen van het Pohorje-gebergte (NKA 51: 62-63)
W. Feekes. Buitenlanden langs Oost- en Westkust van de Zuiderzee, voor en na de afsluiting (NKA 51: 63-67)
J. Meltzer. Duinbosschen (NKA 51: 67-69)
V. Westhoff, W.H. Diemont & G. Sissingh. Het dwergbiezenverbond (Nanocyperion) in Nederland (NKA 51: 69-70)
W. van der Kloot. De landschapsvorming op De Beer (NKA 51: 70-71)

10.  23 March 1941. Utrecht (NKA 52: 303-311)
D.M. de Vries. Over den invloed van jaargetijde en weer op de botanische samenstelling van grasland (NKA 52: 303-307)
G. Sissingh. Graslandtypen om Wageningen (NKA 52: 308-309)
L.A.AE. van Eerde & W. Feekes. De landaanwinning 'Het Noorderleegs Buitenveld', een belangrijke proef met Spartina Townsendii in de Waddenzee (NKA 52: 309; Tijdschr. Kon. Ned. Aardrijksk. Genootsch. 59: 1-23)
W.H. Diemont. Het wintereiken-berkenbosch in Nederland (NKA 52: 309-310)
C. Sipkes. De invloed van konijnen en wateronttrekking op de duinflora (NKA 52: 310-311)

11.  16 November 1941. Utrecht (NKA 52: 311-320)
A.C. Boer. Biezencultuur (NKA 52: 311-314)
J. Vlieger. Experimenteel successieonderzoek (NKA 52: 314)
F.P. Jonker. De flora van het Mindel-Riss-interglaciaal in Nederland (NKA 52: 314-315)
F. Florschütz. Het Cuspidatumveen (NKA 52: 315)
J. Doeksen. Bevolkingsproblemen bij dieren, in het bijzonder bij insecten (NKA 52: 315-317)
V. Westhoff. Onderzoekingen naar de sociologische plaats van de mieren in de Nederlandsche bosschen (NKA 52: 317-320)

12.  3 May 1942. Utrecht (NKA 52: 320-332)
D.M. de Vries, M.L. 't Hart & A.A. Kruijne. Landbouwkundige waardeering van grasland op grond van de plantkundige samenstelling (NKA 52: 320-324)
D.M. de Vries & A.A. Kruijne. De invloed van bemesting met kalk, fosforzuur, kali of stikstof op de plantkundige samenstelling van grasland (NKA 52: 324-327)
E.M. van Zinderen Bakker. De voorgeschiedenis van het Naardermeer (NKA 52: 328-329)
B. Veen. Epiphyten op eikenstronken (NKA 52: 329)
W. Voorbeytel Cannenburg. De oecologie en de sociologie van de 'Fransche Berg' in het nationale Park 'De Hooge Veluwe' (NKA 52: 329-332)

13. 15 November 1942. Utrecht (NKA 54: 264-275)
M.L. 't Hart. Lolium perenne-typen in onze graslanden (NKA 54: 264-267)
M.A.J. Goedewaagen. Het veen in het Zuidelijk deel der Geldersche Vallei (NKA 54: 267-268)
F. Florschütz. De laagterrasflora en het veen op grootere diepte bij Velsen (Tijdschr. Kon. Ned. Aardrijksk. Genootsch. 61 (2): 25-33)
G.L. Funke. Enkele voorbeelden van experimenteele plantensociologie (NKA 54: 269-270; Blumea 5: 281-293, 294-296)
N. Hubbeling. De boschvegetatie der Twentsche beekdalen
M.F. Mörzer Bruyns. Biosociologische onderzoekingen van de molluskenfauna in het IJsseldal (NKA 54: 270-275)

14. 2 May 1943. Utrecht (NKA 54: 276)
J.J. Franssen. De invloed van de fauna op de bodemvorming (Tijdschr. Ned. Heidemaatschappij 54: 25-34, 42-62, 106-115, 138-144, 170-174, 185-190; 55: 39-47, 49-58)
D.M. de Vries. Aanpassing van ons atlantisch grasland aan de meer continentale weersgesteldheid van de laatste jaren (Landbouwk. Tijdschr. 55: 268-274)
W. Feekes. De natuurlijke vegetatie van den Noordoostpolder

15. 21 November 1943. Utrecht. Theme: De Zuiderzee en alle met de verandering in het landschap samenhangende problemen (NK/_ 54: 276-303)
S. Smeding. Een woord ter inleiding (NKA 54: 277-279)
P. Jansen. Het werk der Zuiderzee-commissie (NKA 54: 280-288)
P.J.R. Modderman. Het Zuiderzeegebied historisch-geographisch beschouwd (NKA 54: 288-292)
F. Florschütz. Veenonderzoek in Wieringermeer en Noordoostpolder
A.J. Zuur. De bodem van de Zuiderzee (NKA 54: 292-296)
H.C. Redeke. Over de hydrobiologie van de Zuiderzee en het IJsselmeer (NKA 54: 296-303)

16. 30 April 1944. Utrecht (NKA 56: 21-41)
M.L. 't Hart. Kartering van landbouwkundige graslandtypen in het lage midden van Friesland (NKA 56: 21-22)
D.M. de Vries. Een graslandtypering van landbouwkundige betekenis volgens de combinatie van frequente soorten en de dominantie (NKA 56: 22)
G.M. Castenmiller. Een oriënterend onderzoek naar de vormen van Agrostis in enkele graslandtypen en in handelszaad
F.P. Jonker. Palaeobotanische problemen bij het onderzoek van het West-Nederlandse Holoceen (NKA 56: 23-24)
V. Westhoff. De betekenis van de phaenologie voor het plantensociologisch onderzoek (NKA 56: 24-31)
G. Sissingh. Klimaatsverschillen in Nederland en hun invloed op de vegetatie (NKA 56: 31-38)
M.F. Mörzer Bruyns. Biosociologie en biocoenologie (NKA 56: 38-41)

17. 26 May 1946. Utrecht (NKA 57: 21-33)
E.C. Wassink. Enkele sociologische en plantengeografische opmerkingen in verband met de hoogveenvegetatie der Engbertsdijkvenen bij Vriezenveen (NKA 57: 21-23)
F. Florschütz. Poging tot ouderdomsbenadering van de Gelderse löss langs geologisch-palynologische weg (NKA 57: 23-28)
D.M. de Vries. Graslandtypen en hun ecologie (NKA 57: 28-31)
F.P. Jonker. Botanische bijdragen tot het onderzoek van Holocene transgressie en regressie (NKA 57: 31-33)

18. 15 December 1946. Utrecht (NKA 57: 33-39)
M.F. Mörzer Bruyns. De molluskenfauna van het Naardermeer (NKA 57: 33-35)
L. Tinbergen. Vegetatie en vogelbevolking (NKA 57: 35-37)
E.M. van Zinderen Bakker. Pollenanalyse van Veluwse venen
G. Sissingh. De levensvormen van onze akkeronkruiden en de levensvormen spectra der akkeronkruiden
J. van der Drift. De bodemfauna in bossen (NKA 57: 37-39)

294

19. 17 May 1947. Utrecht (NKA 57: 76-83)
H.Tj. Waterbolk. De natuurlijke bosvegetaties in Drente (NKA 57: 76-79)
A.D. Voûte. Regulerende elementen in de dierlijke samenleving
Th.A. de Boer. Plantensociologische waarnemingen op geïnundeerd grasland (NKA 57: 79-80)
P. Tideman. Flora van de garrigues in de Franse Midi. Oecologische en genetische opmerkingen over enkele van zijn plantengezelschappen in verband met de mogelijkheden tot beter bodemgebruik (NKA 57: 80-82)
H.J. Zwart. De vegetatie van de brongebieden in de Lutte (NKA 57: 82-83)

20. 23 November 1947. Utrecht (NKA 57: 84-91)
J.J. Barkman. Enige epiphytenassociaties in Zwitserland (NKA 57: 84-85)
A.C. Stolk. Het pollenanalytisch onderzoek van het Haaksbergerveen (NKA 57: 85-87)
D. Bakker. De flora en fauna van Walcheren tijdens en na de inundatie (NKA 57: 87-88)
P. den Dulk. De vegetatie van de voormalige blauwgraslanden van het Binnenveld bij Wageningen
V. Westhoff. De plantengroei van de Botshol (bij Abcoude) (NKA 57: 88-91)

21. 25 April 1948. Utrecht. Theme: Bewegingen van bodem en zeeniveau in Nederland tijdens het Holoceen (NKA 57: 123-134)
F.P. Jonker. Het botanisch onderzoek naar het optreden van transgressies en regressies (NKA 57: 123-124)
J.H.F. Umbgrove. Bewegingen van bodem en zeeniveau in ons kustgebied gedurende de laatste 4000 jaren (NKA 57: 124-126)
P.A. Florschütz. De mariene mollusken in de Holocene afzettingen te Velzen (NKA 57: 126-128)
T.K. Huizinga. Inklinking van de grond (NKA 57: 128-132)
W.C. Visser. Bodemkundig aspect van de zeespiegelvariaties (NKA 57: 133-134)

22. 21 November 1948. Utrecht (NKA 57: 134-148)
D. Bakker. De vegetatie van de toekomstige Zuid-Sloepolder (NKA 57: 135-137)
M.J. Adriani. Over vergelijkend waterbalansonderzoek (NKA 57: 137-138)
E. Meijer Drees. Bosassociaties op Timor (NKA 57: 138-139)
J. Bennema. Enkele voorlopige resultaten der veengrondenkartering in West-Nederland (NKA 57: 139-143)
W. van Zeist. Veenonderzoek van het Princehof (Friesland) (NKA 57: 143-144)
W. Meijer. Enkele problemen bij vegetatie-studie in West-Nederlandse venen (NKA 57: 144-145)
J.J. Barkman. Duinvegetaties van Noord-Schotland (NKA 57: 145-148)

23. 24 April 1949. Utrecht (NKA 58: 10-19)
J.G.P. Dirven & D.M. de Vries. Oecologisch graslandonderzoek (NKA 58: 11-12)
Th.A. de Boer. Graslandkartering in Nederland (NKA 58: 12-14)
H.Tj. Waterbolk. Pollenanalytisch onderzoek bij enkele Veluwse tumuli-opgravingen (NKA 58: 14-16)
T. van der Hammen. De Alleröd-oscillatie en de laat-glaciale vegetatie-ontwikkeling in Nederland (NKA 58: 16-17)
H.J. Zwart. Palynologisch veenonderzoek in West-Nederland (NKA 58: 18-19)
V. Westhoff. Synoecologie

24. 20 November 1949. Utrecht (NKA 58: 19-27)
D. Bakker. De natuurlijke vegetatie van de gebieden in de omgeving van het voormalige eiland Urk (NKA 58: 19-21)
C. vanden Berghen. La carthographie phytosociologique en Belgique (NKA 58: 21-22)
W. Meijer. Flora en vegetatie van de Kierse Wiede (NKA 58: 22-23)
H.Tj. Waterbolk. De betekenis van het kruidenpollen voor het palynologisch onderzoek (NKA 58: 24-25)
B.A. de Planque. Palynologisch onderzoek van het Holoceen en Laat-Glaciaal (met Alleröd-schommeling) in Zuid-Oost Friesland (NKA 58: 25-26)
M. Sanders & D.M. de Vries. De soortenrijkdom van graslanden in verband met de standplaats (NKA 58: 26-27)

25.  23 April 1950. Wageningen. Theme: Vegetatiekartering (VB 30: 93-94)
H.J. Venema. Inleidend overzicht
V. Westhoff. Vegetatiekartering in buiten- en binnenland
V. Westhoff. Detailkartering van een deel van het bosgebied van Middachten
Th.A. de Boer. Waardering en kartering van het grasland op botanische grondslag
B. de Jong. Een bosbedrijfsplan op grond van een vegetatiekartering op de Hoge Lutte bij Oldenzaal
C.P. van Goor. Een vegetatiekartering van het landgoed Singraven bij Denekamp
N.P.H.J. Roorda van Eysinga. De geschiedenis van het landschap bij Renkum
G. Sissingh. Een plantensociologische kartering in de omgeving van Didam

26.  19 November 1950. Utrecht (VB 30: 233)
W. Meijer. Het Ilperveld als terrein voor plantensociologisch onderzoek
H.J. Zwart. Palynologisch en geologisch onderzoek van een Holocene sedimentenreeks bij Nootdorp (Z.-H.)
G.C. Ennik & D.M. de Vries. Dominantie en dominantiegezelschappen
I.S. Zonneveld. De vegetatie van de Biesbosch
Ph. Stoutjesdijk & J.P. Kruyt. Opmerkingen naar aanleiding van een vegetatiekartering op de Beer
T. van der Hammen. Een botanische excursie naar Lapland

27.  22 April 1951. Utrecht (VB 31: 116)
D. Bakker. Over de sociologische waardering van Tussilago farfara L.
D. Bakker. De levensvormen van lucerne-onkruiden in de Noordoostpolder
A.C. Boer. De veranderingen in de vegetatie van het buitendijkse randgebied van de Noordoostpolder
D.M. de Vries. Objectieve soortencombinaties en hun oecologie
W.H. Zagwijn. De 'viviers' op het plateau van de Hautes Fagnes
W. van der Zweep. De migratie van plantensoorten naar de drooggevallen terreinen om Urk

28.  18 November 1951. Utrecht (VB 31: 229)
V. Westhoff & J. van Dijk Jr. Experimenteel successie-onderzoek in natuurreservaten
F. Florschütz. Nieuwe inzichten in de aard van de Nederlandse Dryas-flora
R.J. de Wit. Het Caricetum lasiocarpae in het gebied van de Vechtplassen en in N.W. Overijssel
W. Meijer. Hydrospherische spectra van waterplantbegroeiingen
L.G. Kop & E. Stapelveld. Kartering van bosvegetaties in Losser
D.M. de Vries. Dynamiek van vogelbevolkingen
F.P. Jonker. Het veen op grotere diepte in Noord-Friesland

29.  17 April 1952. Utrecht (JB 1953: 10-11)
H.J.W. Schimmel. De vegetatie van het Zwin
W.G. Beeftink. Vegetatie en standplaatsfactoren van de schorren en duinen van de Kaloot
D.M. de Vries & Th.A. de Boer. Oecologische indicatie in cijfers
J. Hendriks & B. Belderok. Palynologie rondom Amsterdam
H. Doing Kraft. De bosvegetatie van de binnenduinrand van Den Haag tot Bergen
K.J. Zandstra. Uitkomsten van een onderzoek naar de akkeronkruidvegetaties van de Wageningse Eng

30.  19 December 1952. Utrecht (JB 1953: 10-11)
D. Burger. Bodemkundige onderzoekingen in het ongestoorde en bewerkte profiel van het droge Querceto-Betuletum
M.F. Mörzer Bruyns. Een biosociologisch onderzoek van een vochtige duinvallei in het buitenduin van Terschelling
Chr.G. van Leeuwen. Een biosociologisch onderzoek van een binnenduinpan op Terschelling
A.W.H. Damman & D.M. de Vries. Toetsing van associaties aan soortencombinaties
V. Westhoff. De plantensociologische tuin in Hannover

31. 17 April 1953. Utrecht (JB 1954: 9-10)
J.H.A. Boerboom. De vegetatie van Meyendel
A.C. Boer. De vegetatie van de Brakman
W.A.E. van Donselaar-ten Bokkel Huinink & J. van Donselaar. De vegetatie van de heide van Duurswoude (Z.O. Friesland)
W. van Zeist. Een palynologisch onderzoek betreffende de grenshorizont van Zuid-Oost Drenthe
A. Coops & G. Slettenhaar. Een vegetatiekartering van Oostereng te Wageningen, proefboswachterij van de Landbouwhogeschool
I.S. Zonneveld. Een oriënterend onderzoek naar de reactie van de vegetatie op de verdichting van de grond ten gevolge van de betreding in een Zwitsers bos

32. 18 December 1953. Utrecht (JB 1954: 9-10)
H. Doing Kraft. Een poging tot herziening van de bosassociaties van het systeem der Frans-Zwitserse school
L.O. Zwillenberg & J. Hendriks. Pollenanalytisch onderzoek naar de Cardium-transgressie in Waterland
D. Bakker. Een prognose van de natuurlijke vegetatie in Oostelijk Flevoland
C. den Hartog. Een systeem voor de algengemeenschappen van onze kust
M.F. Mörzer Bruyns. Broedvogeltellingen in natuurlijke bossen

33. 9 April 1954. Utrecht (JB 1955: 11)
H.M. Heybroek. Standplaatseisen en onderlinge beinvloeding van planten (JB 1955: 25)
J.J. Barkman. Verband tussen epiphytenverspreiding en klimaat in Nederland (JB 1955: 25-26)
A.L. Stoffers. Opmerkingen over de vegetatie van de Benedenwindse Eilanden (JB 1955: 27)
D.M. de Vries. Driedimensionale constellatie van graslandplanten en haar oecologische interpretatie (JB 1955: 27-28)
M. Sonnema. Akkeronkruiden op zandgrond in verband met de zuurgraad (JB 1955: 28-29)

34. 23 December 1954. Utrecht (JB 1955: 11)
V. Westhoff & H.J. Venema. De vegetatie in en buiten de honderd-jarige strikte reservaten in het woud van Fontainebleau: een bijdrage tot het onderzoek van het climax-probleem
W. van Zeist. Pollenanalyse van enige veenvondsten (JB 1955: 29)
I.S. Zonneveld. Skleromorphie als criterium bij levensvormen (JB 1955: 29-30)
D. Bakker. De begrippen annuel en biannuel (JB 1955: 30)
F.P. Jonker. Veenvorming en bosgeschiedenis in Z.O. Friesland (JB 1955: 30-31)

35. 7 April 1955. Utrecht (JB 1956: 10-11)
Th.A. de Boer. Resultaten en mogelijkheden van de graslandkartering
D. Bakker & W. van der Zweep. Het disseminatie-spectrum van de vegetatie in de omgeving van het voormalige eiland Urk, II
E. Stapelveld. De bodemvegetatie van lariksbossen in Drente (JB 1956: 25)
W.G. Beeftink. De invloed van het zoutgehalte van het vloedwater op de vegetatie en molluskenfauna van de slikken en schorren langs Wester- en Oosterschelde (JB 1956: 25-26)
L.G. Kop. Vegetatie, bodem en groeikracht van de eik in enkele Twentse bossen (JB 1956: 26-27)

36. 22 December 1955. Utrecht (JB 1956: 10-11)
W.H. Diemont. Het wetenschappelijk verantwoord beheer van de Zuid-Limburgse bosreservaten (JB 1956: 27)
W. van Zeist. Enige absolute dateringen van een veenprofiel uit Zuidoost-Drente (JB 1956: 27-28)
A.W.H. Damman. Een nieuwe indeling van de heidegezelschappen (JB 1956: 28-29)
F.M. Maas. Samenstelling en synoecologie van bronvegetaties (JB 1956: 29)
H. Doing Kraft. De tegenwoordige opvattingen omtrent het associatiebegrip en de systematiek van plantengezelschappen volgens de methode van Braun-Blanquet (JB 1956: 29-30)

37. 5 April 1956. Utrecht (JB 1957: 13)
J.H.A. Boerboom. Zonatie en overspoeling van halophyten op de Bosplaat (Terschelling)

V. Westhoff. De ethologische synoecologie van de xero-serie op Terschelling
W.H. Zagwijn. Aspecten van de vegetatie bij de overgang van het Plioceen naar het Pleistoceen
F.M. Maas. Samenstelling en synoecologie van bronvegetaties

38.  22 December 1956. Utrecht (JB 1957: 13)
P.J.C. Kuiper. Verlandingsvegetaties in de moerasgebieden van N.W. Overijsel (JB 1957: 30-31)
D.M. de Vries. Indicatie door graslandplanten (JB 1957: 31-32)
W.G. Beeftink. De halophiele vegetatie van Skallingen (Denemarken) (JB 1957: 34-35)
B. Zwart Jr. Enkele gegevens over het veen op grote diepte onder Amsterdam (JB 1957: 33-34)
W.A.E. van Donselaar-ten Bokkel Huinink. Successie-onderzoek van moerasvegetaties door analyse van plantenresten in de bodem (JB 1957: 32-33)
J.J. Barkman. De begrippen vitaliteit, fertiliteit en trouw in de vegetatiekunde

39.  25 April 1957. Utrecht (JB 1958: 12-13)
V. Westhoff, J. Vlieger & J. van Dijk Jr. Een kwart eeuw successie-onderzoek op de permanente quadraten bij IJdoorn (IJsselmeerkust) en in Waalenburg en Dijkmanshuizen (Texel)
F.M. Maas. Verspreiding, systematiek en oecologie van Glycerieto-Sparganion-gezelschappen (JB 1958: 34-36)
D.M. de Vries & J.F.M. van Leeuwen. De begroeiing der dijken in samenhang met de standplaats (JB 1958: 36-37)
W.G. Beeftink. De verwerking van opnametabellen als grondslag voor een typering en indeling van plantengemeenschappen (JB 1958: 38-39)

40.  21 December 1957. Utrecht (JB 1958: 12-13)
F. Florschütz. Over de laat-glaciale en de Holocene vegetatie-geschiedenis in het Noordwesten van Spanje (JB 1958: 39-40)
A.A. Manten. Een palynologisch onderzoek van de Nederlandse Miocene bruinkool bij Haanrade (L.) (JB 1958: 40-41)
J.H.A. Boerboom & H. Doing Kraft. Over de zonering in landschap en plantengroei van de duinen bij Wassenaar en Bloemendaal (JB 1958: 41-43)
I.S. Zonneveld. Iets over continuiteit en discontinuiteit in de vegetatie, en de betekenis daarvan voor de synsystematiek (JB 1958: 43-45)
H.J. Over. Factoren die de broedvogelstand in de Nederlandse bossen bepalen
F.J.J. van Heyst. Voorlopige resultaten van een palynologisch onderzoek in de Franse Jura (Lac de Chalain) (JB 1958: 45-46)

41.  10 April 1958. Utrecht (JB 1959: 11-12)
Ph. Stoutjesdijk. Mikroklimatologische metingen in het algemeen (JB 1959: 41-43)
V. Westhoff, J.J. Barkman, H. Doing Kraft & Chr.G. van Leeuwen. Enige opmerkingen over de terminologie in de vegetatiekunde (JB 1959: 44-46)
E.E. van der Voo & V. Westhoff. Over het verband tussen het voorkomen van enige limnophyten en de waterbeweging in oude rivierlopen (JB 1959: 43-44)
W. van Zeist. Enige opmerkingen over veranderingen in de vegetatie in de tweede helft van het postglaciaal (JB 1959: 48-49)

42.  20 December 1958. Utrecht (JB 1959: 11-12)
B. Polak. Palynologie van het Uddelermeer (JB 1959: 39-41)
D. Bakker. Enige opmerkingen over vegetatie-successie in Oostelijk Flevoland
A.A. Kruijne & D.M. de Vries. Benadering van milieu-eigenschappen van grasland uitgaande van de botanische samenstelling (JB 1959: 37-38)
V. Westhoff & H. Doing Kraft. De plaats van de beuk in het West-Europese bos (JB 1959: 46-48)
W. Meijer. Plantensociologische opnamen in bergbossen van West-Java en Midden-Sumatra (JB 1959: 38-39)
J.P. Schulz. Ecologische studies in het Surinaamse bos

43.  9 April 1959. Utrecht (JB 1960: 13)
D.M. de Vries. Over schommelingen in de vogelstand gedurende het jaar en in de loop der jaren (JB 1960: 40-42)

W.H. Zagwijn. Enkele opmerkingen over veranderingen in samenstelling van jong-Coenophytische makroflorae uit onze streken

Ph. Stoutjesdijk. Hoe moeten we ons de invloed van de vegetatie op het klimaat voorstellen?

P. Leentvaar. Het phytoplankton als maatstaf van de eutrophie van het water van de grote meren (JB 1960: 37-38)

I.S. Zonneveld. Enkele resultaten van een onderzoek van bodem en vegetatie op de Kalmthoutse Heide (JB 1960: 45-46)

44. 19 December 1959. Utrecht (JB 1960: 13)

D.M. de Vries & E.H. Zeiler. Actuele pH-amplituden, bij verschillend hoge frequentie van soortencombinaties en kensoorten in grasland (JB 1960: 42-43)

V. Westhoff. Een vegetatiekundige excursie door Zuid-Zweden (JB 1960: 43-45)

C. den Hartog. Wetmatigheden in het littorale zoneringssysteem (JB 1960: 36-37)

H. Doing Kraft. Begroeiingen in de zeereep der West-Franse duinen, bezien in het licht van het 'retractie-fenomeen' (JB 1960: 35-36)

Chr.G. van Leeuwen. Micropatronen in pioniervegetaties (JB 1960: 38-39)

S. Segal & J.J. Barkman. Enige opmerkingen over dominantie en abundantie bij het opnemen van kwadraten (JB 1960: 39-40)

45. 14 April 1960. (JB 1961: 15-16)

F.P. Jonker. De begroeiing van in kreken liggende rotsblokken in de Emmaketen, Suriname (JB 1961: 56-57)

V. Westhoff. Indrukken van de vegetatie van Quebec en Newfoundland (JB 1961: 61-64)

J. van Donselaar. De vegetatie van oude rivierlopen in de uiterwaarden (JB 1961: 52-54)

J.J. Barkman & P.J. den Boer. Biosociologisch onderzoek in Wijster (JB 1961: 51-52)

46. 17 December 1960. Utrecht (JB 1961: 15-16)

A.J. Havinga. Een pollenanalytisch onderzoek van vegetatieprofielen in dekzand (JB 1961: 54-55)

C. den Hartog. Het zoneringspatroon van geëxponeerde rotskusten langs het Kanaal (JB 1961: 54)

C.R. Janssen. De betekenis van de vegetatiekunde bij de interpretatie van pollendiagrammen (JB 1961: 56)

E. van der Maarel. Een vegetatiekartering van de duinen bij Oostvoorne (JB 1961: 57-59)

I.S. Zonneveld. Enkele (speculatieve) opmerkingen over de resultaten van een gecombineerd bodemkundig en palaeobotanisch onderzoek op het Nederlandse deel van de Kalmthoutse Heide (JB 1961: 64-66)

S. Segal. Vegetaties op oude muren in Nederland (JB 1961: 59-61)

47. 15 April 1961. Utrecht (JB 1962: 13-14)

D.M. de Vries. Indicatie en concurrentie in verband met de chemische samenstelling van graslandplanten (JB 1962: 44-45)

W. Groenman-van Waateringe. Palynologisch onderzoek van drie laat-Neolitische tumuli te St. Walrick bij Overasselt (Gld.) (JB 1962: 42)

J.J. Barkman. Kartering van de epifytenvegetatie van Belgisch Limburg in verband met de industrialisatie (JB 1962: 40-41)

Chr.G. van Leeuwen. De begroeiing van contactgordels en storingsmilieu's (JB 1962: 43)

V. Westhoff. Resultaten internationale conferentie herziening vegetatiesysteem gematigd Europa (JB 1962: 46-47)

48. 30 September 1961. Utrecht (JB 1962: 13-14)

I.S. Zonneveld. Vegetatie, boomgroei en milieu van Pinusbos in eerste generatie (JB 1962: 48-49)

S. van der Werf. Vegetatie-onderzoek van enige dennenbossen op de Veluwe (JB 1962: 45-46)

H. Doing. Nederlandse bossen en struwelen (JB 1962: 41-42)

Ph. Stoutjesdijk. Opmerkingen over een microklimatologische studiereis naar Oostenrijk (JB 1962: 44)

49. 22 December 1961. Utrecht. Theme: Migratie en vestiging (JB 1962: 13-14)
J. Heimans. Inleiding (JB 1962: 49-50)
D. Bakker. Migratie en vestiging (JB 1962: 50-51)
F.P. Jonker. Pleistocene refugia en interglaciale en Holocene migratie en vestiging (JB 1962: 51-52)
Chr.G. van Leeuwen. Ervaringen ten aanzien van de verspreiding van diasporen en vestiging van soorten in de natuur (JB 1962: 52-53)
E. Laarman. Ervaringen betreffende vestiging van planten in het terrein De Wolf (JB 1962: 53-54)

50. 14 April 1962. Utrecht (JB 1963: 12-13)
J.J. Barkman. Een nieuwe associatie op hoogveenturf (JB 1963: 37)
D.M. de Vries & H.N. Leijs. Over vogelbevolking en landschap (JB 1963: 44-45)
J.P. van den Bergh. Concurrentie tussen meerjarige graslandplanten (JB 1963: 38-39)
A.L. Stoffers. Enige opmerkingen over de vegetatie van Bonaire (JB 1963: 44)
V. Westhoff. De internationale plantengeografische excursie door Finland en Noorwegen in 1961 (JB 1963: 45-47)

51. 29 September 1962. Utrecht. Theme: De beuk (JB 1963: 12-13)
F.P. Jonker. De geschiedenis van de beuk in West-Europa (JB 1963: 47)
H. Doing. De oecologie van de beuk (JB 1963: 48-49)
V. Westhoff. De systematiek van de beukenbossen (JB 1963: 49-50)
G. Sissingh. De Deense beukenbossen (JB 1963: 51)
W.J. Reijnders. De mycoflora van beukenbossen (JB 1963: 51)

52. 21 December 1962. Utrecht (JB 1963: 12-13)
S. Segal. Over het Sagineto- en 'Filici'-Bryetum argentei (JB 1963: 41-42)
J.Th. de Smidt. Nederlandse heidegezelschappen (JB 1963: 43)
E. van der Maarel. Een gedetailleerde vegetatiekartering van de duintjes bij 'Weevers' Duin' (JB 1963: 40-41)
W.A. Casparie. Veenvorming en Rhizopoden (JB 1963: 39-40)
J.J. Barkman. Vegetatie-indrukken van hoogvenen in de Harz (JB 1963: 38)

53. 4 April 1963. Utrecht (JB 1964: 13-14)
V. Westhoff. Contactbegroeiingen tussen eutroof en oligotroof milieu in het hoogveengebied van de Peel (JB 1964: 44-45)
T.A. Wijmstra. Jong-Holocene veranderingen van vegetatie en klimaat in de beneden Magdalena (Columbia) (JB 1964: 45)
J.P. van den Bergh. Enkele ervaringen met droogte- en vochtaanwijzers onder de grassen (JB 1964: 37-38)
Ph. Stoutjesdijk. De spectrale samenstelling van het licht in een bos en de oecologische betekenis daarvan (JB 1964: 42-43)
J. van der Burgh. De overblijfselen van houtige planten in de Miocene bruinkool van Haanrade (L.) (JB 1964: 38-39)
I.S. Zonneveld. Bodem en vegetatie in het Speulderbos (JB 1964: 45-47)

54. 28 September 1963. Utrecht. Theme: Oligotrofe venen (JB 1964: 13-14)
F.P. Jonker. Algemene inleiding
V. Westhoff & Chr.G. van Leeuwen. Geografische differentiatie in de hoogvenen van Europa (JB 1964: 32-33)
W.A. Casparie. Stratigrafisch veenonderzoek (JB 1964: 35-36)
P.J. den Boer. Iets over de arthropodenfauna van hoogvenen en oligotrofe 'veentjes' (JB 1964: 35)
J.J. Barkman. Typologie en oecologie van de Nederlandse en Noord-Duitse vlaktehoogvenen (JB 1964: 33-35)
B. Polak. Oligotrofe bosvenen in de tropen (JB 1964: 36-37)

55.  20 December 1963. Utrecht (JB 1964: 13-14)
D.M. de Vries & A.A. Kruijne. Spruitvormen bij grassen (JB 1964: 43-44)
A.J. Havinga. Onderzoek naar corrosiegevoeligheid van stuifmeel (JB 1964: 40)
E.M. Eisma-Donker. Een analyse van het Cirsieto-Molinietum (JB 1964: 39)
F.P. Jonker. Moderne aspecten uit pollenanalytisch onderzoek (JB 1964: 40-41)
E. van der Maarel. Over doel en verwerking van analysegegevens in de vegetatiekunde (JB 1964: 41-42)

56.  18 April 1964. Utrecht (JB 1965: 17)
J.G.P. Dirven. Plantkundig graslandonderzoek in Suriname (JB 1965: 48-49)
Chr.G. van Leeuwen. Over grenzen en grensmilieus (JB 1965: 53-54)
A. Voorrips. Pollenanalytisch onderzoek van de donk te Hillegersberg
V. Westhoff. Inleiding ter discussie over 'De wenselijkheid van naamswijziging der Commissie in verband met haar werkterrein'

57.  26 September 1964. Utrecht (JB 1965: 17)
E. van der Maarel. Het 10e Internationale Botanische Congres te Edinburgh, augustus 1964, sectie Plant Sociology and Phytogeography (JB 1965: 57-58)
V. Westhoff. Het Internationale Symposium over Vegetatiesystematiek te Stolzenau, maart 1964
V. Westhoff. Structuur als diagnostisch criterium in de vegetatiesystematiek
J.J. Barkman. Microgezelschappen en hun systematiek

58.  18 December 1964. Utrecht (JB 1965: 17)
A.H.J. Freijsen. Relaties tussen het voorkomen van Centaurium littorale en het bodemwater (JB 1965: 50-51)
S. Segal. Vegetatie-onderzoek van lemniden (JB 1965: 59-60)
J.Th. de Smidt. Heidegezelschappen in Bretagne en Ierland (JB 1965: 60-61)
P.A. Bakker. Enige botanische indrukken uit het Zwarte Woud (JB 1965: 45-47)

59.  10 April 1965. Utrecht (JB 1966: 13-14)
G. Londo. De vegetaties in het infiltratiegebied der Amsterdamse Waterleidingduinen (JB 1966: 39-41)
W.A. Casparie & A.V. Munaut. Veenvorming, dendrochronologie en C14
D.M. de Vries. Terreinkeus van vogels in winter en broedtijd (JB 1966: 42)
C.R. Janssen. Enige aspecten van bosvegetaties van N.W. Minnesota (JB 1966: 35)
P. Schroevers. Hydrobiologisch onderzoek in het Peelgebied

*Dagen voor het Vegetatie-onderzoek*

60.  21 December 1965. Utrecht. Theme: Zeldzaamheid bij planten en dieren. (In cooperation with the Oecologische Kring of the Nederlandse Dierkundige Vereniging) (JB 1966: 13–14)
K.H. Voous. Beschouwingen over zeldzaamheid van diersoorten
J.J. Barkman. Beschouwingen over zeldzaamheid bij planten (JB 1966: 32-33)
Chr.G. van Leeuwen. Zeldzame planten als uitdrukking van zeldzame toestanden (JB 1966: 36-39)
P.J. den Boer. Zeldzaamheid als dieroecologisch probleem (JB 1966: 33-34)

61.  19 April 1966. Utrecht (JB 1967/1968: 15-16)
W.H. Zagwijn. Subatlantische beukenbossen in het kustgebied van Holland (JB 1967/1968: 49-50)
D.M. de Vries & A.A. Kruijne. Over indelingswijzen van grasland, in het bijzonder een oecologische
A.J. Gottenbos. Successie-onderzoek in de Biesbosch (JB 1967/1968: 33-34)
J.F. Bannink, H.N. Leijs & I.S. Zonneveld. Vochtindicatie en grondwaterstanden op een aantal akkers in het Lollebeekgebied (JB 1967/1968: 32-33)
A.H.J. Freijsen. The germination of Centaurium vulgare Rafn, some observations on field plots (ABN 17: 161-162)

62. 1 October 1966. Utrecht (JB 1967/1968: 15-16)
A.L. Stoffers. Oecologisch onderzoek van de vegetatie in kalk- en diabaasgebied op Curaçao
F.P. Jonker. Second International Conference on Palynology, Utrecht, 1966
Ph. Stoutjesdijk. Vegetatietemperaturen in lage mozaïekvegetaties (JB 1967/1968: 45-46)
J.H.M. Hilgers, W. Colaris & C. van Driel. Populatiestudie in de kalkgraslandvegetatie van de Berghofweide (Zuid-Limburg). Een vegetatiekundig en bodemkundig onderzoek (JB 1967/1968: 35-38)
H.A.R. Velthuis & S. Broekhuizen. Vegetatie-onderzoek op het Asselse veld (JB 1967/1968: 46-47)

63. 20 December 1966. Utrecht. Theme: Historic and dynamic aspects of coastal dune vegetations in the Netherlands (JB 1967/1968: 15-16; ABN 20: 173-174)
S. Jelgersma. An outline of the geological history of the coastal dunes in the Western Netherlands (Geologie en Mijnbouw 48: 335-342)
W.H. Zagwijn. Vegetational history of the coastal dunes in the Western Netherlands (ABN 20: 174-182)
H. Doing & C.J. Doing-Huis in 't Veld. History of landscape and vegetation of coastal dune areas in the province of North Holland (ABN 20: 183–190)
E. van der Maarel & Chr.G. van Leeuwen. Pattern and process in coastal dune vegetations (ABN 20: 191-198)
C.J.M. Sloet van Oldruitenborgh & M.J. Adriani. On the relation between vegetation and soil-development in dune-shrub vegetations (ABN 20: 198-204)

64. 11 March 1967. Utrecht (JB 1967/1968: 59-60)
I.S. Zonneveld, J.F. Bannink & H.N. Leijs. Een vegetatie-indeling voor akkers en naaldbossen t.b.v. praktijk- en landschapsoecologische karteringen (JB 1967/1968: 87-89)
P. Oosterveld, J.H. de Haas, H.C. van der Meulen & H. Nolten. Patroonstudie in duinheidevegetaties van de Berkenvallei, Boschplaat, Terschelling
J. Klein & S.Th.J. Fabius. Patroonstudie in het grensgebied van duinheide en zilte graslanden van De Groede, Boschplaat, Terschelling
S. Segal. Algemene principes bij successie
J.J. Barkman. De vegetatie van jeneverbesstruwelen in Nederland

65. 7 October 1967. Utrecht (JB 1967/1968: 59-60)
G. Grosse-Brauckmann. Möglichkeiten und Ergebnisse einer vegetationskundlichen Auswertung botanischer Torfuntersuchungen (Makrofossil-analysen)
D.M. de Vries. Vogelstand en landschap
F.M. Muller. Seedlings in vegetation types (ABN 18: 396-398)
A.A. Sterk. Variabiliteit en milieu van Spergularia media en S. marina (JB 1967/1968: 84-85)
W.A. Casparie. Bulten en slenken; algemene principes bij hoogveenvorming? (JB 1967/1968: 78-79)

66. 19 December 1967. Utrecht (JB 1967/1968: 59-60)
D. Teunissen. Some aspects of the forest history in Late Glacial and Early Holocene times in the vicinity of Nijmegen (The Netherlands) (ABN 17: 160-161)
E.M. van Zinderen Bakker. Vegetatie en pollenanalytisch onderzoek in het Oost-Afrikaanse hooggebergte
J.J. Barkman. Microgezelschappen in jeneverbesstruwelen en hun paddestoelenflora
F.J.A. Daniëls. Shrub heath communities in south-east Greenland (ABN 18: 483-484)
J.G. de Molenaar. Laag-arctische chionophiele vegetaties op Zuidoost-Groenland

67. 23 March 1968. Utrecht (JB 1969: 13-15)
S.R. Gradstein & J.H. Smittenberg. Vegetaties van beekoevers en moerassen in Kreta (CO 4(2): 11-12)
A.J. Havinga. Selektieve korrosie van stuifmeel (JB 1969: 47-48; CO 4(2): 13)
S. Segal. Zoneringen in het water (JB 1969: 51-52; CO 4(2): 15-16)
C.J.M. Sloet van Oldruitenborgh. Some remarks on vegetation and vegetation-research in South Africa (ABN 17: 330; CO 4(2): 14)

68.   19 October 1968. Utrecht (JB 1969: 13-15)
J. van der Toorn. Ecological differentiation of Phragmites communis Trin. (ABN 18: 484-485)
J. van Donselaar. Een vegetatiekundige verkenning in het Voltzberggebied, Suriname
P.A.I. Oremus. Experimenteel oecologisch onderzoek aan primaire duinplanten op de afsluit-dam in het Brielse Gat
G. Londo. Vegetatie-onderzoek op de oevers van een gegraven duinplas in de Kennemerduinen (JB 1969: 50-51)

69.   20 December 1968. Utrecht (JB 1969: 13-15)
J.H. Willems. Heath communities with Sarothamnus scoparius and Erica cinerea in the eastern part of the Belgian Kempen and the Dutch province of Limburg (ABN 18: 485-486; CO 4(4): 16-17)
H. Doing. Sociological species groups (ABN 18: 398-400; JB 1969: 47; CO 4(4): 20)
D.C.P. Thalen. Soortdiversiteit en variatie in enkele duin- en kweldervegetaties op Schiermon-nikoog in relatie tot variatie in het milieu (JB 1969: 53-54; CO 4(4): 19)
J.G.M. Janssen. Phenological study of Lonicera periclymenum L. and Senecio fuchsii Gmel. in the nature reserve 'De Duivelsberg' near Nijmegen (JB 1969: 48-49; CO 4(4): 18)

70.   11 April 1969. Utrecht (JB 1970: 13-16)
L.F.M. Fresco & H. Meijboom. A quantitative analysis of vegetation boundaries and gradients in a north Drenthe heath (ABN 18: 578)
C.R. Janssen. Atlantische en postatlantische vegetatiegeschiedenis in Noord-Brabant in verband met de geschiedenis van de occupatie door de mens
D.A. Vestergaard, J.G. van der Made & R. Vis. Investigating relations between moth popula-tions (Lepidoptera) and vegetation structure in the isle of Voorne (ABN 18: 578-579)
J.Th. de Smidt, A.M. Cleef, A.G. van Embden, J. Kers, R. Pos, P. Scheijgrond & L.E. Verwey. Nederlandse stuifzandvegetaties

71.   25 October 1969. Utrecht (JB 1970: 13-16)
E. van der Maarel. Enkele begrippen en methoden in de kwantitatieve vegetatiesystematiek (JB 1970: 52-54; CO 6(2): 89-91)
W. Groenman-van Waateringe & M.J. Jansma. L'analyse de diatomées et de pollen de la crique de Vlaardingen. Une interprétation révisée (ABN 19: 112)
J.A. Baars. Ordening van vegetaties met Thelypteris palustris
R. Norde, F.H.F. Oldenburger & H.Th. Riezebos. Savanne-onderzoek aan de Boven-Sipaliwini

72.   19 December 1969. Utrecht. Theme: Plantengroei in kunstmatige bossen (JB 1970: 13-16)
H. Doing. The vegetation of artificial forests (ABN 19: 454-455)
G. Sissingh. De plantengroei in douglasbossen
J.J. Barkman. De betekenis van de mycoflora voor de karakterisering van naaldbossen in Neder-land
I.S. Zonneveld. Toepassing van vegetatiestudies in naaldbos voor groeiplaatsboniteting

73.   10 April 1970. Utrecht (JB 1971: 13-15)
A.H.J. Freijsen. Groei-fysiologie en mineralenhuishouding van Centaurium littorale (Aanpassin-gen aan het oligotroof brak duinmilieu)
J.M.C.P. Schoonen. Flora en vegetatie van de Beerse Overlaat bij Vogelshoek (N.Br.) (CO 6(3/4): 114)
A.J. Havinga. Een palynologisch onderzoek in het zwarte aardegebied van Neder-Oostenrijk (CO 6(3/4): 113)
V. Westhoff. Botanische indrukken uit het noordwesten van de Verenigde Staten

74.   3 October 1970. Utrecht (JB 1971: 13-15)
W.G. Beeftink. Ontwikkeling van de vegetatie in het Veerse Meer
H. Doing. Vegetation formations in Australia (ABN 20: 258-259)
D. van der Laan. Some aspects of vegetational and environmental research of the dune slacks of Voorne (ABN 20: 717-718)
J.C. Lindeman. Indrukken van de vegetatie van Paraná, Zuid-Brazilië

75.  17 December 1970. Wageningen. Theme: Cryptogamenvegetaties (JB 1971: 13-15)
J.J. Barkman. De relatie tussen cryptogamengezelschappen en fytocoenosen (CO 7(1): 7-10)
R.K.J. van Hulst. De relatie epifyt-omringende bosvegetatie in de Hautes Fagnes (België) (CO 7(1): 5-6)
R. Ketner-Oostra. Onderzoek van terrestrische cryptogamenvegetaties in de droge duinen van het Waddendistrict (CO 7(1): 6-7)

76.  22 April 1971. Wageningen (JB 1972: 10-11)
C. den Hartog. Klassificatie van zeegrasgezelschappen (JB 1972: 32-33)
A. Voorrips. Een palynologisch onderzoek van prehistorische bewoningen in het West-Nederlands rivierengebied
C.W.P.M. Blom & A.M. Blom-Steinbusch. Een vegetatiekartering van het natuurmonument 'Quackjeswater' in de duinen van Voorne (CO 7(4): 103-104)
H. Nell. Stikstofhuishouding in verschillende duinvegetaties
S.P. Tjallingii. Op de grens van bos en savanne. Enkele landschapsoecologische opmerkingen naar aanleiding van een verblijf in Ivoorkust

77.  1 October 1971. Wageningen. Theme: Betrekkingen tussen planten- en diergemeenschappen. (In cooperation with the Oecologische Kring of the Nederlandse Dierkundige Vereniging) (JB 1972: 10-11)
E. van der Maarel. Algemene inleiding: Richtingen in het onderzoek van levensgemeenschappen
H. Strijbosch. Biocoenologisch onderzoek aan reptielen en amphibiën (CO 7(4): 104-105)
P. van der Aart. Relatie tussen de verspreiding van wolfsspinnen en karakteristieken van vegetatie en bodem (CO 7(4): 106-114)
J. van der Drift. Produktie en afbraak van organische stof in een eikenbos (CO 6(1): 43-51)
J. Gardeniers. Betrekkingen tussen levensgemeenschappen in stromend water

78.  16 December 1971. Wageningen. Theme: Palynologie en vegetatiekunde. (In cooperation with the Palynologische Kring of the Koninklijk Nederlands Geologisch Mijnbouwkundig Genootschap) (JB 1972: 10-11)
C.R. Janssen. Een overzicht van de methoden ter reconstructie van plantengezelschappen in het verleden door middel van palynologisch onderzoek
Sv.Th. Andersen. The differential pollen productivity of trees and its significance for the interpretation of pollen diagrams from forested regions
D. Teunissen. Een poging tot het ontwerpen van een kwantitatieve methode tot herkenning van plantengemeenschappen uit palynologische gegevens

79.  13 April 1972. Wageningen (JB 1973: 10-11)
M.J.M. Martens & G.H. Boonen. Vergelijkend geobotanisch onderzoek van de Berghofweide (gemeente Wijlre) (CO 8 (3/4): 66)
G. de Groot-Veenbaas & R.W. Tienstra. De bossen van Oud- en Nieuw-Amelisweerd mede gezien in het licht van hun historische ontwikkeling
J.C. Smittenberg. Moerasvegetaties langs het Zuidlaardermeer (CO 8(3/4): 64-65)
F.J.A. Daniëls. Opmerkingen over en indrukken van licheenvegetaties op steen in arctische en alpiene gebieden (JB 1973: 34-35)

80.  6 October 1972. Wageningen (JB 1973: 10-11)
D.A.J. Vogelpoel. Enkele aspecten van de autoecologie van Dicranum scoparium in de droge duinen van Terschelling (CO 9(1): 24-25)
S.J. ter Borg. Het verband tussen habitat en intraspecifieke variabiliteit bij Rhinanthus serotinus (CO 9(1): 27-28)
J. Groot. Seizoensgebonden resultaten van experimenten aan Plantago major, in kas en klimaatkamer
J.H. Willems. Experimenteel botanisch synoecologisch onderzoek in het Gerendal (Z.-Limburg) (CO 9(1): 25-27)

81. 19 December 1972. Wageningen. Theme: Het laagveen (JB 1973: 10-11)
A. Smit. De geschiedenis van het veenlandschap en zijn vegetatie
G. van Wirdum. Het verband tussen de successie en enige veranderingen in de eigenschappen van het water in de Weerribben (JB 1973: 49-50)
A.J. den Held. A comparative study of the vegetation and flora of recently formed peat areas in the fens in the western part of the Netherlands (ABN 22: 264-265)
H. Piek. De beheersproblematiek van de Wieden, in het bijzonder van het Kiersche Wijde

82. 2 May 1973. Utrecht (JB 1974: 10-11)
A. van Haperen. De vegetatie van het Schiepersberg-complex (Z. Limburg), met name de kalkgraslanden (CO 9(4): 91-92)
Ph. Stoutjesdijk. De open schaduw, een interessant micro-klimaat (ABN 23: 125-130)
H. de Boois. Patronen en processen in de Biesbosch voor en na de afsluiting (CO 9(4): 95-99)
G. Londo. Ervaringen met een oecologische proeftuin (CO 9(4): 93-95)

83. 16 October 1973. Wageningen. Theme: De oecologie van (kleine) verwante taxa (JB 1974: 10-11)
G. Zijlstra. Vegetatiekundig onderzoek aan één- en tweejarige vormen van Linum catharticum L. (CO 10(1): 10-14)
I. Koch. Over de oecologie van Scirpus planifolius Grimm en S. rufus (Huds.) Schrad. (JB 1974: 33)
J.H. Neuteboom. Variabiliteit van de grassoort kweek (Elytrigia repens (L.) Desv.) op Nederlandse landbouwgronden (CO 10(1): 15-17)
D.M. Pegtel. Effect of crop rotation on the distribution of two ecotypes of Sonchus arvensis L. in The Netherlands (ABN 23: 349-350)

84. 18 December 1973. Enschede. Theme: De toepassing van luchtfoto's in de vegetatiekunde (JB 1974: 10-11)
P.R.J. Satter. Luchtfotografie (inleiding d.m.v. een film)
I.S. Zonneveld. Algemene aspecten van de luchtfoto-interpretatie voor vegetatiekundig onderzoek (CO 10(1): 17-26)
J. Leemans & B. Verspaandonk. Luchtfoto-interpretatie voor de vegetatiekartering van het Land van Saeftinghe (CO 10(1): 27-28)
J. van der Toorn. Successie-onderzoek in Zuid-Flevoland met gebruik van luchtfoto's (CO 10(1): 28)
D. de Hoop. Toepassing van de luchtfotografie in het watermilieu
D.A. Stellingwerf. Aspecten van het verkrijgen van kwantitatieve bosbouwkundige gegevens via luchtfoto's (CO 10(1): 29)

85. 7 May 1974. Wageningen (JB 1975: 12-13)
J.W.M. Kuipers. De vegetatie van de Beninger en Korendijksche slikken (CO 10(2): 45-48)
C.W.P.M. Blom. De invloed van bodemverdichting en betreding op het voorkomen van enkele plantensoorten (CO 10(3): 17-20)
M.C. Groenhart. Enkele problemen rond scheidingsmaten in de vegetatiekunde (CO 10(3): 20-21)
J.G.M. Janssen. Simulatie en de ontkieming van winterannuellen (CO 10(2): 48-54)

86. 2 October 1974. Wageningen. Theme: Wegbermen en sloten (JB 1975: 12-13)
P. Zonderwijk. Mogelijkheden tot het herstel van de flora langs wegen door wijzigingen van het beheer (CO 10(4): 2-4)
M. Hoogerkamp. De aanleg van weinig productieve bermen en het maai-onderhoud van bestaande bermen (CO 10(4): 5-9)
C.J.M. Sloet van Oldruitenborgh & J.M. Gleichman. Over milieu, vegetatie en beheer van wegbermen op voedselarme zandgronden (CO 10(4): 10-17)
A.S.N. Liem, H. Doornebal & A. Hendriks. Enige wegbermexperimenten in de Bijlmermeer, Amsterdam (CO 10(4): 18-24)
J.C.J. van Zon. De waarde en het beheer van slootvegetaties (CO 10(4): 25-28)

87. 17 December 1974. Wageningen. Theme: Geomorfologisch, vegetatiekundig, bodemkundig en palaeobotanisch onderzoek van de Centrale Vogezen. (In cooperation with the Palynologische Kring of the Koninklijk Nederlands Geologisch Mijnbouwkundig Genootschap) (JB 1975: 12-13)

A.I. Salomé. Geomorfologisch overzicht van de Hoge Vogezen met enkele opmerkingen over het klimaat

R. Carbiener. Overzicht van de vegetatie van de Centrale Vogezen

A.J. Kalis, A.A.M.L. Verbeek-Reuvers, L.H. Batenburg & E.J. de Valk. Montane en subalpine bosgezelschappen van de Centrale Vogezen

H. Bick, J.H.J. Krüger & P. van der Knaap. De vegetatie van noord en noordoost geëxponeerde kaarwanden

C.R. Janssen. Overzicht van het palaeobotanisch onderzoek in de Vogezen

G. Tamboer-van de Heuvel. De recente pollen neerslag in de subalpiene heiden en beukenstruwelen

E.J. de Valk. Palynologische opmerkingen over de boomgrens in het Hohneck-Kastelberg massief

H. Edelman. De huidige en vroegere veenvegetatie van het Feigne d'Artimont

88. 9 April 1975. Wageningen (JB 1976: 10)

H.A.M. van Gils. De syntaxonomie van vegetatietypen zonder associatie-kentaxa aan voorbeelden uit het Geranion sanguinei (CO 12(1): 19-23)

J.L.J. Hendriks. De invloed van voormalig agrarisch gebruik op de vegetatie van loofbossen op jonge voedselrijke gronden (Alno-Padion) in het Fluviatiel en Kempens District

J. Heyink. Bodem, vegetatie en beheer van het landschapsreservaat Cranendonck (N.Br.)

B. Spiers. Vegetatie en ecologie van 'Dehesa'-landschappen bij Merida (Z.W.-Spanje)

L. de Lange. Ecological aspects of stands of macrophytes in ditches (ABN 24: 358)

89. 30 September 1975. Wageningen. Theme: Dynamics of algal vegetations in the Netherlands (JB 1976: 10)

H. Hillebrand. Periodicity of multicellular green algae (ABN 25: 117)

P.F.M. Coesel. Succession of desmid communities in the broads area of NW Overijssel (ABN 25: 118-119)

J. Simons. Dynamics of algal vegetations dominated by Vaucheria species (ABN 25: 119-120)

P.H. Nienhuis. Dynamics of environment and algal vegetations in salt marshes in the S.W.-Netherlands (ABN 25: 120-121)

P.J.G. Polderman. Dynamics of the algal vegetation of saltmarshes in the Wadden Sea (ABN 25: 121-122)

F. Colijn, H. Nienhuis, V.N. de Jong & R.P.T. Koeman. Distribution and seasonal periodicity of sediment inhabiting diatoms in the Waddensea and the Ems-Dollard estuary (ABN 25: 122-123)

90. 19 December 1975. Wageningen (JB 1976: 10)

M.J. Jansma. Kwantitatieve diatomeeënanalyse als hulpmiddel bij prehistorisch onderzoek

J.G. Vermeer & H. Weijs. Wrakelberg 1968-1973: de ontwikkeling van een kalkgraslandvegetatie met maaien als beheersmaatregel

C. Daan. Vegetatie-onderzoek langs de Oude Maas

J.J. Barkman. Transplantatieproeven met cryptogamen in Drenthe

H. Doing. Kustvegetaties in New South Wales (Australia)

91. 6 April 1976. Wageningen (JB 1977: 10-11)

B. Korf. Vegetatiekartering ten behoeve van de planologie in de gemeente Zaanstad (CO 12(3): 66-68)

N.J.M. Gremmen. De vegetatie van het subantarctische eiland Marion

M.G.C. Schouten & M.J. Nooren. Coastal vegetation types and soil features in South-East Ireland (ABN 26: 357-358)

J.G.H.M. Eijsink & G.A. Ellenbroek. Droge en half-droge graslanden in het Weinviertel van Nieder-Österreich

92. 30 September 1976. Wageningen. Theme: Beheer van vegetaties (JB 1977: 10-11)
G. Londo. Uitgangspunten en ideeën betreffende het natuurbeheer (CO 12(4): 77-81)
J.P. Bakker. Tussentijdse resultaten van een aantal beheersexperimenten in de madelanden van het stroomdallandschap Drentsche Aa (CO 12(4): 81-92)
M.J.M. Oomes. Cutting regime experiments on extensively used grasslands (ABN 26: 265-266; CO 12(4): 92-99)
P. Oosterveld. Integratie van voormalige landbouwgronden in natuurterreinen door middel van een begrazingsbeheer met IJslandse ponies in de Baronie Cranendonck (CO 12(4): 99-109)
H.P.G. Helsper. Vegetatie en milieu van het Beuven, een door sluipende eutrophiëring bedreigd voedselarm ven (CO 12(4): 109-110)

93. 17 December 1976. Wageningen (JB 1977: 10-11)
A.K. Masselink. De synsystematiek van gagelstruwelen
M.J.P.W. van Sambeek & A.M.F.C. van Pruissen. Vegetatiekundig onderzoek van enige vennen op het Hoogterras van de Maas in de oostelijke Belgische Kempen
H.M. van der Steeg. De vegetatie-ontwikkeling op de dijkverzwaringsproefvakken onder Ewijk
J. Haeck & R. Hengeveld. Het verband tussen oecologische indicatiewaarde van soorten en de positie in hun areaal
A.J.M. Roozen, S.H.M. Hochstenbach, M.J.M. van Mansfeld & J.M. Groenendael. Vegetatie en substraat van de kuststrook van Connemara, West-Ierland, als gradiëntzone tussen zee en spreihoogveen

94. 23 March 1977. Wageningen (JB 1978: 11-12)
M.M. Kwak. Hommelbestuiving van Rhinanthus serotinus en R. minor; hybridisatie en isolatie (ABN 26: 97-107)
J. van Andel. Levenscyclus en minerale voeding van enkele plantensoorten op kap- en storm-vlakten; experimenteel onderzoek naar de oorzaken van successie
P.B.Ph.M. Bogaers, J. Prins & J. Wiertz. Typologie en kartering van het Hol en de Suikerpot (gemeente 's Graveland)
I. Koch. De sociologische en oecologische plaats van Scirpus planifolius binnen Europa
G.J.R. Allersma. Aantal en type van koeieplakken als indicatoren van beweidingsdruk op kwel-dervegetaties (CO 13(3): 44-51)

95. 22 September 1977. Wageningen. Theme: Kieming en vestiging (JB 1978: 11-12)
M.J.M. Oomes. Verschillen in kieming en vestiging en de oecologische konsekwenties ervan
A.H.J. Freijsen. De ontkieming van de kalkplant Cynoglossum officinale L. op substraten met een verschillende stikstof-toestand
C.W.P.M. Blom. De invloed van bodemdichtheid en betreding op de vestiging van Plantago kiemplanten
J.P.C.M. Breek. Kiemingsoecologie van Juniperus communis in struwelen van Drente en Overijssel
G. Londo. Ervaringen betreffende kieming en vestiging van plantensoorten in permanente kwa-draten

96. 16 December 1977. Wageningen (JB 1978: 11-12)
W.G. Braakhekke. Stabiele evenwichten tussen graslandplanten
P.J. van Loenhoud & J.C.P.M. van de Sande. De 'Braun-Blanquet'-methode in een onderzoek naar de litorale zonering op de Nederlandse Antillen
S. van der Werf. Het voorkomen van beuk en linde bezien vanuit de toponymie
A. Barendregt. Vegetaties van het Zwanenwater (Gem. Callandsoog, N.H.)
J. Rozema. Oecologie van halophyten op de strandvlakte van Schiermonnikoog; beschrijvend en experimenteel onderzoek aan zonering en successie

97. 11 April 1978. Wageningen
R.M. Mooij. Onderzoek naar vegetatiesuccessie in het Molenven (gem. Saesveld, Overijssel)
H. Sprangers. Samenstelling en structuur van het droge bostype van Zuid-Oost India
J.C. de Ruyter. Effecten van vijf jaar beweiding op de Oosterkwelder van Schiermonnikoog
J. Schouw & J. Wolf. Onkruidgezelschappen op een biologisch-dynamisch landbouwbedrijf
M.J.A. Werger. Vegetatiestructuur en substraat van de 'Great Dyke' in Rhodesië

98. 27 September 1978. Wageningen. Theme: Produktie-onderzoek in graslandvegetaties

D.C.P. Thalen. Het meten en schatten van produktie voor het karteren en evalueren van vegetaties

M.J.M. Oomes. Produktie van marginale graslanden bij verschillende gebruiksvormen

S.E. van Wieren & A. Vreugdenhil. De invloed van beweiden, maaien en plaggen op de snelheid van mineralenafvoer in relatie tot veranderingen in produktiviteit en samenstelling van de vegetatie

A. Grootjans. Enkele opmerkingen over de relaties tussen grondwaterstand, stikstofmineralisatie en 'standing crop' in natte hooilanden

J.H. Willems. Bovengrondse biomassa en soortsdiversiteit in kalkgrasland

99. 14 December 1978. Wageningen

H. Doing. Landschapskartering op vegetatiekundige grondslag in Nederlandse duingebieden

E. van der Maarel. Ontwikkeling van graslandvegetaties in een voormalige boomgaard onder invloed van verschillende beheersmaatregelen

M.C. Groenhart. Over het gebruik van abundantie-dominantie schattingsschalen in de vegetatiekunde en oecologie

G. van Wirdum. De paludisfeer; oecologische samenhang van de wereld der moerassen

M. Rijken. De ruimtelijke relaties tussen een aantal bosgemeenschappen in het oerbos van Bialowieza (Polen)

100. 23 March 1979. Wageningen

(See contents of this book)


## Some observations on the list

It is interesting to outline briefly a few observations which are immediately apparent from the list. Though the task of the Commission of the Royal Botanical Society of the Netherlands which organized the hundred meetings documented in the list remained the same throughout its period of existence, its name changed once, while the designation of the meetings changed twice. From 1933 till 1965 the name of the Commission was 'Commission for Biosociology and Peat Research of the Netherlands' and subsequently became 'Commission for the Study of Vegetation'. From 1933 till 1939 the meetings were called 'Dutch Meetings for Phytosociology and Palaeobotany of the Holocene', which then changed to 'Dutch Meetings for Biosociology and Palaeobotany'. In 1965 the name changed again, now to 'Meetings on the Study of Vegetation'. These name changes apparently were attempts to improve on the flag covering the cargo, rather than a change in the field covered by the meetings.

Until about 1966 the speakers on the meetings remained restricted to a fairly small group of scientists. This contrasts sharply with the strong proliferation in speakers' names during the last twelve years. Very many young scientists addressed the meetings during that period, though this observation cannot be concluded from this list.

As to the topics presented at the meetings, it is remarkable that tropical ecology is rarely discussed, even in the early years of these get-togethers when the Netherlands still were a colonial power. Much attention was paid, however, to the ecological effects of the damming of the Zuiderzee and the subsequent reclamations in the IJselmeer. These large projects offered unique opportunities for succession and colonization studies. Conspicuous among the topics discussed at the meetings are also the frequent occurrences of the famous grassland studies of D.M. de Vries.

The meetings always have included some lectures on experimental ecological research, but particularly during the last decade aspects of this branch of ecology figured regularly on the agenda. By far most lectures dealt with descriptive ecology, however, which is understandable as this approach always has been the one that was most practised by ecologists. It is not possible to draw conclusions from the list on the change in and refinement of methods over the period of time covered, but such changes obviously have occurred and are discussed elsewhere in this book.

Finally, it is interesting to note that, though on the meetings there have always been lectures on applied ecological research next to those of a purely scientific character, the aims of the applied research discussed here seem to have changed. While in the early years applied ecology evidently aimed at agriculture, since about 1970 it definitely concentrates on nature conservation and management. In this the severely increased needs of nature conservation in the Netherlands since the last war are clearly reflected.

### INDEX TO THE LECTURES

## OFFICE-BEARERS OF THE COMMISSION FOR THE STUDY OF VEGETATION 1933-1979*

| Year | Chairman | Secretary | | | | | |
|---|---|---|---|---|---|---|---|
| 1933 | A.A. Pulle | J.W. van Dieren | J. Vlieger | F. Florschütz | B. Polak | P. van Oije | D.M. de Vries |
| 1934 | A.A. Pulle | J.W. van Dieren | J. Vlieger | F. Florschütz | B. Polak | P. van Oije | D.M. de Vries |
| 1935 | A.A. Pulle | A. Scheygrond | J. Vlieger | F. Florschütz | B. Polak | P. van Oije | D.M. de Vries |
| 1936 | J. Jeswiet | A. Scheygrond | M.J. Adriani | E.C. Wassink | B. Polak | P. van Oije | D.M. de Vries |
| 1937 | J. Jeswiet | A. Scheygrond | M.J. Adriani | E.C. Wassink | F. Florschütz | W. Feekes | G.W. Harmsen |
| 1938 | J. Jeswiet | D.M. de Vries | M.J. Adriani | E.C. Wassink | F. Florschütz | W. Feekes | G.W. Harmsen |
| 1939 | Th. Weevers | W. Feekes | D.M. de Vries | B. Polak | F. Florschütz | W.C. de Leeuw | G.W. Harmsen |
| 1940 | Th. Weevers | D.M. de Vries | W.H. Diemont | A.W. Moll | E.C. Wassink | W.C. de Leeuw | G. Sissingh |
| 1941 | Th. Weevers | A. Scheygrond | W.H. Diemont | F. Florschütz | E.C. Wassink | P. van Oije | G. Sissingh |
| 1942 | A.A. Pulle | A. Scheygrond | W.H. Diemont | F. Florschütz | E.C. Wassink | P. van Oije | G. Sissingh |
| 1943 | A.A. Pulle | A. Scheygrond | W. Feekes | F. Florschütz | F.P. Jonker | P. van Oije | V. Westhoff |
| 1944 | A.A. Pulle | D.M. de Vries | W. Feekes | F. Florschütz | F.P. Jonker | W.G. van der Kloot | V. Westhoff |
| 1945 | A.A. Pulle | D.M. de Vries | W. Feekes | E.C. Wassink | F.P. Jonker | W.G. van der Kloot | V. Westhoff |
| 1946 | A. Scheygrond | D.M. de Vries | W. Feekes | E.C. Wassink | F.P. Jonker | W.G. van der Kloot | V. Westhoff |
| 1947 | A. Scheygrond | D.M. de Vries | F. Florschütz | E.C. Wassink | J. Vlieger | W.G. van der Kloot | M.F. Mörzer Bruyns |
| 1948 | A. Scheygrond | F.P. Jonker | F. Florschütz | H.T. Waterbolk | J. Vlieger | P.R. den Dulk | M.F. Mörzer Bruyns |
| 1949 | J. Heimans | F.P. Jonker | F. Florschütz | H.T. Waterbolk | J. Vlieger | P.R. den Dulk | M.F. Mörzer Bruyns |
| 1950 | J. Heimans | F.P. Jonker | J. Lanjouw | H.T. Waterbolk | T. van der Hammen | P.R. den Dulk | V. Westhoff |
| 1951 | J. Heimans | D. Bakker | F. Florschütz | D.M. de Vries | T. van der Hammen | J.J. Barkman | V. Westhoff |
| 1952 | J. Heimans | D. Bakker | F. Florschütz | D.M. de Vries | T. van der Hammen | J.J. Barkman | V. Westhoff |
| 1953 | J. Lanjouw | D. Bakker | F. Florschütz | D.M. de Vries | J. Venema | J.J. Barkman | M.F. Mörzer Bruyns |
| 1954 | D.M. de Vries | F.P. Jonker | H.T. Waterbolk | I.S. Zonneveld | J. Venema | W. van Zeist | M.F. Mörzer Bruyns |
| 1955 | J. Heimans | F.P. Jonker | H.T. Waterbolk | I.S. Zonneveld | J. Venema | W. van Zeist | M.F. Mörzer Bruyns |
| 1956 | J. Heimans | F.P. Jonker | H.T. Waterbolk | I.S. Zonneveld | W.H. Zagwijn | W. van Zeist | V. Westhoff |
| 1957 | J. Heimans | H.T. Waterbolk | B. Polak | B. Polak | W.H. Zagwijn | J. Vlieger | V. Westhoff |
| 1958 | D.M. de Vries | W. van Zeist | H. Doing Kraft | B. Polak | W.H. Zagwijn | J. Vlieger | V. Westhoff |
| 1959 | J. Heimans | W. van Zeist | H. Doing Kraft | B. Polak | P. Tideman | J. Vlieger | W.H. Zagwijn |
| 1960 | J. Heimans | W. van Zeist | H. Doing Kraft | T. van der Hammen | P. Tideman | J.J. Barkman | W.H. Zagwijn |
| 1961 | J. Heimans | J.J. Barkman | E. Stapelveld | T. van der Hammen | P. Tideman | F.P. Jonker | W.H. Zagwijn |
| 1962 | F.P. Jonker | E. van der Maarel | E. Stapelveld | T. van der Hammen | B. Polak | D. Bakker | V. Westhoff |
| 1963 | F.P. Jonker | E. van der Maarel | E. Stapelveld | W.H. Zagwijn | B. Polak | D. Bakker | V. Westhoff |
| 1964 | J.J. Barkman | E. van der Maarel | S. Segal | W.H. Zagwijn | B. Polak | D. Bakker | V. Westhoff |
| 1965 | J.J. Barkman | W.A.E. van Donselaar-ten Bokkel Huinink | S. Segal | W.H. Zagwijn | M.J. Adriani | C.G. van Leeuwen | A.J. Havinga |
| 1966 | J.J. Barkman | W.A.E. van Donselaar-ten Bokkel Huinink | S. Segal | J.Th. de Smidt | M.J. Adriani | C.G. van Leeuwen | A.J. Havinga |
| 1967 | V. Westhoff | W.A.E. van Donselaar-ten Bokkel Huinink | H.N. Leys | J.Th. de Smidt | M.J. Adriani | C.G. van Leeuwen | A.J. Havinga |
| 1968 | V. Westhoff | W.A. Casparie | H.N. Leys | J.Th. de Smidt | I.S. Zonneveld | C.R. Janssen | C.J.M. Sloet van Oldruitenborgh |
| 1969 | E. van der Maarel | W.A. Casparie | H.N. Leys | W.H. Zagwijn | I.S. Zonneveld | C.R. Janssen | C.J.M. Sloet van Oldruitenborgh |
| 1970 | E. van der Maarel | W.A.E. van Donselaar-ten Bokkel Huinink | P. Stoutjesdijk | W.H. Zagwijn | I.S. Zonneveld | C.R. Janssen | C.J.M. Sloet van Oldruitenborgh |
| 1971 | E. van der Maarel | W.A.E. van Donselaar-ten Bokkel Huinink | P. Stoutjesdijk | W.H. Zagwijn | G. Londo | L. Kop | W. Groenman-van Waateringe |
| 1972 | V. Westhoff | A.H.J. Freijsen | P. Stoutjesdijk | T.A. Wijmstra | G. Londo | L. Kop | W. Groenman-van Waateringe |
| 1973 | J. van Donselaar | A.H.J. Freijsen | L.F.M. Fresco | T.A. Wijmstra | G. Londo | L. Kop | W. Groenman-van Waateringe |
| 1974 | J. van Donselaar | A.H.J. Freijsen | L.F.M. Fresco | T.A. Wijmstra | C.C. Bakels | J.H. Smittenberg | J.P. van den Bergh |
| 1975 | J. van Donselaar | J.H. Willems | L.F.M. Fresco | D.M. Pegtel | C.C. Bakels | J.H. Smittenberg | J.P. van den Bergh |
| 1976 | M.J.A. Werger | J.H. Willems | P.A. Bakker | D.M. Pegtel | C.C. Bakels | J.H. Smittenberg | J.P. van den Bergh |
| 1977 | M.J.A. Werger | J.H. Willems | P.A. Bakker | D.M. Pegtel | M.J. Oomes | A. de Wit | W.H.E. Gremmen |
| 1978 | M.J.A. Werger | J. Wiegers | H. Piek | E. van der Meijden | M.J. Oomes | A. de Wit | W.H.E. Gremmen |
| 1979 | I.S. Zonneveld | J. Wiegers | H. Piek | E. van der Meijden | M.J. Oomes | A. de Wit | W.H.E. Gremmen |

* List compiled by Jaap Wiegers.